ADVANCED ERROR CONTROL TECHNIQUES FOR DATA STORAGE SYSTEMS

ADVANCED ERROR CONTROL TECHNIQUES FOR DATA STORAGE SYSTEMS

EDITED BY

Erozan M. Kurtas

Seagate Technology
Pittsburgh, PA

Bane Vasic

University of Arizona
Tucson, AZ

Taylor & Francis
Taylor & Francis Group
Boca Raton   London   New York

A CRC title, part of the Taylor & Francis imprint, a member of the
Taylor & Francis Group, the academic division of T&F Informa plc.

This material was previously published in *Coding and Signal Processing for Magnetic Recording Systems* © CRC Press LLC 2005.

Published in 2006 by
CRC Press
Taylor & Francis Group
6000 Broken Sound Parkway NW, Suite 300
Boca Raton, FL 33487-2742

© 2006 by Taylor & Francis Group, LLC
CRC Press is an imprint of Taylor & Francis Group

No claim to original U.S. Government works
Printed in the United States of America on acid-free paper
10 9 8 7 6 5 4 3 2 1

International Standard Book Number-10: 0-8493-9547-X (Hardcover)
International Standard Book Number-13: 978-0-8493-9547-5 (Hardcover)
Library of Congress Card Number 2005050632

Library of Congress Cataloging-in-Publication Data

Kurtas, Erozan.
    Advanced error control techniques for data storage systems / Erozan M. Kurtas, Bane Vasic.
        p. cm.
    Includes bibliographical references and index.
    ISBN 0-8493-9547-X (alk. paper)
    1. Coding theory. 2. Error-correcting codes (Information theory) I. Title: Error control and correction methods for data storage systems. II. Vasic, Bane. III. Title.

TK5102.92.K87 2005
003'.54--dc22   2005050632

Taylor & Francis Group
is the Academic Division of Informa plc.

Visit the Taylor & Francis Web site at
http://www.taylorandfrancis.com

and the CRC Press Web site at
http://www.crcpress.com

Preface

We live in an age that is defined by the massive production and use of digital information. Everyday we send and receive e-mails, exchange pictures of our loved ones, download music and video, and transact business on-line. The digital revolution has become such a fundamental pillar of our existence it is very difficult to imagine a life without it. According to one recent study, data was being created at a rate of 1.4 Gigabytes/person/year in the United States in 2004. Every year 75 Petabytes or 100 Million hours of new broadcast content is created. We exchange 400 Petabytes of e-mail and purchase 488 Petabytes or about 814 M Cds worth (only in the United States) audio files yearly. Although these numbers are amazing, they are growing even in a more phenomenal pace. Clearly, we need a reliable way to store, protect and exchange all these data.

This book focuses on explaining the state of the art error control and correction methods used in data storage devices such as hard disk drives. The advanced error control and correction methods ensure the reliable transmission and storage of digital data. This in turn enables us to create, use and manipulate data as we wish. The reliability levels that are required by storage devices are extremely high since unlike communication systems generally no retransmission is possible. We expect to save our data and be able to retrieve it perfectly at any future time. It is the art of error control and correction that makes this possible.

We start the book with an introduction on error control codes. Chapter 1 explains the finite fields, and defines linear codes over these fields. The relationship between minimum distance and error correction capability of a code is subsequently discussed. We give a detailed description of Reed-Solomon codes, which are widely employed for error control and correction, and describe their encoding and decoding.

Chapter 2 and Chapter 3 introduce modulation codes that are widely used in data storage devices. Modulation codes serve a number of useful purposes, the main one being the elimination of sequences, which cause problems for signal processing, synchronization or coding purposes. The most common constraint in use is the run-length one, and it is imposed on encoded sequences in order to bound the minimal, and/or maximal lengths of consecutive like channel symbols. Chapter 4 investigates maximum transition run (MTR) codes. These codes limit the maximum number of consecutive transitions and improve the minimum distance properties of recorded sequences. Chapter 5 introduces the spectrum shaping codes, which are deployed widely by recording systems. For example, in optical recording systems these codes are used to circumvent the interference between data tracks and servo tracking system.

Chapter 6 and Chapter 7 study the very important relationship between constrained codes and error control and correction systems. They do this by looking at both code design techniques for different constraint and the architectural options such as reverse concatenation. Chapter 8 takes the analysis of previous two chapters one step further by introducing encoding and decoding techniques based on convolutional codes for partial response channels which represent a wide variety of recording systems. Chapter 9 introduces the channel model and capacity definitions, and designs new classes of codes that approach the information theoretical performance limits. Chapter 10 takes us into another direction by investigating the coding and detection methods for multitrack systems. Multitrack systems envision a new architecture where multiple-read elements are employed both to record and retrieve the data. The data rate and access time benefits of such a configuration is clear. Moreover, as Chapter 10 shows, with properly designed codes one can improve the data density as well.

The last part of the book focuses on introducing the most recent developments in error control and correction systems as applied to data storage channels. Chapter 11 is about one of the most important developments in information theory in the last few decades, namely Turbo Codes. This chapter explains both the basic principles of turbo coding and decoding techniques as well as their applications to data storage systems. Soft decoding, turbo equalization, parallel and serially concatenated codes are among the important concepts that are investigated in this chapter. The next chapter, Chapter 12, takes the investigation of codes that employ soft decoding techniques one step further by giving a very good tutorial on Low-Density Parity Check Codes. Gallager Codes, McKay Codes and Finite Geometry Codes are among the Low-Density Parity Check Codes that are studied in this chapter. This chapter concludes with the explanation of the sum-product decoding algorithm.

Chapter 13 focuses on a special class of turbo codes called Turbo Product Codes. These codes are constructed by partitioning the data into rows and columns, and then encoding them separately. During decoding phase soft-information exchange between rows and columns is used to increase the reliability of each decoded information bit. This chapter uses both analytical and numerical techniques to evaluate the performance of Turbo Product Codes. The next chapter, Chapter 14, continues in a sense in the same direction as the pervious one by studying the Structured Low-Density Parity-Check Codes. The classical construction for Low-Density Parity-Check Codes is based on large, sparse and randomly generated parity check matrices. Clearly such a construction creates a number of implementation issues as well as analysis ones. By designing and studying Low-Density Parity-Check Codes constructed from combinatorial objects we can come up with better ways of analyzing and implementing them. Chapter 15, the final chapter of this book, expands the turbo coding techniques to previously multitrack systems.

We believe this book covers all the relevant developments in error control and correction techniques as applied to data storage channels. From this perspective we hope that it will be useful to both seasoned practitioners as a reference material, and to interested readers who are new to this growing area of information theory.

Bane Vasic
Erozan M. Kurtas

Contributors

Mario Blaum
Hitachi Global Storage
  Technologies
San Jose, CA

Barrett J. Brickner
Bermai, Inc.
Minnetonka, MN

Willem A. Clarke
Department of Electrical and
  Electronic Engineering
Rand Afrikaans University
Auckland Park, South Africa

Stojan Denic
School of Information
  Technology and Engineering
University of Ottawa
Ottawa, ON

Miroslav Despotović
Department of Electrical
  Engineering and
  Computer Science
Faculty of Engineering
University of Novi Sad,
Serbia and Montenegro

Tolga M. Duman
Department of Electrical
  Engineering
Arizona State University
Tempe, AZ

John L. Fan
Flarion Technologies
Bedminster, NJ

Hendrik C. Ferreira
Department of Electrical and
  Electronic Engineering
Rand Afrikaans University
Auckland Park, South Africa

Mark A. Herro
Department of Electrical and
  Electronic Engineering
University College
Dublin, Ireland

Kees A. Schouhamer Immink
University of Essen
Essen, Germany
Turing Machines Inc.
Rotterdam, Netherlands

Aleksandar Kavčić
Division of Engineering and
  Applied Sciences
Harvard University
Boston, MA

Mustafa Kaynak
Department of Electrical
  Engineering
Arizona State University
Tempe, AZ

Erozan M. Kurtas
Seagate Technology
Pittsburgh, PA

Alexander Kuznetsov
Seagate Technology
Pittsburgh, PA

Jing Li
Department of Electrical and
  Computer Engineering
Lehigh University
Bethlehem, PA

Xiao Ma
Department of Electronic
  Engineering
City University of Hong Kong
Kowloon, Hong Kong

Brian Marcus
Department of Mathematics
University of British Columbia
Vancouver, BC

Olgica Milenkovic
Electrical and Computer
  Engineering Department
University of Colorado
Boulder, CO

Krishna R. Narayanan
Texas A&M University
College Park, TX

Travis Oenning
IBM Corporation,
Rochester, MI

William E. Ryan
Department of Electrical and
  Computer Engineering
University of Arizona
Tucson, AZ

Vojin Šenk
Department of Electrical
    Engineering and
    Computer Science
Faculty of Engineering
University of Novi Sad
Novi Sad, Yugoslavia

Emina Soljanin
Bell Laboratories
Lucent Technologies
Murray Hill, NJ

Bartolomeu F. Uchôa-Filho
Department of Electrical
    Engineering
Federal University of Santa
    Catarina
Florianopolis, Brazil

Nedeljko Varnica
Division of Engineering and
    Applied Sciences
Harvard University
Boston, MA

Bane Vasic
Department of Electrical and
    Computer Engineering
Department of Mathematics
University of Arizona
Tucson, AZ

Zheng Zhang
Department of Electrical
    Engineering
Arizona State University
Tempe, AZ

Contents

1

An Introduction to Error-Correcting Codes

Mario Blaum

Hitachi Global Storage Technologies
San Jose, CA

1.1 Introduction

When digital data are transmitted over a noisy channel, it is important to have a mechanism allowing recovery against a limited number of errors. Normally, a user string of 0s and 1s, called bits, is encoded by adding a number of redundant bits to it. When the receiver attempts to reconstruct the original message sent, it starts by examining a possibly corrupted version of the encoded message, and then makes a decision. This process is called the *decoding*.

The set of all possible encoded messages is called an error-correcting code. The field was started in the late 1940s by the work of Shannon and Hamming, and since then thousands of papers on the subject have been published. There are also several very good books touching different aspects of error-correcting codes, for instance, [1, 3, 4, 5, 7, 8], to mention just a few.

The purpose of this chapter is giving an introduction to the theory and practice of error-correcting codes. In particular, it will be shown how to encode and decode the most widely used codes, Reed Solomon codes.

In principle, we will assume that our information symbols are bits, that is, 0s and 1s. The set $\{0, 1\}$ has a field structure under the exclusive-OR ($\oplus$) and product operations. We denote this field $GF(2)$, which means Galois field of order 2.

Roughly, there are two types of error-correcting codes: codes of block type and codes of convolutional type. Codes of block type encode a fixed number of bits, say k bits, into a vector of length n. So, the information string is divided into blocks of k bits each. Convolutional codes take the string of information bits globally and slide a window over the data in order to encode. A certain amount of memory is needed by the encoder. However, in this chapter we concentrate on block codes only. For more on convolutional codes, see [3, 8].

As said above, we encode k information bits into n bits. So, we have a 1-1 function f,

$$f : GF(2)^k \rightarrow GF(2)^n$$

The function f defines the encoding procedure. The set of 2^k encoded vectors of length n is called a code of *length* n and *dimension* k, and we denote it as an $[n, k]$ code. We call codewords the elements of the code while we call words the vectors of length n in general. The ratio k/n is called the *rate* of the code.

The error-correcting power of a code is characterized by a parameter called the minimum (Hamming) distance of the code. Formally:

Definition 1.1   Given two vectors of length n, say a and b, we call the Hamming distance between a and b the number of coordinates in which they differ (notation, $d_H(a, b)$). Given a code C of length n and dimension k, let

$$d = \min\{d_H(a, b) : a \neq b, \, a, b, \in C\}$$

We call d the minimum (Hamming) distance of the code C and we say that C is an $[n, k, d]$ code.

It is easy to verify that $d_H(a, b)$ verifies the axioms of distance, that is,

1. $d_H(a, b) = d_H(b, a)$
2. $d_H(a, b) = 0$ if and only if $a = b$
3. $d_H(a, c) \leq d_H(a, b) + d_H(b, c)$

We call a sphere of radius r and center a the set of vectors that are at distance at most r from a. The relation between d and the maximum number of errors that code C can correct is given by the following lemma:

Lemma 1.1   *The maximum number of errors that an $[n, k, d]$ code can correct is $\lfloor (d - 1)/2 \rfloor$, where $\lfloor x \rfloor$ denotes the largest integer smaller or equal to x.*

Proof 1.1   Assume that vector a was transmitted but a possibly corrupted version of a, say r, was received. Moreover, assume that no more than $\lfloor (d - 1)/2 \rfloor$ errors have occurred.

Consider the set of 2^k spheres of radius $\lfloor (d - 1)/2 \rfloor$ whose centers are the codewords in C. By the definition of d, all these spheres are disjoint. Hence, r belongs to one and only one sphere: the one whose center is codeword a. So, the decoder looks for the sphere in which r belongs, and outputs the center of that sphere as the decoded vector. As we see, whenever the number of errors is at most $\lfloor (d - 1)/2 \rfloor$, this procedure will give the correct answer.

Moreover, $\lfloor (d-1)/2 \rfloor$ is the maximum number of errors that the code can correct. Let $a, b, \in C$ such that $d_H(a, b) = d$. Let u be a vector such that $d_H(a, u) = 1 + \lfloor (d-1)/2 \rfloor$ and $d_H(b, u) = d - 1 - \lfloor (d-1)/2 \rfloor$. We easily verify that $d_H(b, u) \leq d_H(a, u)$, so, if a is transmitted and u is received (i.e., $1 + \lfloor (d - 1)/2 \rfloor$ errors have occurred), the decoder cannot decide that the transmitted codeword was a, since codeword b is at least as close to u as a.   □

Example 1.1

Consider the following 1-1 relationship between $GF(2)^2$ and $GF(2)^5$ defining the encoding:

00	↔	00000
10	↔	00111
01	↔	11100
11	↔	11011

The four vectors in $GF(2)^5$ constitute a [5, 2, 3] code $\mathcal{C}$. From Lemma 1.1, $\mathcal{C}$ can correct one error. For instance, assume that we receive the vector $\underline{r} = 10100$. The decoder looks into the four spheres of radius 1 (each sphere has six elements!) around each codeword, finding that $\underline{r}$ belongs in the sphere with center 11100. If we look at the table above, the final output of the decoder is the information block 01.

Example 1.1 shows that the decoder has to make at most 24 checks before arriving to the correct decision. When large codes are involved, as is the case in applications, this decoding procedure is not practical, since it amounts to an exhaustive search over a huge set of vectors. One of the goals in the theory of error-correcting codes is finding codes with high rate and minimum distance as large as possible. The possibility of finding codes with the right properties is often limited by bounds that constrain the choice of parameters n, k and d. We give some of these bounds in the next section.

Let us point out that error-correcting codes can be used for detection instead of correction of errors. The simplest example of an error-detecting code is given by a parity code: a parity is added to a string of bits in such a way that the total number of bits is even (a more sophisticated way of saying this, is that the sum modulo-2 of the bits has to be 0). For example, 0100 is encoded as 01001. If an error occurs, or, more generally, an odd number of errors, these errors will be detected since the sum modulo 2 of the received bits will be 1. Notice that two errors will be undetected. In general, if an $[n, k, d]$ code is used for detection only, the decoder checks whether the received vector is in the code or not. If it is not, then errors are detected. It is easy to see that an $[n, k, d]$ code can detect up to $d - 1$ errors. Also, we can choose to correct less than $\lfloor (d - 1)/2 \rfloor$ errors, say s errors, by taking disjoint spheres of radius s around codewords, and using the remaining capacity to detect errors. In other words, we want to correct up to s errors or detect up to $s + t$ errors when more than s errors occur.

Another application of error-correcting codes is in erasure correction. An erased bit is a bit that cannot be read, so the decoder has to decide if it was a 0 or a 1. An erasure is normally denoted with the symbol ?. For instance, 01?0 means that we cannot read the third symbol. Obviously, it is easier to correct erasures than to correct errors, since in the case of erasures we already know the location, we simply have to find what the erased bit was. It is not hard to prove that an $[n, k, d]$ code can correct up to $d - 1$ erasures. We may also want to simultaneously correct errors and erasures. In fact, a code $\mathcal{C}$ with minimum distance d can correct s errors together with t erasures whenever $2s + t \leq d - 1$.

1.2   Linear Codes

We have seen in the previous section that a binary code of length n is a subset of $GF(2)^n$. Notice that, being $GF(2)$ a field, $GF(2)^n$ has a structure of vector space over $GF(2)$. We say that a code $\mathcal{C}$ is linear if it is a subspace of $GF(2)^n$, that is,

1. $\underline{0} \in \mathcal{C}$
2. $\forall\, \underline{a}, \underline{b} \in \mathcal{C}, \underline{a} \oplus \underline{b} \in \mathcal{C}$

The symbol $\underline{0}$ denotes the all-zero vector. In general, vectors will be denoted with underlined letters, otherwise letters denote scalars.

There are many interesting combinatorial questions regarding nonlinear codes. Probably, the most important question is the following: given the length n and the minimum distance d, what is the maximum number of codewords that a code can have? For more about nonlinear codes, the reader is referred to [4]. From now on, we assume that all codes are linear. Linear codes are in general easier to encode and decode than their nonlinear counterparts, hence they are more suitable for implementation in applications.

In order to find the minimum distance of a linear code, it is enough to find its minimum *weight*. We say that the (Hamming) weight of a vector $\underline{u}$ is the distance between $\underline{u}$ and the zero vector. In other words, the weight of $\underline{u}$, denoted $w_H(\underline{u})$, is the number of nonzero coordinates of the vector $\underline{u}$. The minimum weight of a code is the minimum between all the weights of the nonzero codewords. The proof of the following lemma is left as an exercise.

Lemma 1.2   *Let C be a linear $[n,k,d]$ code. Then, the minimum distance and the minimum weight of C are the same.*

Next, we introduce two important matrices that define a linear error-correcting code. Since a code C is now a subspace, the dimension k of C is the cardinality of a basis of C. Consider then an $[n,k,d]$ code C. We say that a $k \times n$ matrix G is a *generator* matrix of a code C if the rows of G are a basis of C. Given a generator matrix, the encoding process is simple.

Explicitly, let $\underline{u}$ be an information vector of length k and G a $k \times n$ generator matrix, then $\underline{u}$ is encoded into the n-vector $\underline{v}$ given by

$$\underline{v} = \underline{u}G \tag{1.1}$$

Example 1.2

Let G be the 2×5 matrix

$$G = \begin{pmatrix} 0 & 0 & 1 & 1 & 1 \\ 1 & 1 & 1 & 0 & 0 \end{pmatrix}$$

It is easy to see that G is a generator matrix of the $[5,2,3]$ code described in Example 1.1.

Notice that, although a code may have many generator matrices, the encoding depends on the particular matrix chosen, according to Equation 1.1. We say that G is a *systematic* generator matrix if G can be written as

$$G = (I_k \mid V) \tag{1.2}$$

where I_k is the $k \times k$ identity matrix and V is a $k \times (n-k)$ matrix. A systematic generator matrix has the following advantage: given an information vector $\underline{u}$ of length k, the encoding given by Equation 1.1 outputs a codeword $(\underline{u}, \underline{w})$, where $\underline{w}$ has length $n-k$. In other words, a systematic encoder adds $n-k$ redundant bits to the k information bits, so information and redundancy are clearly separated. This also simplifies the decoding process, since, after decoding, the redundant bits are simply discarded. For that reason, most encoders used in applications are systematic.

A permutation of the columns of a generator matrix gives a new generator matrix defining a new code. The codewords of the new code are permutations of the coordinates of the codewords of the original code. We then say that the two codes are *equivalent*. Notice that equivalent codes have the same distance properties, so their error correcting capabilities are exactly the same.

By permuting the columns of the generator matrix in Example 1.2, we obtain the following generator matrix G':

$$G' = \begin{pmatrix} 1 & 0 & 0 & 1 & 1 \\ 0 & 1 & 1 & 1 & 0 \end{pmatrix} \tag{1.3}$$

The matrix G' defines a systematic encoder for a code that is equivalent to the one given in Example 1.1. For instance, the information vector 11 is encoded into 11 101.

The second important matrix related to a code is the so called *parity check* matrix. We say that an $(n-k) \times n$ matrix H is a parity check matrix of an $[n,k]$ code C if and only if, for any $\underline{c} \in C$,

$$\underline{c}H^T = \underline{0} \tag{1.4}$$

where H^T denotes the transpose of matrix H and $\underline{0}$ is a zero vector of length $n - k$. We say that the parity check matrix H is in systematic form if

$$H = (W \mid I_{n-k}) \tag{1.5}$$

where I_{n-k} is the $(n - k) \times (n - k)$ identity matrix and W is an $(n - k) \times k$ matrix.

Given a systematic generator matrix G of a code $\mathcal{C}$, it is easy to find the systematic parity check matrix H (and conversely). Explicitly, if G is given by Equation 1.2, H is given by

$$H = (V^T \mid I_{n-k}) \tag{1.6}$$

We leave the proof of this fact to the reader.

For example, the systematic parity check matrix of the code whose systematic generator matrix is given by Equation 1.3, is

$$H = \begin{pmatrix} 0 & 1 & 1 & 0 & 0 \\ 1 & 1 & 0 & 1 & 0 \\ 1 & 0 & 0 & 0 & 1 \end{pmatrix} \tag{1.7}$$

We state now an important property of parity check matrices.

Lemma 1.3 *Let $\mathcal{C}$ be a linear $[n, k, d]$ code and H a parity-check matrix. Then, any $d - 1$ columns of H are linearly independent.*

Proof 1.2 Numerate the columns of H from 0 to $n - 1$. Assume that columns $0 \leq i_1 < i_2 < \cdots < i_m \leq n - 1$ are linearly dependent, where $m \leq d - 1$. Without loss of generality, we may assume that the sum of these columns is equal to the column vector zero. Let $\underline{v}$ be a vector of length n whose non-zero coordinates are in locations $i_1, i_2, \ldots, i_m$. Then, we have

$$\underline{v} H^T = \underline{0}$$

hence $\underline{v}$ is in $\mathcal{C}$. But $\underline{v}$ has weight $m \leq d - 1$, contradicting the fact that $\mathcal{C}$ has minimum distance d. $\square$

Corollary 1.1 For any linear $[n, k, d]$ code, the minimum distance d is the smallest number m such that there is a subset of m linearly dependent columns.

Proof 1.3 It follows immediately from Lemma 1.3. $\square$

Corollary 1.2 (Singleton Bound) For any linear $[n, k, d]$ code,

$$d \leq n - k + 1$$

Proof 1.4 Notice that, since H is an $(n - k) \times n$ matrix, any $n - k + 1$ columns are going to be linearly dependent, so if $d > n - k + 1$, we would contradict Corollary 1.1. $\square$

Codes meeting the Singleton bound are called maximum distance separable (MDS). In fact, except for trivial cases, binary codes are not MDS. In order to obtain MDS codes, we will define codes over larger fields, like the so-called Reed Solomon codes, to be described later in the chapter.

We also give a second bound relating the redundancy and the minimum distance of an $[n, k, d]$ code: the so-called Hamming or volume bound. Let us denote by $V(r)$ the number of elements in a sphere of radius r whose center is an element in $GF(2)^n$. It is easy to verify that

$$V(r) = \sum_{i=0}^{r} \binom{n}{i} \tag{1.8}$$

We then have:

Lemma 1.4 (Hamming bound)   *Let C be a linear $[n, k, d]$ code, then*

$$n - k \geq \log_2 V(\lfloor (d - 1)/2 \rfloor) \tag{1.9}$$

Proof **1.5**   Notice that the 2^k spheres with the 2^k codewords as centers and radius $\lfloor (d - 1)/2 \rfloor$ are disjoint. The total number of vectors contained in these spheres is $2^k V(\lfloor (d - 1)/2 \rfloor)$. This number has to be smaller than or equal to the total number of vectors in the space, that is,

$$2^n \geq 2^k V(\lfloor (d - 1)/2 \rfloor) \tag{1.10}$$

Inequality 1.9 follows immediately from Inequality 1.10.   □

A *perfect* code is a code for which Inequality 1.9 is in effect equality. Geometrically, a perfect code is a code for which the 2^k spheres of radius $\lfloor (d - 1)/2 \rfloor$ and the codewords as centers cover the whole space.

There are not many perfect codes. In the binary case, the only nontrivial linear perfect codes are the Hamming codes (to be presented in the next section) and the $[23, 12, 7]$ Golay code (see [4]).

1.3   Syndrome Decoding, Hamming Codes, and Capacity of the Channel

In this section, we study the first important family of codes, the so called Hamming codes. As we will see, Hamming codes can correct up to one error.

Let C be an $[n, k, d]$ code with parity check matrix H. Let $\underline{u}$ be a transmitted vector and $\underline{r}$ a possibly corrupted received version of $\underline{u}$. We say that the syndrome of $\underline{r}$ is the vector $\underline{s}$ of length $n - k$ given by

$$\underline{s} = \underline{r} H^T \tag{1.11}$$

Notice that, if no errors occurred, the syndrome of $\underline{r}$ is the zero vector. The syndrome, however, tells us more than a vector being in the code or not. Say, as before, that $\underline{u}$ was transmitted and $\underline{r}$ was received, where $\underline{r} = \underline{u} \oplus \underline{e}$, $\underline{e}$ an error vector. Notice that,

$$\underline{s} = \underline{r} H^T = (\underline{u} \oplus \underline{e}) H^T = \underline{u} H^T \oplus \underline{e} H^T = \underline{e} H^T$$

since $\underline{u}$ is in C. Hence, the syndrome does not depend on the received vector but on the error vector. In the next lemma, we show that to every error vector of weight $\leq (d - 1)/2$ corresponds a unique syndrome.

Lemma 1.5   *Let C be a linear $[n, k, d]$ code with parity check matrix H. Then, there is a 1-1 correspondence between errors of weight $\leq (d - 1)/2$ and syndromes.*

Proof **1.6**   Let $\underline{e}_1$ and $\underline{e}_2$ be two distinct error vectors of weight $\leq (d - 1)/2$ with syndromes $\underline{s}_1 = \underline{e}_1 H^T$ and $\underline{s}_2 = \underline{e}_2 H^T$. If $\underline{s}_1 = \underline{s}_2$, then $\underline{s} = (\underline{e}_1 \oplus \underline{e}_2) H^T = \underline{s}_1 \oplus \underline{s}_2 = \underline{0}$, hence $\underline{e}_1 \oplus \underline{e}_2 \in C$. But $\underline{e}_1 \oplus \underline{e}_2$ has weight $\leq d - 1$, a contradiction.   □

Lemma 1.5 gives the key for a decoding method that is more efficient than exhaustive search. We can construct a table with the 1-1 correspondence between syndromes and error patterns of weight $\leq (d-1)/2$ and decode by look-up table. In other words, given a received vector, we first find its syndrome and then we look in the table to which error pattern it corresponds. Once we obtain the error pattern, we add it to the received vector, retrieving the original information. This procedure may be efficient for small codes, but it is still too complex for large codes.

Example 1.3

Consider the code whose parity matrix H is given by (7). We have seen that this is a $[5, 2, 3]$ code. We have 6 error patterns of weight ≤ 1. The 1-1 correspondence between these error patterns and the syndromes, can be immediately verified to be

$$
\begin{array}{ccc}
00000 & \leftrightarrow & 000 \\
10000 & \leftrightarrow & 011 \\
01000 & \leftrightarrow & 110 \\
00100 & \leftrightarrow & 100 \\
00010 & \leftrightarrow & 010 \\
00001 & \leftrightarrow & 001
\end{array}
$$

For instance, assume that we receive the vector $\underline{r} = 10111$. We obtain the syndrome $\underline{s} = \underline{r}H^T = 100$. Looking at the table above, we see that this syndrome corresponds to the error pattern $\underline{e} = 00100$. Adding this error pattern to the received vector, we conclude that the transmitted vector was $\underline{r} \oplus \underline{e} = 10011$.

Given a number r or redundant bits, we say that a $[2^r - 1, 2^r - r - 1, 3]$ Hamming code is a code having an $r \times (2^r - 1)$ parity check matrix H such that its columns are all the different nonzero vectors of length r.

A Hamming code has minimum distance 3. This follows from its definition and Corollary 1.1: notice that any two columns in H, being different, are linearly independent. Also, if we take any two different columns and their sum, these three columns are linearly dependent, proving our assertion.

A natural way of writing the columns of H in a Hamming code, is by considering them as binary numbers on base 2 in increasing order. This means, the first column is 1 on base 2, the second columns is 2 and so on. The last column is $2^r - 1$ on base 2, that is, $(1, 1, \ldots, 1)^T$. This parity check matrix, although nonsystematic, makes the decoding very simple.

In effect, let $\underline{r}$ be a received vector such that $\underline{r} = \underline{v} \oplus \underline{e}$, where $\underline{v}$ was the transmitted codeword and $\underline{e}$ is an error vector of weight 1. Then, the syndrome is $\underline{s} = \underline{e}H^T$, which gives the column corresponding to the location in error. This column, as a number on base 2, tells us exactly where the error has occurred, so the received vector can be corrected.

Example 1.4

Consider the $[7, 4, 3]$ Hamming code C with parity check matrix

$$
H = \begin{pmatrix} 0 & 0 & 0 & 1 & 1 & 1 & 1 \\ 0 & 1 & 1 & 0 & 0 & 1 & 1 \\ 1 & 0 & 1 & 0 & 1 & 0 & 1 \end{pmatrix} \tag{1.12}
$$

Assume that vector $\underline{r} = 1100101$ is received. The syndrome is $\underline{s} = \underline{r}H^T = 001$, which is the binary representation of the number 1. Hence, the first location is in error, so the decoder estimates that the transmitted vector was $\underline{v} = 0100101$.

We can obtain 1-error correcting codes of any length simply by shortening a Hamming code. This procedure works as follows: Assume that we want to encode k information bits into a 1-error correcting code. Let r be the smallest number such that $k \leq 2^r - r - 1$. Let H be the parity-check matrix of a $[2^r - 1, 2^r - r - 1, 3]$ Hamming code. Then construct a matrix H' by eliminating some $2^r - r - 1 - k$ columns from H. The code whose parity-check matrix is H' is a $[k + r, k, d]$ code with $d \geq 3$, hence it can correct one error. We call it a shortened Hamming code. For instance, the $[5, 2, 3]$ code whose parity-check matrix is given by Equation 1.7, is a shortened Hamming code.

In general, if H is the parity-check matrix of a code $\mathcal{C}$, H' is a matrix obtained by eliminating a certain number of columns from H and $\mathcal{C}'$ is the code with parity-check matrix H', we say that $\mathcal{C}'$ is obtained by shortening $\mathcal{C}$.

A $[2^r - 1, 2^r - r - 1, 3]$ Hamming code can be extended to a $[2^r, 2^r - r - 1, 4]$ Hamming code by adding to each codeword a parity bit that is the exclusive-OR of the first $2^r - 1$ bits. The new code is called an extended Hamming code.

So far we have not talked about probabilities of errors. Assume that we have a binary symmetric channel (BSC), that is, the probability of a 1 becoming a 0 or of a 0 becoming a 1 is $p < .5$. Let P_{err} be the probability of error after decoding using a code, that is, the output of the decoder does not correspond to the originally transmitted information vector. A fundamental question is the following: given a BSC with bit error probability p, does it exist a code of high rate that can arbitrarily lower P_{err}? The answer, due to Shannon, is yes, provided that the code has rate below a parameter called the capacity of the channel, as defined next:

Definition 1.2   Given a BSC with probability of bit error p, we say that the capacity of the channel is

$$\mathcal{C}(p) = 1 + p \log_2 p + (1 - p) \log_2 (1 - p) \tag{1.13}$$

Theorem 1.1 (Shannon)   *For any $\epsilon > 0$ and $R < \mathcal{C}(p)$, there is an $[n, k]$ binary code of rate $k/n \geq R$ with $P_{err} < \epsilon$.*

For a proof of Theorem 1.1 and some of its generalizations, the reader is referred to [5], or even to Shannon's original paper [6].

Theorem 1.1 has enormous theoretical importance: it shows that reliable communication is not limited in the presence of noise, only the rate of communication is. For instance, if $p = 0.01$, the capacity of the channel is $\mathcal{C}(0.01) = 0.9192$. Hence, there are codes of rate ≥ 0.9 with P_{err} arbitrarily small. It also tells us not to look for codes with rate 0.92 making P_{err} arbitrarily small.

The proof of Theorem 1.1, though, is based on probabilistic methods and the assumption of arbitrarily large values of n. In practical applications, n cannot be too large. The theorem does not tell us how to construct efficient codes, it just asserts their existence. Moreover, when we construct codes, we want them to have efficient encoding and decoding algorithms. In the last few years, coding methods approaching the Shannon limit have been developed, the so called *turbo codes*. Although great progress has been made towards practical implementations of turbo codes, in applications like magnetic recording their complexity is still a problem. A description of turbo codes is beyond the scope of this introduction. We refer the reader to [2].

1.4   Codes over Bytes and Finite Fields

So far, we have considered linear codes over bits. Next we want to introduce codes over larger symbols, mainly over bytes. A byte of size v is a vector of v bits. Mathematically, bytes are vectors in $GF(2)^v$. Typical cases in magnetic and optical recording involve 8-bit bytes. Most of the general results in the previous sections for codes over bits easily extend to codes over bytes. It is trivial to multiply bits, but we need a method to multiply bytes. To this end, the theory of finite fields has been developed. Next we give a brief introduction to the theory of finite fields.

We know how to add two binary vectors: we simply exclusive-OR them componentwise. What we need now is a rule that allows us to multiply bytes while preserving associative, distributive, and multiplicative inverse properties, that is, a product that gives to the set of bytes of length v the structure of a field. To this end, we will define a multiplication between vectors that satisfies the associative and commutative properties, it has a 1 element, each nonzero element is invertible, and it is distributive with respect to the sum operation.

Recall the definition of the ring Z_m of integers modulo m: Z_m is the set $\{0, 1, 2, \ldots, m - 1\}$, with a sum and product of any two elements defined as the residue of dividing by m the usual sum or product. It is not difficult to prove that Z_m is a field if and only if m is a prime number. Using this analogy, we will give to $(GF(2))^\nu$ the structure of a field.

Consider the vector space $(GF(2))^\nu$ over the field $GF(2)$. We can view each vector as a polynomial of degree $\leq \nu - 1$ as follows: the vector $a = (a_0, a_1, \ldots, a_{\nu-1})$ corresponds to the polynomial $a(\alpha) = a_0 + a_1\alpha + \cdots + a_{\nu-1}\alpha^{\nu-1}$.

Our goal is to give to $(GF(2))^\nu$ the structure of a field. We will denote such a field by $GF(2^\nu)$. The sum in $GF(2^\nu)$ is the usual sum of vectors in $(GF(2))^\nu$. We need now to define a product.

Let $f(x)$ be an irreducible polynomial (i.e., it cannot be expressed as the product of two polynomials of smaller degree) of degree ν whose coefficients are in $GF(2)$. Let $a(\alpha)$ and $b(\alpha)$ be two elements of $GF(2^\nu)$. We define the product between $a(\alpha)$ and $b(\alpha)$ in $GF(2^\nu)$ as the unique polynomial $c(\alpha)$ of degree $\leq \nu - 1$ such that $c(\alpha)$ is the residue of dividing the product $a(\alpha)b(\alpha)$ by $f(\alpha)$ (the notation $g(x) \equiv h(x) \pmod{f(x)}$ means that $g(x)$ and $h(x)$ have the same residue after dividing by $f(x)$, i.e., $g(\alpha) = h(\alpha)$).

The sum and product operations defined above give to $GF(2^\nu)$ a field structure. The role of the irreducible polynomial $f(x)$ is the same as the prime number m when Z_m is a field. In effect, the proof that $GF(2^\nu)$ is a field when m is irreducible is essentially the same as the proof that Z_m is a field when m is prime. From now on, we denote the elements in $GF(2^\nu)$ as polynomials in α of degree $\leq \nu - 1$ with coefficients in $GF(2)$. Given two polynomials $a(x)$ and $b(x)$ with coefficients in $GF(2)$, $a(\alpha)b(\alpha)$ denotes the product in $GF(2^\nu)$, while $a(x)b(x)$ denotes the regular product of polynomials. Notice that, for the irreducible polynomial $f(x)$, in particular, $f(\alpha) = 0$ in $GF(2^\nu)$, since $f(x) \equiv 0 \pmod{f(x)}$.

So, the set $GF(2^\nu)$ given by the irreducible polynomial $f(x)$ of degree ν, is the set of polynomials of degree $\leq \nu - 1$, where the sum operation is the regular sum of polynomials, and the product operation is the residue of dividing by $f(x)$ the regular product of two polynomials.

Example 1.5

Let us construct the field $GF(8)$. Consider the polynomials of degree ≤ 2 over $GF(2)$. Let $f(x) = 1 + x + x^3$. Since $f(x)$ has no roots over $GF(2)$, it is irreducible (notice that such an assessment can be made only for polynomials of degree 2 or 3). Let us consider the powers of α modulo $f(\alpha)$. Notice that $\alpha^3 = \alpha^3 + f(\alpha) = 1 + \alpha$. Also, $\alpha^4 = \alpha\alpha^3 = \alpha(1 + \alpha) = \alpha + \alpha^2$. Similarly, we obtain $\alpha^5 = \alpha\alpha^4 = \alpha(\alpha + \alpha^2) = \alpha^2 + \alpha^3 = 1 + \alpha + \alpha^2$, and $\alpha^6 = \alpha\alpha^5 = \alpha + \alpha^2 + \alpha^3 = 1 + \alpha^2$. Finally, $\alpha^7 = \alpha\alpha^6 = \alpha + \alpha^3 = 1$.

As we can see, every nonzero element in $GF(8)$ can be obtained as a power of the element α. In this case, α is called a *primitive* element and the irreducible polynomial $f(x)$ that defines the field is called a *primitive* polynomial. It can be proven that it is always the case that the multiplicative group of a finite field is cyclic, so there is always a primitive element. A convenient description of $GF(8)$ is given in Table 1.1.

The first column in Table 1.1 describes the element of the field in vector form, the second one as a polynomial in α of degree ≤ 2, the third one as a power of α, and the last one gives the logarithm (also called Zech logarithm): it simply indicates the corresponding power of α. As a convention, we denote by $-\infty$ the logarithm corresponding to the element 0.

TABLE 1.1   The Finite Field $GF(8)$ Generated by $1 + x + x^3$

Vector	Polynomial	Power of α	Logarithm
000	0	0	$-\infty$
100	1	1	0
010	α	α	1
001	α^2	α^2	2
110	$1 + \alpha$	α^3	3
011	$\alpha + \alpha^2$	α^4	4
111	$1 + \alpha + \alpha^2$	α^5	5
101	$1 + \alpha^2$	α^6	6

It is often convenient to express the elements in a finite field as powers of α: when we multiply two of them, we obtain a new power of α whose exponent is the sum of the two exponents modulo $2^\nu - 1$. Explicitly, if i and j are the logarithms of two elements, in $GF(2^\nu)$, then their product has logarithm $i + j \pmod{2^\nu - 1}$. In the example, if we want to multiply the vectors 101 and 111, we first look at their logarithms. They are 6 and 5, respectively, so the logarithm of the product is $6 + 5 \pmod 7 = 4$, corresponding to the vector 011.

In order to add vectors, the best way is to express them in vector form and add coordinate to coordinate in the usual way.

1.5   Cyclic Codes

In the same way we defined codes over the binary field $GF(2)$, we can define codes over any finite field $GF(2^\nu)$. Now, a code of length n is a subset of $(GF(2^\nu))^n$, but since we study only linear codes, we require that such a subset is a vector space. Similarly, we define the minimum (Hamming) distance and the generator and parity-check matrices of a code. Some properties of binary linear codes, like the Singleton bound, remain the same in the general case. Others, like the Hamming bound, require some modifications.

Consider a linear code $\mathcal{C}$ over $GF(2^\nu)$ of length n. We say that $\mathcal{C}$ is cyclic if, for any codeword $(c_0, c_1, \ldots, c_{n-1}) \in \mathcal{C}$, then $(c_{n-1}, c_0, c_1, \ldots, c_{n-2}) \in \mathcal{C}$. In other words, the code is invariant under cyclic shifts to the right.

If we write the codewords as polynomials of degree $<n$ with coefficients in $GF(2^\nu)$, this is equivalent to say that if $c(x) \in \mathcal{C}$, then $xc(x) \bmod (x^n - 1) \in \mathcal{C}$. Hence, if $c(x) \in \mathcal{C}$, then, given any polynomial $w(x)$, the residue of dividing $w(x)c(x)$ by $x^n - 1$ is in $\mathcal{C}$. In particular, if the degree of $w(x)c(x)$ is smaller than n, then $w(x)c(x) \in \mathcal{C}$.

From now on, we write the elements of a cyclic code $\mathcal{C}$ as polynomials modulo $x^n - 1$.

Theorem 1.2 $\mathcal{C}$ *is an* $[n, k]$ *cyclic code over* $GF(2^\nu)$ *if and only if there is a (monic) polynomial* $g(x)$ *of degree* $n - k$ *such that* $g(x)$ *divides* $x^n - 1$ *and each* $c(x) \in \mathcal{C}$ *is a multiple of* $g(x)$, *that is,* $c(x) \in \mathcal{C}$ *if and only if* $c(x) = w(x)g(x), \deg(w) < k$. *We call* $g(x)$ *a generator polynomial of* $\mathcal{C}$.

Proof 1.7   Let $g(x)$ be a monic (i.e., lead coefficient is 1) polynomial in $\mathcal{C}$ such that $g(x)$ has minimal degree. If $\deg(g) = 0$ (i.e., $g = 1$), then $\mathcal{C}$ is the whole space $(GF(2^\nu))^n$, so assume $\deg(g) \geq 1$. Let $c(x)$ be any element in $\mathcal{C}$. We can write $c(x) = w(x)g(x) + r(x)$, where $\deg(r) < \deg(g)$. Since $\deg(wg) < n, g \in \mathcal{C}$ and $\mathcal{C}$ is cyclic, in particular, $w(x)g(x) \in \mathcal{C}$. Hence, $r(x) = c(x) - w(x)g(x) \in \mathcal{C}$. If $r \neq 0$, we would contradict the fact that $g(x)$ has minimal degree, hence, $r = 0$ and $c(x)$ is a multiple of $g(x)$.

Similarly, we can prove that $g(x)$ divides $x^n - 1$. Let $x^n - 1 = h(x)g(x) + r(x)$, where $\deg(r) < \deg(g)$. In particular, $h(x)g(x) \equiv -r(x) \bmod (x^n - 1)$, hence, $r(x) \in \mathcal{C}$. Since $g(x)$ has minimal degree, $r = 0$, so $g(x)$ divides $x^n - 1$.

Conversely, assume that every element in $\mathcal{C}$ is a multiple of $g(x)$ and g divides $x^n - 1$. It is immediate that the code is linear and that it has dimension k. Let $c(x) \in \mathcal{C}$, hence, $c(x) = w(x)g(x)$ with $\deg(w) < k$. Also, since $g(x)$ divides $x^n - 1, x^n - 1 = h(x)g(x)$. Assume that $c(x) = c_0 + c_1 x + c_2 x^2 + \cdots + c_{n-1}x^{n-1}$, then, $xc(x) \equiv c_{n-1} + c_0 x + \cdots + c_{n-2}x^{n-1} \pmod{x^n - 1}$. We have to prove that $c_{n-1} + c_0 x + \cdots + c_{n-2}x^{n-1} = q(x)g(x)$, where $q(x)$ has degree $\leq k - 1$. Notice that

$$c_{n-1} + c_0 x + \cdots + c_{n-2}x^{n-1} = c_{n-1} + c_0 x + \cdots + c_{n-2}x^{n-1} + c_{n-1}x^n - c_{n-1}x^n$$
$$= c_0 x + \cdots + c_{n-2}x^{n-1} + c_{n-1}x^n - c_{n-1}(x^n - 1)$$
$$= xc(x) - c_{n-1}(x^n - 1)$$
$$= xw(x)g(x) - c_{n-1}h(x)g(x)$$
$$= (xw(x) - c_{n-1}h(x))g(x)$$

proving that the element is in the code.   □

Theorem 1.1 gives a method to find all cyclic codes of length n: simply take all the (monic) factors of $x^n - 1$. Each one of them is the generator polynomial of a cyclic code.

Example 1.6

Consider the $[7, 4]$ cyclic code over $GF(2)$ generated by $g(x) = 1 + x + x^3$. We can verify that $x^7 - 1 = g(x)(1 + x)(1 + x^2 + x^3)$, hence, $g(x)$ indeed generates a cyclic code.

In order to encode an information polynomial over $GF(2)$ of degree ≤ 3 into a codeword, we multiply it by $g(x)$.

Say that we want to encode $\underline{u} = (1, 0, 0, 1)$, which in polynomial form is $u(x) = 1 + x^3$. Hence, the encoding gives $c(x) = u(x)g(x) = 1 + x + x^4 + x^6$. In vector form, this gives $\underline{c} = (1 1 0 0 1 0 1)$.

It can be easily verified that the $[7, 4]$ code given in this example has minimum distance 3 and is equivalent to the Hamming code of Example 1.4. In other words, the codewords of the code given in this example are permutations of the codewords of the $[7,4,3]$ Hamming code given in Example 1.4.

The encoding method of a cyclic code with generator polynomial g is then very simple: we multiply the information polynomial by g. However, this encoder is not systematic. A systematic encoder of a cyclic code is given by the following algorithm:

Algorithm 1.1 (Systematic Encoding Algorithm for Cyclic Codes)

Let C be a cyclic $[n, k]$ code over $GF(2^v)$ with generator polynomial $g(x)$. Let $u(x)$ be an information polynomial, $\deg(u) < k$. Let $r(x)$ be the residue of dividing $x^{n-k}u(x)$ by $g(x)$. Then, $u(x)$ is encoded into the polynomial $c(x) = u(x) - x^k r(x)$.

We leave as an exercise proving that Algorithm 1.2 produces indeed a codeword in C.

Example 1.7

Consider the $[7, 4]$ cyclic code over $GF(2)$ of Example 1.6. If we want to encode systematically the information vector $\underline{u} = (1, 0, 0, 1)$(or $u(x) = 1 + x^3$), we have to obtain first the residue of dividing $x^3 u(x) = x^3 + x^6$ by $g(x)$. This residue is $r(x) = x + x^2$. Hence, the output of the encoder is $c(x) = u(x) - x^4 r(x) = 1 + x^3 + x^5 + x^6$. In vector form, this gives $\underline{c} = (1 0 0 1 0 1 1)$.

1.6 Reed Solomon Codes

Throughout this section, the codes considered are over the field $GF(2^v)$. Let α be a primitive element in $GF(2^v)$, that is, $\alpha^{2^v-1} = 1, \alpha^i \neq 1$ for $i \neq 0 \bmod 2^v - 1$. A Reed Solomon (RS) code of length $n = 2^v - 1$ and dimension k is the cyclic code generated by

$$g(x) = (x - \alpha)(x - \alpha^2) \cdots (x - \alpha^{n-k-1})(x - \alpha^{n-k})$$

Since each α^i is a root of unity, $x - \alpha^i$ divides $x^n - 1$, hence g divides $x^n - 1$ and the code is cyclic.

An equivalent way of describing a RS code, is as the set of polynomials over $GF(2^v)$ of degree $\leq n - 1$ with roots $\alpha, \alpha^2, \ldots, \alpha^{n-k}$, that is, F is in the code if and only if $\deg(F) \leq n - 1$ and $F(\alpha) = F(\alpha^2) = \cdots = F(\alpha^{n-k}) = 0$.

This property allows us to find a parity check matrix for a RS code. Say that $F(x) = F_0 + F_1 x + \cdots + F_{n-1}x^{n-1}$ is in the code. Let $1 \leq i \leq n - k$, then

$$F(\alpha^i) = F_0 + F_1\alpha^i + \cdots + F_{n-1}\alpha^{i(n-1)} = 0 \tag{1.14}$$

In other words, Equation 5.14 tells us that codeword $(F_0, F_1, \ldots, F_{n-1})$ is orthogonal to the vectors $(1, \alpha^i, \alpha^{2i}, \ldots, \alpha^{i(n-1)}), 1 \leq i \leq n - k$. Hence these vectors are the rows of a parity check matrix for the

RS code. A parity check matrix of an $[n, k]$ RS code over $GF(2^v)$ is then

$$H = \begin{pmatrix} 1 & \alpha & \alpha^2 & \cdots & \alpha^{n-1} \\ 1 & \alpha^2 & \alpha^4 & \cdots & \alpha^{2(n-1)} \\ \vdots & \vdots & \vdots & \ddots & \vdots \\ 1 & \alpha^{n-k} & \alpha^{(n-k)2} & \cdots & \alpha^{(n-k)(n-1)} \end{pmatrix} \qquad (1.15)$$

In order to show that H is in fact a parity check matrix, we need to prove that the rows of H are linearly independent. The next lemma provides an even stronger result.

Lemma 1.6 *Any set of $n - k$ columns in matrix H defined by Equation 1.15 is linearly independent.*

Proof 1.8 Take a set $0 \le i_1 < i_2 < \cdots < i_{n-k} \le n-1$ of columns of H. Denote α^{i_j} by α_j, $1 \le j \le n-k$. Columns $i_1, i_2, \ldots, i_{n-k}$ are linearly independent if and only if their determinant is nonzero, that is, if and only if

$$\det \begin{pmatrix} \alpha_1 & \alpha_2 & \cdots & \alpha_{n-k} \\ (\alpha_1)^2 & (\alpha_2)^2 & \cdots & (\alpha_{n-k})^2 \\ \vdots & \vdots & \ddots & \vdots \\ (\alpha_1)^{n-k} & (\alpha_2)^{n-k} & \cdots & (\alpha_{n-k})^{n-k} \end{pmatrix} \ne 0 \qquad (1.16)$$

Let

$$V(\alpha_1, \alpha_2, \ldots, \alpha_{n-k}) = \det \begin{pmatrix} 1 & 1 & \cdots & 1 \\ \alpha_1 & \alpha_2 & \cdots & \alpha_{n-k} \\ \vdots & \vdots & \ddots & \vdots \\ (\alpha_1)^{n-k-1} & (\alpha_2)^{n-k-1} & \cdots & (\alpha_{n-k})^{n-k-1} \end{pmatrix} \qquad (1.17)$$

We call the determinant $V(\alpha_1, \alpha_2, \ldots, \alpha_{n-k})$ a *Vandermonde determinant*: it is the determinant of an $(n-k) \times (n-k)$ matrix whose rows are the powers of vector $\alpha_1, \alpha_2, \ldots, \alpha_{n-k}$, the powers running from 0 to $n - k - 1$. By properties of determinants, if we consider the determinant in Equation 1.16, we have

$$\det \begin{pmatrix} \alpha_1 & \alpha_2 & \cdots & \alpha_{n-k} \\ (\alpha_1)^2 & (\alpha_2)^2 & \cdots & (\alpha_{n-k})^2 \\ \vdots & \vdots & \ddots & \vdots \\ (\alpha_1)^{n-k} & (\alpha_2)^{n-k} & \cdots & (\alpha_{n-k})^{n-k} \end{pmatrix} = \alpha_1 \alpha_2 \ldots \alpha_{n-k} V(\alpha_1, \alpha_2, \ldots, \alpha_{n-k}). \qquad (1.18)$$

Hence, by Equation 1.16 and Equation 1.18, since the α_j's are nonzero, it is enough to prove that $V(\alpha_1, \alpha_2, \ldots, \alpha_{n-k}) \ne 0$. A well known result in literature states that

$$V(\alpha_1, \alpha_2, \ldots, \alpha_{n-k}) = \prod_{1 \le i < j \le n-k} (\alpha_j - \alpha_i) \qquad (1.19)$$

Since α is a primitive element in $GF(2^v)$, its powers $\alpha^l, 0 \le l \le n - 1$ are distinct. In particular, the α_i's, $l \le i \le n - k$ are distinct, hence, the product in the right hand side of Equation 1.19 nonzero. □

Corollary 1.3 *An $[n, k]$ RS code has minimum distance $d = n - k + 1$.*

Proof 1.9 Let H be the parity check matrix of the RS code defined by Equation 1.15. Notice that, since *any* $n - k$ columns in H are linearly independent, $d \geq n - k + 1$ by Lemma 1.3.

On the other hand, $d \leq n - k + 1$ by the Singleton bound (Corollary 1.2), so we have equality. □

Since RS codes meet the Singleton bound with equality, they are MDS (see Section 1.2).

Example 1.8

Consider the $[7, 3, 5]$ RS code over $GF(8)$, where $GF(8)$ is given by Table 1.1. The generator polynomial is

$$g(x) = (x - \alpha)(x - \alpha^2)(x - \alpha^3)(x - \alpha^4) = \alpha^3 + \alpha x + x^2 + \alpha^3 x^3 + x^4$$

Assume that we want to encode the 3 byte vector $\underline{u} = 101\,001\,111$. Writing the bytes as powers of α in polynomial form, we have $u(x) = \alpha^6 + \alpha^2 x + \alpha^5 x^2$.

In order to encode $u(x)$, we perform

$$u(x)g(x) = \alpha^2 + \alpha^4 x + \alpha^2 x^2 + \alpha^6 x^3 + \alpha^6 x^4 + \alpha^4 x^5 + \alpha^5 x^6$$

In vector form the output of the encoder is given by $001\,011\,001\,101\,101\,011\,111$. If we encode $u(x)$ using a systematic encoder (Algorithm 1.1), then the output of the encoder is

$$\alpha^6 + \alpha^2 x + \alpha^5 x^2 + \alpha^6 x^3 + \alpha^5 x^4 + \alpha^4 x^5 + \alpha^4 x^6$$

which in vector form is $101\,001\,111\,101\,111\,011\,011$.

Next we make some observations:

- The definition given above for an $[n, k]$ Reed Solomon code states that $F(x)$ is in the code if and only if it has as roots the powers $\alpha, \alpha^2, \ldots, \alpha^{n-k}$ of a primitive element α. However, it is enough to state that F has as roots a set of *consecutive* powers of α, say, $\alpha^m, \alpha^{m+1}, \ldots, \alpha^{m+n-k-1}$, where $0 \leq m \leq n - 1$. Although our definition (i.e., $m = 1$) gives the most usual setting for RS codes, often engineering reasons may determine different choices of m. It is easy to verify that with the more general definition of RS codes, the minimum distance remains $n - k + 1$.
- Given an $[n, k]$ RS code, there is an easy way to shorten it and obtain an $[n - l, k - l]$ code for $l < k$. In effect, if we have only $k - l$ bytes of information, we add l zeroes in order to obtain an information string of length k. We then find the $n - k$ redundant bytes using a systematic encoder. When writing, of course, the l zeroes are not written, so we have an $[n - l, k - l]$ code, called a shortened RS code. It is immediately verified that shortened RS codes are also MDS.

We have defined RS codes, proven that they are MDS and showed how to encode them systematically. The next step, to be developed in the next sections, is decoding them.

1.7 Decoding of RS Codes: The Key Equation

Through this section C denotes an $[n, k]$ RS code (unless otherwise stated). Assume that a codeword $F(x) = \sum_{i=0}^{n-1} F_i x^i$ in C is transmitted and a word $R(x) = \sum_{i=0}^{n-1} R_i x^i$ is received; hence, F and R are related by an error vector $E(x) = \sum_{i=0}^{n-1} E_i x^i$, where $R(x) = F(x) + E(x)$. The decoder will attempt to find $E(x)$.

Let us start by computing the syndromes. For $1 \leq j \leq n - k$, we have

$$S_j = R(\alpha^j) = \sum_{i=0}^{n-1} R_i \alpha^{ij} = \sum_{i=0}^{n-1} E_i \alpha^{ij} \tag{1.20}$$

Before proceeding further, consider Equation 1.20 in a particular case.

Take the $[n, n - 2]$ 1-byte correcting RS code. In this case, we have two syndromes S_1 and S_2, so, if exactly one error has occurrred, say in location i, by Equation 1.20, we have

$$S_1 = E_i\alpha^i \quad \text{and} \quad S_2 = E_i\alpha^{2i} \tag{1.21}$$

Hence, $\alpha^i = S_2/S_1$, so we can determine the location i in error. The error value is $E_i = (S_1)^2/S_2$.

Example 1.9

Consider the $[7, 5, 3]$ RS code over $GF(8)$, where $GF(8)$ is given by Table 1.1. Assume that we want to decode the received vector.

$$\underline{r} = (101\,001\,110\,001\,011\,010\,100)$$

which in polynomial form is

$$R(x) = \alpha^6 + \alpha^2 x + \alpha^3 x^2 + \alpha^2 x^3 + \alpha^4 x^4 + \alpha x^5 + x^6$$

Evaluating the syndromes, we obtain $S_1 = R(\alpha) = \alpha^2$ and $S_2 = R(\alpha^2) = \alpha^4$. Thus, $S_2/S_1 = \alpha^2$, meaning that location 2 is in error. The error value is $E_2 = (S_1)^2/S_2 = (\alpha^2)^2/\alpha^4 = 1$, which in vector form is 100. The output of the decoder is then

$$\underline{c} = (101\,001\,010\,001\,011\,010\,100)$$

which in polynomial form is

$$C(x) = \alpha^6 + \alpha^2 x + \alpha x^2 + \alpha^2 x^3 + \alpha^4 x^4 + \alpha x^5 + x^6$$

Let $\mathcal{E}$ be the subset of $\{0, 1, \ldots, n - 1\}$ of locations in error, that is, $\mathcal{E} = \{l : E_l \neq 0\}$. With this notation, Equation 1.20 becomes

$$S_j = \sum_{i \in \mathcal{E}} E_i\alpha^{ij}, \quad 1 \leq j \leq n - k \tag{1.22}$$

The decoder will find the error set $\mathcal{E}$ and the error values E_i when the error correcting capability of the code is not exceeded. Thus, if s is the number of errors and $2s \leq n - k$, the system of equations given by Equation 1.22 has a unique solution. However, this is a nonlinear system, and it is very difficult to solve it directly.

In order to find the set of locations in error $\mathcal{E}$ and the corresponding error values $\{E_i : i \in \mathcal{E}\}$, we define two polynomials. The first one is called the *error locator polynomial*, which is the polynomial that has as roots the values α^{-i}, where $i \in \mathcal{E}$. We denote this polynomial by $\sigma(x)$. Explicitly,

$$\sigma(x) = \prod_{i \in \mathcal{E}} (x - \alpha^{-i}) \tag{1.23}$$

If somehow we can determine the polynomial $\sigma(x)$, by finding its roots, we can obtain the set $\mathcal{E}$ of locations in error. Once we have the set of locations in error, we need to find the errors themselves. We define a second polynomial, called the *error evaluator polynomial* and denoted by $w(x)$, as follows:

$$w(x) = \sum_{i \in \mathcal{E}} E_i \prod_{\substack{l \in \mathcal{E} \\ l \neq i}} (x - \alpha^{-l}) \tag{1.24}$$

Since an $[n, k]$ RS code corrects at most $(n - k)/2$ errors, we assume that $|\mathcal{E}| = \deg(\sigma) \leq (n - k)/2$. Notice also that $\deg(w) \leq |\mathcal{E}| - 1$, since w is a sum of polynomials of degree $|\mathcal{E}| - 1$. Given a polynomial $f(x) = a_0 + a_1 x + \cdots + a_m x^m$ with coefficients over a field F, we define the (formal) derivative of f, denoted f', as the polynomial

$$f'(x) = a_1 + 2a_2 x + \cdots + m a_m x^{m-1}$$

For instance, over $GF(8)$, if $f(x) = \alpha + \alpha^3 x + \alpha^4 x^2$, then $f'(x) = \alpha^3$ (since $2 = 0$ over $GF(2)$). The formal derivative has several properties similar to the traditional derivative, like the derivative of a product, $(fg)' = f'g + fg'$. Back to the error locator and error evaluator polynomials, we have the following relationship between the two:

$$E_i = \frac{w(\alpha^{-i})}{\sigma'(\alpha^{-i})} \tag{1.25}$$

Let us prove some of these facts in the following lemma:

Lemma 1.7 *The polynomials $\sigma(x)$ and $w(x)$ are relatively prime, and the error values E_i are given by Equation 1.25*

Proof 1.10 In order to show that $\sigma(x)$ and $w(x)$ are relatively prime, it is enough to observe that they have no roots in common. In effect, if α^{-j} is a root of $\sigma(x)$, then $j \in \mathcal{E}$. By Equation 1.24,

$$w(\alpha^{-j}) = \sum_{i \in \mathcal{E}} E_i \prod_{\substack{l \in \mathcal{E} \\ l \neq i}} (\alpha^{-j} - \alpha^{-l}) = E_j \prod_{\substack{l \in \mathcal{E} \\ l \neq j}} (\alpha^{-j} - \alpha^{-l}) \neq 0 \tag{1.26}$$

Hence, $\sigma(x)$ and $w(x)$ are relatively prime.

In order to prove Equation 1.25, notice that

$$\sigma'(x) = \sum_{i \in \mathcal{E}} \prod_{\substack{l \in \mathcal{E} \\ l \neq i}} (x - \alpha^{-l})$$

hence,

$$\sigma'(\alpha^{-j}) = \prod_{\substack{l \in \mathcal{E} \\ l \neq j}} (\alpha^{-j} - \alpha^{-l}) \tag{1.27}$$

By Equation 1.26 and Equation 1.27, Equation 1.25 follows. $\square$

The decoding methods of RS codes are based on finding the error locator and the error evaluator polynomials. By finding the roots of the error locator polynomial, we determine the locations in error, while the errors themselves can be found using Equation 1.25. We will establish a relationship between $\sigma(x)$ and $w(x)$, but first we need to define a third polynomial, the syndrome polynomial. We define the syndrome polynomial as the polynomial of degree $\leq n - k - 1$ whose coefficients are the $n - k$ syndromes. Explicitly,

$$S(x) = S_1 + S_2 x + S_3 x^2 + \cdots + S_{n-k} x^{n-k-1} = \sum_{j=0}^{n-k-1} S_{j+1} x^j \tag{1.28}$$

Notice that $R(x)$ is in $\mathcal{C}$ if and only if $S(x) = 0$.

The next theorem gives the so called *key equation* for decoding RS codes, and it establishes a fundamental relationship between $\sigma(x), w(x)$ and $S(x)$.

Theorem 1.3 *There is a polynomial $\mu(x)$ such that the error locator, the error evaluator, and the syndrome polynomials verify the following equation:*

$$\sigma(x)S(x) = -w(x) + \mu(x)x^{n-k} \tag{1.29}$$

Alternatively, Equation 1.29 can be written as a congruence as follows:

$$\sigma(x)S(x) \equiv -w(x) \pmod{x^{n-k}} \tag{1.30}$$

Proof 1.11   By Equation 1.28 and Equation 1.22, we have

$$
\begin{aligned}
S(x) &= \sum_{j=0}^{n-k-1} S_{j+1}x^j \\
&= \sum_{j=0}^{n-k-1} \left(\sum_{i \in \mathcal{E}} E_i \alpha^{i(j+1)} \right) x^j \\
&= \sum_{i \in \mathcal{E}} E_i \alpha^i \sum_{j=0}^{n-k-1} (\alpha^i x)^j \\
&= \sum_{i \in \mathcal{E}} E_i \alpha^i \frac{(\alpha^i x)^{n-k} - 1}{\alpha^i x - 1} \\
&= \sum_{i \in \mathcal{E}} E_i \frac{(\alpha^i x)^{n-k} - 1}{x - \alpha^{-i}}
\end{aligned}
\tag{1.31}
$$

since $\sum_{l=0}^m a^l = (a^{m+1} - 1)/(a-1)$ for $a \neq 1$. Multiplying both sides of Equation 1.31 by $\sigma(x)$, where $\sigma(x)$ is given by Equation 1.23, we obtain

$$
\sigma(x)S(x) = \sum_{i \in \mathcal{E}} E_i((\alpha^i x)^{n-k} - 1) \prod_{\substack{l \in \mathcal{E} \\ l \neq i}} (x - \alpha^{-l})
$$

$$
= -\sum_{i \in \mathcal{E}} E_i \prod_{\substack{l \in \mathcal{E} \\ l \neq i}} (x - \alpha^{-l}) + \left(\sum_{i \in \mathcal{E}} E_i \alpha^{i(n-k)} \prod_{\substack{l \in \mathcal{E} \\ l \neq i}} (x - \alpha^{-l}) \right) x^{n-k}
$$

$$
= -\omega(x) + \mu(x) x^{n-k}
$$

since $\omega(x)$ is given by Equation 1.24. This completes the proof.   □

The decoding methods for RS codes concentrate on solving the key equation. In the next section we describe an efficient decoder based on Euclid's algorithm for polynomials. Another efficient decoding algorithm is the so-called Berlekamp-Massey decoding algorithm [1].

1.8   Decoding RS Codes with Euclid's Algorithm

Given two polynomials or integers A and B, Euclid's algorithm provides a recursive procedure to find the greatest common divisor C between A and B, denoted $C = \gcd(A, B)$. Moreover, the algorithm also finds two polynomials or integers S and T such that $C = SA + TB$.

Recall that we want to solve the key equation

$$\mu(x)x^{n-k} + \sigma(x)S(x) = -\omega(x)$$

In the recursion, x^{n-k} will play the role of A and $S(x)$ the role of B; $\sigma(x)$ and $\omega(x)$ will be obtained at a certain step of the recursion.

Let us describe Euclid's algorithm for integers or polynomials. Consider A and B such that $A \geq B$ if they are integers and $\deg(A) \geq \deg(B)$ if they are polynomials. We start from the initial conditions $r_{-1} = A$ and $r_0 = B$.

We perform a recursion in steps $1, 2, \ldots, i, \ldots$. At step i of the recursion, we obtain r_i as the residue of dividing r_{i-2} by r_{i-1}, that is , $r_{i-2} = q_i r_{i-1} + r_i$, where $r_i < r_{i-1}$ for integers and $\deg(r_i) < \deg(r_{i-1})$ for polynomials. The recursion is then given by

$$r_i = r_{i-2} - q_i r_{i-1} \tag{1.32}$$

We also obtain values s_i and t_i such that $r_i = s_i A + t_i B$. Hence, the same recursion is valid for s_i and t_i as well:

$$s_i = s_{i-2} - q_i s_{i-1} \tag{1.33}$$

$$t_i = t_{i-2} - q_i t_{i-1} \tag{1.34}$$

Since $r_{-1} = A = (1)A + (0)B$ and $r_0 = B = (0)A + (1)B$, we set the initial conditions $s_{-1} = 1, t_{-1} = 0, s_0 = 0$ and $t_0 = 1$.

Let us illustrate the process with $A = 124$ and $B = 46$. We will find $\gcd(124, 46)$. The idea is to divide recursively by the residues of the division until obtaining a last residue 0. Then, the last divisor is the gcd. The procedure works as follows:

$$
\begin{aligned}
124 &= \quad (1)124 + \quad (0)46 \\
46 &= \quad (0)124 + \quad (1)46 \\
32 &= \quad (1)124 + (-2)46 \\
14 &= (-1)124 + \quad (3)46 \\
4 &= \quad (3)124 + (-8)46 \\
2 &= (-10)124 + \quad (27)46
\end{aligned}
$$

Since 2 divides 4, 2 is the greatest common divisor between 124 and 46.

The best way to develop the process above, is to construct a table for r_i, q_i, s_i and t_i using the initial conditions and recursions from Equation 1.32 through Equation 1.34.

Let us do it again for 124 and 46.

i	r_i	q_i	$s_i = s_{i-2} - q_i s_{i-1}$	$t_i = t_{i-2} - q_i t_{i-1}$
-1	124		1	0
0	46		0	1
1	32	2	1	-2
2	14	1	-1	3
3	4	2	3	-8
4	2	3	-10	27
5	0	2	23	-62

From now on, let us concentrate on Euclid's algorithm for polynomials. If we want to solve the key equation

$$\mu(x)x^{n-k} + \sigma(x)S(x) = -\omega(x)$$

and the error correcting capability of the code has not been exceeded, then applying Euclid's algorithm to x^{n-k} and to $S(x)$, at a certain point of the recursion, we obtain

$$r_i(x) = s_i(x)x^{n-k} + t_i(x)S(x)$$

where $\deg(r_i) \leq \lfloor (n - k)/2 \rfloor - 1$, and i is the first with this property. Then, $\omega(x) = -\lambda r_i(x)$ and $\sigma(x) = \lambda t_i(x)$, where λ is a constant that makes $\sigma(x)$ monic. For a proof that Euclid's algorithm gives the right solution, see [1] or [5].

We illustrate the decoding of RS codes using Euclid's algorithm with an example. Notice that we are interested in $r_i(x)$ and $t_i(x)$ only.

Example 1.10

Consider the [7, 3, 5] RS code over $GF(8)$, and assume that we want to decode the received vector

$$\underline{r} = (011\ 101\ 111\ 111\ 111\ 101\ 010)$$

which in polynomial form is

$$R(x) = \alpha^4 + \alpha^6 x + \alpha^5 x^2 + \alpha^5 x^3 + \alpha^5 x^4 + \alpha^6 x^5 + \alpha x^6$$

Evaluating the syndromes, we obtain

$$S_1 = R(\alpha) = \alpha^5$$
$$S_2 = R(\alpha^2) = \alpha$$
$$S_3 = R(\alpha^3) = 0$$
$$S_4 = R(\alpha^4) = \alpha^3$$

Therefore, the syndrome polynomial is $S(x) = \alpha^5 + \alpha x + \alpha^3 x^3$.

Next, we apply Euclid's algorithm with respect to x^4 and to $S(x)$. When we find the first i for which $r_i(x)$ has degree ≤ 1, we stop the algorithm and we obtain $\omega(x)$ and $\sigma(x)$. The process is tabulated below.

i	$r_i = r_{i-2} - q_i r_{i-1}$	q_i	$t_i = t_{i-2} - q_i t_{i-1}$
-1	x^4		0
0	$\alpha^5 + \alpha x + \alpha^3 x^3$		1
1	$\alpha^2 x + \alpha^5 x^2$	$\alpha^4 x$	$\alpha^4 x$
2	$\alpha^5 + \alpha^2 x$	$\alpha^2 + \alpha^5 x$	$1 + \alpha^6 x + \alpha^2 x^2$

So, for $i = 2$, we obtain a polynomial $r_2(x) = \alpha^5 + \alpha^2 x$ of degree 1. Now, multiplying both $r_2(x)$ and $t_2(x)$ by $\lambda = \alpha^5$, we obtain $\omega(x) = \alpha^3 + x$ and $\sigma(x) = \alpha^5 + \alpha^4 x + x^2$.

Searching the roots of $\sigma(x)$, we verify that these roots are $\alpha^0 = 1$ and α^5; hence, the errors are in locations 0 and 2. The derivative of $\sigma(x)$ is $\sigma'(x) = \alpha^4$. By Equation 1.25, we obtain $E_0 = \omega(1)/\sigma'(1) = \alpha^4$ and $E_2 = \omega(\alpha^5)/\sigma'(\alpha^5) = \alpha^5$. Adding E_0 and E_2 to the received locations 0 and 2, the decoder concludes that the transmitted polynomial was

$$F(x) = \alpha^6 x + \alpha^5 x^3 + \alpha^5 x^4 + \alpha^6 x^5 + \alpha x^6$$

which in vector form is

$$\underline{c} = (000\ 101\ 000\ 111\ 111\ 101\ 010)$$

If the information is carried in the first 3 bytes, then the output of the decoder is

$$\underline{u} = (000\ 101\ 000)$$

1.9 Applications: Burst and Random Error Correction

In the previous sections we have studied how to encode and decode Reed-Solomon codes. In this section, we will briefly examine how they are used in applications, mainly for correction of bursts of errors. The two main methods for burst and combined burst and random error correction are interleaving and product codes.

In practice, errors often come in bursts. A burst of length l is a vector whose nonzero entries are among l consecutive (cyclically) entries, the first and last of them being nonzero. We consider binary bursts, and we use the elements of larger fields (bytes) to correct them. Below are some examples of bursts of length 4 in vectors of length 15:

$$
\begin{array}{ccccccccccccccc}
0 & 0 & 0 & 1 & 0 & 1 & 1 & 0 & 0 & 0 & 0 & 0 & 0 & 0 & 0 \\
0 & 0 & 0 & 0 & 0 & 0 & 1 & 1 & 1 & 1 & 0 & 0 & 0 & 0 & 0 \\
1 & 0 & 0 & 0 & 0 & 0 & 0 & 0 & 0 & 0 & 0 & 0 & 1 & 0 & 0
\end{array}
$$

Errors tend to come in bursts not only because the channel is bursty. Normally, both in optical and magnetic recording, data is encoded using a so called modulation code, which attempts to match the data to the characteristics of the channel. In general, the ECC is applied first to the random data and then the encoded data is modulated using modulation codes (see the chapter on modulation codes in this book). At the decoding, the order is reversed: when data exits the channel, it is first demodulated and then corrected using the ECC. Now, the demodulator tends to propagate errors, even single-bit errors. Although most modulation codes used in practice tend to control error propagation, nevertheless errors have a bursty character. For that reason, we need to implement a burst-correcting scheme, as we will see next.

A well-known relationship between the burst-correcting capability of a code and its redundancy is given by the Reiger bound, to be presented next, and whose proof is left as an exercise.

Theorem 1.4 (Reiger Bound)   *Let C be an $[n,k]$ linear code over a field $GF(2^v)$ that can correct all bursts of length up to l. Then $2l \leq n - k$.*

Cyclic binary codes that can correct bursts were obtained by computer search. A well known family of burst-correcting codes are the so called Fire codes. Here, we concentrate on the use of RS codes for burst correction. There are good reasons for this. One of them is that, although good burst-correcting codes have been found by computer search, there are no known general constructions giving cyclic codes that approach the Reiger bound. Interleaving of RS codes on the other hand, to be described below, provides a burst-correcting code whose redundancy, asymptotically, approaches the Reiger bound. The longer the burst we want to correct, the more efficient interleaving of RS codes is. The second reason for choosing interleaving of RS codes, and probably the most important one, is that, by increasing the error-correcting capability of the individual RS codes, we can correct multiple bursts, as we will see. The known binary cyclic codes are designed, in general, to correct only one burst. Let us start with the use of regular RS codes for correction of bursts. Let C be an $[n,k]$ RS code over $GF(2^b)$ (i.e., b-bit bytes). If this code can correct s bytes, in particular, it can correct a burst of length up to $(s-1)b + 1$ bits. In effect, a burst of length $(s-1)b + 2$ bits may affect $s+1$ consecutive bytes, exceeding the byte-correcting capability of the code. This happens when the burst of length $(s-1)b + 2$ bits starts in the last bit of a byte. How good are the PRS codes as burst-correcting codes? Given a binary $[n,k]$ code that can correct bursts of length up to l, we define a parameter, called the *burst-correcting efficiency* of the code, as follows:

$$
e_l = \frac{2l}{n-k} \tag{1.35}
$$

Notice that, by the Reiger bound, $e_l \leq 1$. The closer e_l is to 1, the more efficient the code is for correction of bursts. Going back to our $[n,k]$ RS code over $GF(2^b)$, it can be regarded as an $[nb, kb]$ binary code. Assuming that the code can correct s bytes and its redundancy is $n - k = 2s$, its burst-correcting efficiency is

$$
e_{(s-1)b+1} = \frac{(s-1)b + 1}{bs}
$$

Notice that, for $s \to \infty$, $e_{(s-1)b+1} \to 1$, justifying our assertion that for long bursts, RS codes are efficient as burst-correcting codes (as a comparison, the efficiency of Fire codes, asymptotically, tends to $2/3$). However, when s is large, there is a problem regarding complexity. It may not be practical to implement a RS code with too much redundancy. Moreover, the length of a RS code is limited; in the case of 8-bit bytes, it cannot be more than 256 (when extended). An alternative would be to implement a 1-byte correcting RS code interleaved s times.

$c_{0,0}$	$c_{0,1}$	$c_{0,2}$	$\cdots$	$c_{0,m-1}$
$c_{1,0}$	$c_{1,1}$	$c_{1,2}$	$\cdots$	$c_{1,m-1}$
$\vdots$	$\vdots$	$\vdots$	$\ddots$	$\vdots$
$c_{k-1,0}$	$c_{k-1,1}$	$c_{k-1,2}$	$\cdots$	$c_{k-1,m-1}$
$c_{k,0}$	$c_{k,1}$	$c_{k,2}$	$\cdots$	$c_{k,m-1}$
$\vdots$	$\vdots$	$\vdots$	$\ddots$	$\vdots$
$c_{n-1,0}$	$c_{n-1,1}$	$c_{n-1,2}$	$\cdots$	$c_{n-1,m-1}$

FIGURE 1.1   Interleaving m times of code $\mathcal{C}$.

An $[n,k]$ code interleaved m times is illustrated in Figure 1.1. Each column $c_{0,j}, \ldots, c_{n-1,j}$ is a codeword in an $[n,k]$ code. In general, each symbol $c_{i,j}$ is a byte and the code is a RS code. The first k bytes carry information bytes and the last $n-k$ bytes are redundant bytes. The bytes are read in row order, and the parameter m is called the depth of interleaving. If each of the individual codes can correct up to s errors, then the interleaved scheme can correct up to s bursts of length up to m bytes each, or $(m-1)b+1$ bits each. This occurs because a burst of length up to m bytes is distributed among m different codewords. Intuitively, interleaving "randomizes" a burst.

The drawback of interleaving is delay: notice that we need to read most of the information bytes before we are able to calculate and write the redundant bytes. Thus, we need enough buffer space to accomplish this.

Interleaving of RS codes has been widely used in magnetic recording. For instance, in a disk, the data are written in concentric tracks, and each track contains a number of information sectors. Typically, a sector consists of 512 information 8-bit bytes (although the latest trends tend to larger sectors). A typical embodiment would consist in dividing the 512 bytes into four codewords, each one containing 128 information bytes and 6 redundant bytes (i.e., each interleaved shortened RS codeword can correct up to 3 bytes). Therefore, this scheme can correct up to three bursts of length up to 25 bits each.

A natural generalization of the interleaved scheme described above is product codes. In effect, we may consider that both rows and columns are encoded into error-correcting codes. The product of an $[n_1, k_1]$ code $\mathcal{C}_1$ with an $[n_2, k_2]$ code $\mathcal{C}_2$, denoted $\mathcal{C}_1 \times \mathcal{C}_2$, is illustrated in Figure 1.2. If $\mathcal{C}_1$ has minimum distance d_1 and $\mathcal{C}_2$ has minimum distance d_2, it is easy to see that $\mathcal{C}_1 \times \mathcal{C}_2$ has minimum distance $d_1 d_2$.

In general, the symbols are read out in row order (although other readouts, like diagonal readouts, are also possible). For encoding, first the column redundant symbols are obtained, and then the row redundant symbols. For obtaining the checks on checks $c_{i,j}, k_1 \leq i \leq n_1 - 1, k_2 \leq j \leq n_2 - 1$, it is easy to see that it is irrelevant if we encode on columns or on rows first. If the symbols are read in row order, normally $\mathcal{C}_1$ is called the outer code and $\mathcal{C}_2$ the inner code. For decoding, there are many possible procedures. The idea

$c_{0,0}$	$c_{0,1}$	$c_{0,2}$	$\cdots$	c_{0,k_2-1}	c_{0,k_2}	c_{0,k_2+1}	$\cdots$	c_{0,n_2-1}
$c_{1,0}$	$c_{1,1}$	$c_{1,2}$	$\cdots$	c_{1,k_2-1}	c_{1,k_2}	c_{1,k_2+1}	$\cdots$	c_{1,n_2-1}
$\vdots$	$\vdots$	$\vdots$	$\ddots$	$\vdots$	$\vdots$	$\vdots$	$\ddots$	$\vdots$
$c_{k_1-1,0}$	$c_{k_1-1,1}$	$c_{k_1-1,2}$	$\cdots$	c_{k_1-1,k_2-1}	c_{k_1-1,k_2}	c_{k_1-1,k_2+1}	$\cdots$	c_{k_1-1,n_2-1}
$c_{k_1,0}$	$c_{k_1,1}$	$c_{k_1,2}$	$\cdots$	c_{k_1,k_2-1}	c_{k_1,k_2}	c_{k_1,k_2+1}	$\cdots$	c_{k_1,n_2-1}
$\vdots$	$\vdots$	$\vdots$	$\ddots$	$\vdots$	$\vdots$	$\vdots$	$\ddots$	$\vdots$
$c_{n_1-1,0}$	$c_{n_1-1,1}$	$c_{n_1-1,2}$	$\cdots$	c_{n_1-1,k_2-1}	c_{n_1-1,k_2}	c_{n_1-1,k_2+1}	$\cdots$	c_{n_1-1,n_2-1}

FIGURE 1.2   Product code $\mathcal{C}_1 \times \mathcal{C}_2$.

is to correct long bursts together with random errors. The inner code C_2 corrects first. In that case, two events may happen when its error-correcting capability is exceeded: either the code will detect the error event or it will miscorrect. If the code detects an error event (that may well have been caused by a long burst), one alternative is to declare an erasure in the whole row, which will be communicated to the outer code C_1. The other event is a miscorrection, that cannot be detected. In this case, we expect that the errors will be corrected by the error-erasure decoder of the outer code.

Product codes are important in practical applications. For instance, the code used in the DVD (digital video disk) is a product code where C_1 is a $[208, 192, 17]$ RS code and C_2 is a $[182, 172, 11]$ RS code. Both RS codes are defined over $GF(256)$, where $GF(256)$ is generated by the primitive polynomial $1 + x^2 + x^3 + x^4 + x^8$.

References

[1] R. E. Blahut, *Theory and Practice of Error Control Codes*, Addison Wesley, Reading 1983.

[2] C. Heegard and S. B. Wicker, *Turbo Coding*, Kluwer Academic Publishers, Dordrecht, 1999.

[3] S. Lin and D. J. Costello, *Error Control Coding: Fundamentals and Applications*, Prentice Hall, New York, 1983.

[4] F. J. MacWilliams and N. J. A. Sloane, *The Theory of Error-Correcting Codes*, North-Holland, Amsterdam, 1978.

[5] R. J. McEliece, *The Theory of Information and Coding*, Addison-Wesley, Reading 1977.

[6] C. E. Shannon, A mathematical theory of communication, *Bell Syst. Tech. J.*, 27, pp. 379–423 and 623–656, 1948.

[7] W. Wesley Peterson and E. J. Weldon, *Error-Correcting Codes*, MIT Press, Cambridge, 2nd ed., 1984.

[8] S. Wicker, *Error Control Systems for Digital Communications and Storage*, Prentice Hall, New York, 1995.

2

Modulation Codes for Storage Systems

Brian Marcus
University of British Columbia
Vancouver, BC

Emina Soljanin
Lucent Technologies
Murray Hill, NJ

2.1 Introduction

Modulation codes are used to constrain the individual sequences that are recorded in data storage channels, such as magnetic or optical disk or tape drives. The constraints are imposed in order to improve the detection capabilities of the system. Perhaps the most widely known constraints are the runlength limited ($\mathrm{RLL}(d, k)$) constraints, in which 1s are required to be separated by at least d and no more than k 0s. Such constraints are useful in data recording channels that employ peak detection: waveform peaks, corresponding to data ones, are detected independently of one another. The d-constraint helps to increase linear density while mitigating intersymbol interference, and the k-constraint helps to provide feedback for timing and gain control.

Peak detection was widely used until the early 1990s. While it is still used today in some magnetic tape drives and some optical recording devices, most high density magnetic disk drives now use a form of maximum likelihood (Viterbi) sequence detection. The data recording channel is modeled as a linear, discrete-time, communications channel with intersymbol interference (ISI), described by its transfer function and white Gaussian noise. The transfer function is often given by $h(D) = (1 - D)(1 + D)^N$, where N depends on and increases with the linear recording density.

Broadly speaking, two classes of constraints are of interest in today's high density recording channels: (1) constraints for improving timing and gain control and simplifying the design of the Viterbi detector for the channel, and (2) constraints for improving noise immunity. Some constraints serve both purposes.

Constraints in the first class usually take the form of a PRML (G, I) constraint: the maximum run of 0s is G and the maximum run of 0s, within each of the two substrings defined by the even indices and odd indices, is I. The G-constraint plays the same role as the k-constraint in peak detection, while the I-constraint enables the Viterbi detector to work well within practical limits of memory.

Constraints in the second class eliminate some of the possible recorded sequences in order to increase the minimum distance between those that remain or eliminate the possibility of certain dominant error events. This general goal does not specify how the constraints should be defined, but many such constraints have been constructed [20] and the references therein for a variety of examples. Bounds on the capacities of constraints that avoid a given set of error events have been given in [26].

Until recently, the only known constraints of this type were the matched-spectral-null (MSN) constraints. They describe sequences whose spectral nulls match those of the channel and therefore increase its minimum distance. For example, a set of DC-balanced sequences (i.e., sequences of ± 1 whose accumulated digital sums are bounded) is an MSN constraint for the channel with transfer function $h(D) = 1 - D$, which doubles its minimum distance [18].

During the past few years, significant progress has been made in defining high capacity distance enhancing constraints for high density magnetic recording channels. One of the earliest examples of such a constraint is the maximum transition run (MTR) constraint [28], which constrains the maximum run of 1s. We explain the main idea behind this type of distance-enhancing codes in Section 2.3.

Another approach to eliminating problematic error events is that of parity coding. Here, a few bits of parity are appended to (or inserted in) each block of some large size, typically 100 bits. For some of the most common error events, any single occurrence in each block can be eliminated. In this way, a more limited immunity against noise can be achieved with less coding overhead [5].

Coding for more realistic recording channel models that include colored noise and intertrack interference are discussed in Section 2.4. We point out that different constraints which avoid the same prescribed set of differences may have different performance on more realistic channels. This makes some of them more attractive for implementation.

For a more complete introduction to this subject, we refer the reader to any one of the many expository treatments, such as [16], [17], or [24].

2.2   Constrained Systems and Codes

Modulation codes used in almost all contemporary storage products belong to the class of constrained codes. These codes encode random input sequences to sequences that obey the constraint of a labeled directed graph with a finite number of states and edges. The set of corresponding constrained sequences is obtained by reading the labels of paths through the graph. Sets of such sequences are called constrained systems or constraints. Figure 2.1 and Figure 2.2 depict graph representations of an RLL constraint and a DC-balanced constraint.

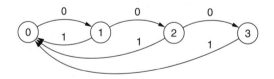

FIGURE 2.1   RLL $(1, 3)$ constraint.

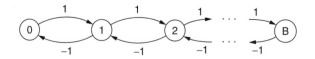

FIGURE 2.2   DC-balanced constraint.

Of special interest are those constraints that do not contain (globally or at certain positions) a finite number of finite length strings. These systems are called systems of finite type (FT). An FT system X over alphabet $\mathcal{A}$ can always be characterized by a finite list of forbidden strings $\mathcal{F} = \{w_1, \ldots, w_N\}$ of symbols in $\mathcal{A}$. Defined this way, FT systems will be denoted by $X_{\mathcal{F}}^{\mathcal{A}}$. The RLL constraints form a prominent class of FT constraints, while DC-balanced constraints are typically not FT.

Design of constrained codes begins with identifying constraints, such as those described in Section 2.1, that achieve certain objectives. Once the system of constrained sequences is specified, information bits are translated into sequences that obey the constraints via an *encoder*, which usually has the form of a finite-state machine. The actual set of sequences produced by the encoder is called a constrained code and is often denoted $\mathcal{C}$. A *decoder* recovers user sequences from constrained sequences. While the decoder is also implemented as a finite-state machine, it is usually required to have a stronger property, called sliding-block decodablility, which controls error propagation [24].

The maximum rate of a constrained code is determined by *Shannon capacity*. The Shannon capacity or simply *capacity* of a constrained system, denoted by C, is defined as

$$C = \lim_{n \to \infty} \frac{\log_2 N(n)}{n}$$

where $N(n)$ is the number of sequences of length n. The capacity of a constrained system represented by a graph G can be easily computed from the *adjacency matrix* (or *state transition matrix*) of G (provided that the labeling of G satisfies some mildly innocent properties). The adjacency matrix of G with r states and a_{ij} edges from state i to state j, $1 \leq i, j \leq r$, is the $r \times r$ matrix $A = A(G) = \{a_{ij}\}_{r \times r}$. The Shannon capacity of the constraint is given by

$$C = \log_2 \lambda(A)$$

where $\lambda(A)$ is the largest real eigenvalue of A.

The *state-splitting algorithm* [1] (see also [24]) gives a general procedure for constructing constrained codes at any rate up to capacity. In this algorithm, one starts with a graph representation of the desired constraint and then transforms it into an encoder via various graph-theoretic operations including splitting and merging of states. Given a desired constraint and a desired rate $p/q \leq C$, one or more rounds of state splitting are performed; the determination of which states to split and how to split them is governed by an approximate eigenvector, that is, a vector $\mathbf{x}$ satisfying $A^q x \geq 2^p x$.

There are many other very important and interesting approaches to constrained code construction — far too many to mention here. One approach combines state-splitting with look-ahead encoding to obtain a very powerful technique which yields competent codes [14]. Another approach involves variable-length and time-varying variations of these techniques [2, 13]. Many other effective coding constructions are described in the monograph [17].

For high capacity constraints, graph transforming techniques, such as the state-splitting algorithm, may result in encoder/decoder architectures with formidable complexity. Fortunately, a block encoder/decoder architecture with acceptable implementation complexity for many constraints can be designed by well-known enumerative [6], and other combinatorial [32] as well as heuristic techniques [25].

Translation of constrained sequences into the channel sequences depends on the modulation method. Saturation recording of binary information on magnetic medium is accomplished by converting an input stream of data into a spatial stream of bit cells along a track where each cell is fully magnetized in one of two possible directions, denoted by 0 and 1. There are two important modulation methods commonly used on magnetic recording channels: *non-return-to-zero* (NRZ) and *modified non-return-to-zero* (NRZI). In NRZ modulation, the binary digits 0 and 1 in the input data stream correspond to 0 and 1 directions of cell magnetizations, respectively. In NRZI modulation, the binary digit 1 corresponds to a magnetic transition between two bit cells, and the binary digit 0 corresponds to no transition. For example, the channel constraint which forbids transitions in two neighboring bit-cells, can be accomplished by either $\mathcal{F} = \{11\}$ NRZI constraint or $\mathcal{F} = \{101, 010\}$ NRZ constraint. The graph representation of these two constraints is shown in Figure 2.3. The NRZI representation is in this case simpler.

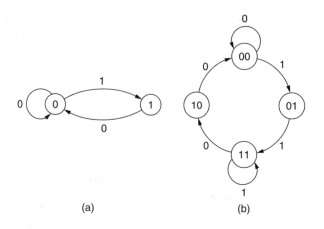

FIGURE 2.3   Two equivalent constraints: (a) $\mathcal{F} = \{11\}$ NRZI, and (b) $\mathcal{F} = \{101, 010\}$ NRZ.

2.3   Constraints for ISI Channels

We discuss a class of codes known as *codes which avoid specified differences*. This is the only class of distance enhancing codes used in commercial magnetic recording systems. There are two main reasons for this: these codes simplify the channel detectors relative to the uncoded channel and even high rate codes in this class can be realized by low complexity encoders and decoders.

2.3.1   Requirements

A number of papers have proposed using constrained codes to provide coding gain on channels with high ISI (see [4, 10, 20, 28]). The main idea of this approach can be described as follows [20]. Consider a discrete-time model for the magnetic recording channel with possibly constrained input $a = \{a_n\} \in \mathcal{C} \subseteq \{-1, 1\}^\infty$, impulse response $\{h_n\}$, and output $y = \{y_n\}$ given by

$$y_n = \sum_m a_m h_{n-m} + \eta_n \tag{2.1}$$

where $h(D) = \sum_n h_n D^n = (1 - D)(1 + D)^3$ (E^{2}PR4) or $h(D) = \sum_n h_n = (1 - D)(1 + D)^4$ (E^{3}PR4), η_n are independent Gaussian random variables with zero mean and variance σ^2. The quantity $1/\sigma^2$ is referred to as the signal-to-noise ratio (SNR). The minimum distance of the uncoded channel (Equation 2.1) is

$$d_{\min}^2 = \min_{\epsilon(D) \neq 0} \|h(D)\epsilon(D)\|^2$$

where $\epsilon(D) = \sum_{i=0}^{l-1} \epsilon_i D^i, (\epsilon_i \in \{-1, 0, 1\}, \epsilon_0 = 1, \epsilon_{l-1} \neq 0)$ is the polynomial corresponding to a normalized input error sequence $\epsilon = \{\epsilon_i\}_{i=0}^{l-1}$ of length l, and the squared norm of a polynomial is defined as the sum of its squared coefficients. The minimum distance is bounded from above by $\|h(D)\|^2$, denoted by

$$d_{\text{MFB}}^2 = \|h(D)\|^2 \tag{2.2}$$

This bound is known as the *matched-filter bound* (MFB), and is achieved when the error sequence of length $l = 1$, that is, $\epsilon(D) = 1$, is in the set

$$\arg\min_{\epsilon(D) \neq 0} \|h(D)\epsilon(D)\|^2 \tag{2.3}$$

For channels that fail to achieve the MFB, that is, for which $d^2_{min} < \|h(D)\|^2$, any error sequences $\epsilon(D)$ for which

$$d^2_{min} \leq \|h(D)\epsilon(D)\|^2 < \|h(D)\|^2 \tag{2.4}$$

are of length $l \geq 2$ and may belong to a constrained system $X_{\mathcal{L}}^{\{-1,0,1\}}$, where $\mathcal{L}$ is an appropriately chosen finite list of forbidden strings.

For code $\mathcal{C}$, we write the set of all admissible nonzero error sequences as

$$\mathcal{E}(\mathcal{C}) = \{\epsilon \in \{-1,0,1\}^{\infty} |$$
$$\epsilon \neq 0, \epsilon = (a - b)/2, \ a, b \in \mathcal{C}\}$$

Given the condition $\mathcal{E}(\mathcal{C}) \subseteq X_{\mathcal{L}}^{\{-1,0,1\}}$, we seek to identify the least restrictive finite collection $\mathcal{F}$ of blocks over the alphabet $\{0, 1\}$ so that

$$\mathcal{C} \subseteq X_{\mathcal{F}}^{\{0,1\}} \implies \mathcal{E}(\mathcal{C}) \subseteq X_{\mathcal{L}}^{\{-1,0,1\}} \tag{2.5}$$

2.3.2 Definitions

A constrained code is defined by specifying $\mathcal{F}$, the list of forbidden strings for code sequences. Prior to that one needs to first characterize error sequences that satisfy Equation 2.4 and then specify $\mathcal{L}$, the list of forbidden strings for error sequences. Error event characterization can be done by using any of the methods described by Karabed et al. [20]. Specification of $\mathcal{L}$ is usually straightforward.

A natural way to construct a collection $\mathcal{F}$ of blocks forbidden in code sequences based on the collection $\mathcal{L}$ of blocks forbidden in error sequences is the following. From the above definition of error sequences $\epsilon = \{\epsilon_i\}$ we see that $\epsilon_i = 1$ requires $a_i = 1$ and $\epsilon_i = -1$ requires $a_i = 0$, that is, $a_i = (1 + \epsilon_i)/2$. For each block $w_{\mathcal{E}} \in \mathcal{L}$, construct a list $\mathcal{F}_{w_{\mathcal{E}}}$ of blocks of the same length l according to the rule:

$$\mathcal{F}_{w_{\mathcal{E}}} = \{w_{\mathcal{C}} \in \{-1,1\}^l |$$
$$w_{\mathcal{C}}^i = (1 + w_{\mathcal{E}}^i)/2 \text{ for all } i \text{ for which } w_{\mathcal{E}}^i \neq 0\}$$

Then the collection $\mathcal{F}$ obtained as $\mathcal{F} = \cup_{w_{\mathcal{E}} \in \mathcal{L}} \mathcal{F}_{w_{\mathcal{E}}}$ satisfies requirement (Equation 2.5). However, the constrained system $X_{\mathcal{F}}^{\{0,1\}}$ obtained this way may not be the most efficient. (Bounds on the achievable rates of codes which avoid specified differences were found recently in [26].)

We illustrate the above ideas on the example of the $E^2 PR4$ channel. Its transfer function is $h(D) = (1 - D)(1 + D)^3$, and its MFB is $\|(1 - D)(1 + D)^3 \cdot 1\|^2 = 10$. The error polynomial $\epsilon(D) = 1 - D + D^2$ is the unique error polynomial for which $\|(1 - D)(1 + D)^3 \epsilon(D)\|^2 = 6$, and the error polynomials $\epsilon(D) = 1 - D + D^2 + D^5 - D^6 + D^7$ and $\epsilon(D) = \sum_{i=0}^{l-1}(-1)^i D^i$ for $l \geq 4$ are the only polynomials for which $\|(1 - D)(1 + D)^3 \epsilon(D)\|^2 = 8$ (see [20]).

It is easy to show that these error events are not in the constrained error set defined by the list of forbidden error strings $\mathcal{L} = \{+-+00, +-+-\}$, where $+$ denotes 1 and $-$ denotes -1. To see that, note that an error sequence that does not contain the string $+-+00$ cannot have error polynomials $\epsilon(D) = 1 - D + D^2$ or $\epsilon(D) = 1 - D + D^2 + D^5 - D^6 + D^7$, while an error sequence that does not contain string $+-+-$ cannot have an error polynomial of the form $\epsilon(D) = \sum_{i=0}^{l-1}(-1)^i D^i$ for $l \geq 4$. Therefore, by the above procedure of defining the list of forbidden code strings, we obtain the $\mathcal{F} = \{+-+\}$ NRZ constraint. Its capacity is about 0.81, and a rate 4/5 code into the constraint was first given in [19].

In [20], the following approach was used to obtain several higher rate constraints. For each of error strings in $\mathcal{L}$, we write all pairs of channel strings whose difference is the error string. To define $\mathcal{F}$, we look for the longest string(s) appearing in at least one of the strings in each channel pair. For the example above and the $+-+00$ error string, a case-by-case analysis of channel pairs is depicted in Figure 2.4. We can distinguish two types (denoted by A and B in the figure) of pairs of code sequences involved in forming an error event. In a pair of type A, at least one of the sequences has a transition run of length 4. In a pair of type B, both sequences have transition runs of length 3, but for one of them the run starts at an even

```
        A              B              A              A
a: 0 | 1 0 1 | 0 0  1 | 1 0 1 | 0 0  0 | 1 0 1 | 0 1  1 | 1 0 1 | 0 1
b: 0 | 0 1 0 | 0 0  1 | 0 1 0 | 0 0  0 | 0 1 0 | 0 1  1 | 0 1 0 | 0 1


        A              A              B              A
a: 0 | 1 0 1 | 1 0  1 | 1 0 1 | 1 0  0 | 1 0 1 | 1 1  1 | 1 0 1 | 1 1
b: 0 | 0 1 0 | 1 0  1 | 0 1 0 | 1 0  0 | 0 1 0 | 1 1  1 | 0 1 0 | 1 1
```

FIGURE 2.4 Possible pairs of sequences for which error event $+ - +00$ may occur.

position and for the other at an odd position. This implies that an NRZI constrained system that limits the run of 1s to 3 when it starts at an odd position, and to 2 when it starts at an even position, eliminates all possibilities shown bold-faced in Figure 2.4. In addition, this constraint eliminates all error sequences containing the string $+-+-$. The capacity of the constraint is about 0.916, and rate 8/9 block codes with this constraint has been implemented in several commercial read channel chips. More about the constraint and the codes can be found in [4, 10, 20, 28].

2.4 Channels with Colored Noise and Intertrack Interference

Magnetic recording systems always operate in the presence of colored noise intertrack interference, and data dependent noise. Codes for these more realistic channel models are studied in [27]. Below, we briefly outline the problem.

Data recording and retrieval process is usually modeled as a linear, continuous-time, communications channel described by its Lorentzian step response and additive white Gaussian noise. The most common discrete-time channel model is given by Equation 2.1. Magnetic recording systems employ channel equalization to the most closely matching transfer function $h(D) = \sum_n h_n D^n$ of the form $h(D) = (1 - D)(1 + D)^N$. This equalization alters the spectral density of the noise, and a better channel model assumes that the η_n in Equation 2.1 are identically distributed, Gaussian random variables with zero mean, variance σ^2, and normalized cross-correlation $E\{\eta_n \eta_k\}/\sigma^2 = \rho_{n-k}$.

In practice, there is always intertrack interference (ITI), that is, the read head picks up magnetization from an adjacent track. Therefore, the channel output is given by

$$y_n = \sum_m a_m h_{n-m} + \sum_m x_m g_{n-m} + \eta_n \tag{2.6}$$

where $\{g_n\}$ is the discrete-time impulse response of the head to the adjacent track, and $x = \{x_n\} \in C$ is the sequence recorded on that track. We assume that the noise is white.

In the ideal case (Equation 2.1), the probability of detecting b given that a was recorded is equal to $Q(d(\epsilon)/\sigma)$, where $d(\epsilon)$ is the distance between a and b given by

$$d^2(\epsilon) = \sum_n \left(\sum_m \epsilon_m h_{n-m} \right)^2 \tag{2.7}$$

Therefore, a lower bound, and a close approximation for small σ, to the minimum probability of an error-event in the system is given by $Q(d_{\min,C}/\sigma)$, where

$$d_{\min,C} = \min_{\epsilon \in \mathcal{E}_C} d(\epsilon)$$

is the channel minimum distance of code C. We refer to

$$d_{\min} = \min_{\epsilon \in \{-1,0,1\}^\infty} d(\epsilon) \tag{2.8}$$

as the minimum distance of the uncoded channel, and to the ratio $d_{\min,C}/d_{\min}$ as the gain in distance of code C over the uncoded channel.

In the case of colored noise, the probability of detecting b given that a was recorded equals to $Q(\Delta(\epsilon)/\sigma)$, where $\Delta(\epsilon)$ is the distance between a and b given by

$$\Delta^2(\epsilon) = \frac{\left[\sum_n \left(\sum_m \epsilon_m h_{n-m} \right)^2 \right]^2}{\sum_n \sum_k \left(\sum_m \epsilon_m h_{n-m} \right) \rho_{n-k} \left(\sum_m \epsilon_m h_{k-m} \right)}$$

Therefore, a lower bound to the minimum probability of an error-event in the system is given by $Q(\Delta_{\min,C}/\sigma)$, where

$$\Delta_{\min,C} = \min_{\epsilon \in \mathcal{E}_C} \Delta(\epsilon)$$

In the case of ITI (Equation 2.6), we are interested in the probability of detecting sequence b given that sequence a was recorded on the track being read and sequence x was recorded on an adjacent track. This probability is

$$Q(\delta(\epsilon, x)/\sigma),$$

where $\delta(\epsilon, x)$ is the distance between a and b in the *presence* of x given by [30]

$$\delta^2(\epsilon, x) = \frac{1}{\left[\sum_n \left(\sum_m \epsilon_m h_{n-m} \right)^2 \right]} \left[\sum_n \left(\sum_m \epsilon_m h_{n-m} \right)^2 + \sum_n \left(\sum_m x_m g_{n-m} \right) \left(\sum_m \epsilon_m h_{n-m} \right) \right]^2$$

Therefore, a lower bound to the minimum probability of an error-event in the system is proportional to $Q(\delta_{\min,C}/\sigma)$, where

$$\delta_{\min,C} = \min_{\epsilon \neq 0, x \in C} \delta(\epsilon, x)$$

Distance $\delta_{\min,C}$ can be bounded as follows [30]:

$$\delta_{\min,C} \geq (1 - M) d_{\min,C} \tag{2.9}$$

where $M = \max_{n, x \in C} \sum_m x_m g_{n-m}$, that is, M is the maximum absolute value of the interference. Note that $M = \sum_n |g_n|$. We will assume that $M < 1$. The bound is achieved if and only if there exists an ϵ, $d(\epsilon) = d_{\min,C}$, for which $\sum_m \epsilon_m h_{n-m} \in \{-1, 0, 1\}$ for all n, and there exists an $x \in C$ such that $\sum_m x_m g_{n-m} = \mp M$ whenever $\sum_m \epsilon_m h_{n-m} = \pm 1$.

2.5 An Example

There are codes that provide gain in minimum distance on channels with ITI and colored noise, but not on the AWGN channel with the same transfer function. This is best illustrated using the example of the partial response channel with the transfer function $h(D) = (1 - D)(1 + D)^2$ known as EPR4. It is well known that for the EPR4 channel $d_{\min}^2 = 4$. Moreover, as discussed in Section 2.3, the following result holds:

Proposition 2.1 Error events $\epsilon(D)$ such that

$$d^2(\epsilon) = d_{\min}^2 = 4$$

take one of the following two forms:

$$\epsilon(D) = \sum_{j=0}^{k-1} D^{2j}, \quad k \geq 1$$

or

$$\epsilon(D) = \sum_{i=0}^{l-1} (-1)^i D^i, \quad l \geq 3$$

Therefore, an improvement of error-probability performance can be accomplished by codes which elimi-nate the error sequences ϵ containing the strings $-1 + 1 - 1$ and $+1 - 1 + 1$. Such codes were extensively studied in [20].

In the case of ITI Equation 2.6, we assume that the impulse response to the reading head from an adjacent track is described by $g(D) = \alpha H(D)$, where the parameter α depends on the track to head distance. Under this assumption, the bound (Equation 2.9) gives $\delta_{min}^2 \geq d_{min}^2 (1 - 4\alpha)^2$. The following result was shown in [30]:

Proposition 2.2   Error events $\epsilon(D)$ such that

$$\min_{x \in C} \delta^2(\epsilon, x) = \delta_{min}^2 = d_{min}^2 (1 - 4\alpha)^2 = 4(1 - 4\alpha)^2$$

take the following form:

$$\epsilon(D) = \sum_{i=0}^{l-1} (-1)^i D^i, \quad l \geq 5$$

For all other error sequences for which $d^2(\epsilon) = 4$, we have $\min_{x \in C} \delta^2(\epsilon, x) = 4(1 - 3\alpha)^2$.

Therefore, an improvement in error-probability performance of this channel can be accomplished by limiting the length of strings of alternating symbols in code sequences to four. For the NRZI type of recording, this can be achieved by a code that limits the runs of successive ones to three. Note that the set of minimum distance error events is smaller than in the case with no ITI. Thus performance improvement can be accomplished by higher rate codes which would not provide any gain on the ideal channel.

Channel equalization to the EPR4 target introduces cross-correlation among noise samples for a range of current linear recording densities (see [27] and references therein). The following result was obtained in [27]:

Proposition 2.3   Error events $\epsilon(D)$ such that

$$\Delta^2(\epsilon) = \Delta_{min}^2$$

take the following form:

$$\epsilon(D) = \sum_{i=0}^{l-1} (-1)^i D^i, \quad l \geq 3, \quad l \text{ odd}$$

Again, the set of minimum distance error events is smaller than in the ideal case (white noise), and performance improvement can be provided by codes which would not give any gain on the ideal channel. For example, since all minimum distance error events have odd parity, a single parity check code can be used.

2.6 Future Directions

2.6.1 Soft-Output Decoding of Modulation Codes

Detection and decoding in magnetic recording systems is organized as a concatenation of a channel detector, an inner decoder, and an outer decoder, and as such should benefit from techniques known as erasure and list decoding. To declare erasures or generate lists, the inner decoder (or channel detector) needs to assess symbol/sequence reliabilities. Although the information required for this is the same one necessary for producing a single estimate, some additional complexity is usually required. So far, the predicted gains for erasure and list decoding of magnetic recording channels with additive white Gaussian noise were not sufficient to justify the increasing complexity of the channel detector and inner and outer decoder. However, this is not the case for systems employing new magneto-resistive reading heads, for which an important noise source, thermal asperities, is to be handled by passing erasure flags from the inner to the outer decoder.

In recent years, one more reason for developing simple soft-output channel detectors has surfaced. The success of turbo-like coding schemes on memoryless channels has sparked the interest in using them as modulation codes for ISI channels. Several recent results show that the improvements in performance that turbo codes offer when applied to magnetic recording channels at moderate linear densities are even more dramatic than in the memoryless case [12, 29]. The decoders for turbo and low density parity check codes (LDPC) either require or perform much better with soft input information which has to be supplied by the channel detector as its soft output. The decoders provide soft outputs which can then be utilized by the outer Reed-Solomon (RS) decoder [22]. A general soft-output sequence detection was introduced in [11], and it is possible to get information on symbol reliabilities by extending those techniques [21, 31].

2.6.2 Reversed Concatenation

Typically, the modulation encoder is the inner encoder, that is, it is placed downstream of an error-correction encoder (ECC) such as an RS encoder; this configuration is known as standard concatenation (Figure 2.5). This is natural since otherwise the ECC encoder might well destroy the modulation properties before passing across the channel. However, this scheme has the disadvantage that the modulation decoder, which must come before the ECC decoder, may propagate channel errors before they can be corrected. This is particularly problematic for modulation encoders of very high rate, based on very long block size. For this reason, a good deal of attention has recently focused on a reversed concatenation scheme, where the encoders are concatenated in the reversed order (Figure 2.6). Special arrangements must be made

FIGURE 2.5   Standard concatenation.

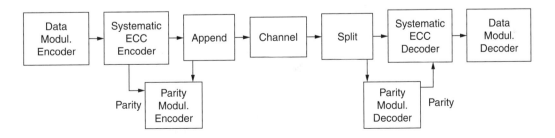

FIGURE 2.6   Reversed concatenation.

to ensure that the output of the ECC encoder satisfies the modulation constraints. Typically, this is done by insisting that this encoder be systematic and then reencoding the parity information using a second modulation encoder (the "parity modulation encoder"), whose corresponding decoder is designed to limit error propagation; the encoded parity is then appended to the modulation-encoded data stream (typically a few merging bits may need to be inserted in between the two streams in order to ensure that the entire stream satisfies the constraint). In this scheme, after passing through the channel the modulation-encoded data stream is split from the modulation-encoded parity stream, and the latter is then decoded via the parity modulation decoder before being passed on to the ECC decoder. In this way, many channel errors can be corrected before the data modulation decoder, thereby mitigating the problem of error propagation. Moreover, if the data modulation encoder has high rate, then the overall scheme will also have high rate because the parity stream is relatively small.

Reversed concatenation was introduced in [3] and later in [23]. Recent interest in the subject has been spurred on by the introduction of a lossless compression scheme, which improves the efficiency of reversed concatenation [15], and an analysis demonstrating the benefits in terms of reduced levels of interleaving [8]; see also [9]. Research on fitting soft decision detection into reversed concatenation can be found in [7, 33].

References

[1] R. Adler, D. Coppersmith, and M. Hassner, Algorithms for sliding-block codes, *IEEE Trans. Inform. Theory*, 29, no. 1, 5–22, January 1983.

[2] J. Ashley and B. Marcus, Time-varying encoders for constrained systems: an approach to limiting error propagation, *IEEE Trans. Inform. Theory*, 46, 1038–1043, 2000.

[3] W. G. Bliss, Circuitry for performing error correction calculations on baseband encoded data to eliminate error propagation, *IBM Tech. Discl. Bull.*, 23, 4633–4634, 1981.

[4] W. G. Bliss, An 8/9 rate time-varying trellis code for high density magnetic recording, *IEEE Trans. Magn.*, 33, no. 5, 2746–2748, September 1997.

[5] T. Conway, A new target response with parity coding for high density magnetic recording, *IEEE Trans. Magnetics*, 34, 2382–2386, 1998.

[6] T. Cover, Enumerative source encoding, *IEEE Trans. Inform. Theory*, 73–77, January 1973.

[7] J. Fan, Constrained coding and soft iterative decoding for storage, Ph.D. Dissertation, Stanford University, 1999.

[8] J. Fan and R. Calderbank, A modified concatenated coding scheme, with applications to magnetic data storage, *IEEE Trans. Inform. Theory*, 44, 1565–1574, 1998.

[9] J. Fan, B. Marcus, and R. Roth, Lossless sliding-block compression of constrained systems, *IEEE Trans. Inform. Theory*, 46, 624–633, 2000.

[10] K. Knudson Fitzpatrick and C. S. Modlin, Time-varying MTR codes for high density magnetic recording, *Proc. 1997 IEEE Global Telecommun. Conf. (GLOBECOM '97)*, Phoenix, AZ, 1250–1253, November 1997.

[11] J. Hagenauer and P. Hoeher, A Viterbi algorithm with soft-decision outputs and its applications, *Proc. 1989 IEEE Global Telecommun. Conf. (GLOBECOM '89)*, Dallas, TX, 1680–16867, November 1989.

[12] C. Heegard, Turbo coding for magnetic recording, *Proc. 1998 Inform. Theory Workshop*, San Diego, CA, 18–19, February 8–11, 1998.

[13] C. D. Heegard, B. H. Marcus, and P. H. Siegel, Variable-length state splitting with applications to average runlength-constrained (ARC) codes, *IEEE Trans. Inform. Theory*, 37, 759–777, 1991.

[14] H. D. L. Hollmann, On the construction of bounded-delay encodable codes for constrained systems, *IEEE Trans. Inform. Theory*, 41, 1354–1378, 1995.

[15] K. A. Schouhamer Immink, A practical method for approaching the channel capacity of constrained channels, *IEEE Trans. Inform. Theory*, 43, 1389–1399, 1997.

[16] K. A. Schouhamer Immink, P. H. Siegel, and J. K. Wolf, Codes for digital recorders, *IEEE Trans. Inform. Theory*, 44, 2260–2299, October 1998.

[17] K. A. Schouhamer Immink, *Codes for Mass Data Storage*, Shannon Foundation Publishers, The Netherlands, 1999.

[18] R. Karabed and P. H. Siegel, Matched spectral null codes for partial response channels, *IEEE Trans. Inform. Theory*, 37, 818–855, 1991.

[19] R. Karabed and P. H. Siegel, Coding for higher order partial response channels, *Proc. 1995 SPIE Int. Symp. on Voice, Video, and Data Communications*, Philadelphia, PA, 2605, 115–126, October 1995.

[20] R. Karabed, P. H. Siegel, and E. Soljanin, Constrained coding for binary channels with high inter-symbol interference, *IEEE Trans. Inform. Theory*, 45, 1777–1797, September 1999.

[21] K. J. Knudson, J. K. Wolf, and L. B. Milstein, Producing soft-decision information on the output of a class IV partial response Viterbi detector, *Proc. 1991 IEEE Int. Conf. Commn. (ICC '91)*, Denver, CO, 26.5.1.–26.5.5, June 1991.

[22] R. Koetter and A. Vardy, preprint 2000.

[23] M. Mansuripur, Enumerative modulation coding with arbitrary constraints and post-modulation error correction coding and data storage systems, *Proc. SPIE*, 1499, 72–86, 1991.

[24] B. Marcus, R. Roth, and P. Siegel, Constrained systems and coding for recording channels, *Handbook of Coding Theory*, V. Pless, C. Huffman, Eds., Elsevier, Amsterdam, 1998, Chap. 20.

[25] D. Modha and B. Marcus, Art of constructing low complexity encoders/decoders for constrained block codes, *IEEE J. Sel. Areas Comm.*, 2001, to appear.

[26] B. E. Moision, A. Orlitsky, and P. H. Siegel, On codes that avoid specified differences, *IEEE Trans. Inform. Theory*, 47, 433–441, January 2001.

[27] B. E. Moision, P. H. Siegel, and E. Soljanin, Distance enhancing codes for high-density magnetic recording channel, *IEEE Trans. Magn.*, January 2001, submitted.

[28] J. Moon and B. Brickner, Maximum transition run codes for data storage systems, *IEEE Trans. Magn.*, 32, 3992–3994, September 1996.

[29] W. Ryan, L. McPheters, and S.W. McLaughlin, Combined turbo coding and turbo equalization for PR4-equalized Lorentzian channels, *Proc. 22nd Annual Conf. Inform. Sciences and Systems*, Princeton, NJ, March 1998.

[30] E. Soljanin, On-track and off-track distance properties of Class 4 partial response channels, *Proc. 1995 SPIE Int. Symp. on Voice, Video, and Data Communications*, Philadelphia, PA, 2605, 92–102, October 1995.

[31] E. Soljanin, Simple soft-output detection for magnetic recording channels, *1998 IEEE Int. Symp. Inform. Theory (ISIT'00)*, Sorrento, Italy, June 2000.

[32] A. J. van Wijngaarden and K. A. Schouhamer Immink, Combinatorial construction of high rate runlength-limited codes, *Proc. 1996 IEEE Global Telecommun. Conf. (GLOBECOM '96)*, London, U.K., 343–347, November 1996.

[33] A. J. van Wijngaarden and K. A. Schouhamer Immink, Maximum run-length limited codes with error control properties, *IEEE J. Select. Areas Commn.*, 19, April 2001.

[34] A. J. van Wijngaarden and E. Soljanin, A combinatorial technique for constructing high rate MTR–RLL codes, *IEEE J. Select. Areas Commn.*, 19, April 2001.

3

Runlength Limited Sequences

Kees A. Schouhamer Immink
Turing Machines Inc.
Rotterdam, Netherlands
and
University of Essen
Essen, Germany

3.1 Introduction

Codes based on runlength-limited sequences have been the state of the art corner stone of current disc recorders whether their nature is magnetic or optical. This chapter provides a detailed description of various properties of runlength-limited sequences and the next section gives a comprehensive review of the code construction methods, ad hoc as well as systematic, that are available.

The length of time usually expressed in channel bits between consecutive transitions is known as the *runlength*. For instance, the runlengths in the word

$$0111100111000000$$

are of length 1, 4, 2, 3, and 6. Runlength-limited (RLL) sequences are characterized by two parameters, $(d+1)$ and $(k+1)$, which stipulate the minimum (with the exception of the very first and last runlength) and maximum runlength, respectively, that may occur in the sequence. The parameter d controls the highest transition frequency and thus has a bearing on intersymbol interference when the sequence is transmitted over a bandwidth-limited channel. In the transmission of binary data it is generally desirable that the received signal is self-synchronizing or self-clocking. Timing is commonly recovered with a phase-locked loop. The maximum runlength parameter k ensures adequate frequency of transitions for synchronization of the read clock.

Recording codes that are based on RLL sequences have found almost universal application in disc recording practice. In consumer electronics, we have the EFM code (rate $= 8/17, d = 2, k = 10$), which is employed in the Compact Disc (CD), and the EFMPlus code (rate $= 8/16, d = 2, k = 10$) used in the DVD.

A *dk*-limited binary sequence, in short, (dk) sequence, satisfies simultaneously the following two conditions:

1. d constraint — two logical 1s are separated by a run of consecutive 0s of length at least d.
2. k constraint — any run of consecutive 0s is of length at most k.

If only proviso (1.) is satisfied, the sequence is said to be d-limited (with $k = \infty$), and will be termed (d) sequence. In general, a (dk) sequence is not employed in optical or magnetic recording without a simple coding step. A (dk) sequence is converted to a runlength-limited channel sequence in the following way. Let the channel signals be represented by a bipolar sequence $\{y_i\}$, $y_i \in \{-1, 1\}$. The channel signals represent the positive or negative magnetization of the recording medium, or pits or lands when dealing with optical recording. The logical 1s in the (dk) sequence indicate the positions of a transition $1 \rightarrow -1$ or $-1 \rightarrow 1$ of the corresponding RLL sequence. The (dk) sequence

$$0 \;\; 1 \;\; 0 \;\; 0 \;\; 0 \;\; 1 \;\; 0 \;\; 0 \;\; 1 \;\; 0 \;\; 0 \;\; 0 \;\; 1 \;\; 1 \;\; 0 \;\; 1 \ldots$$

would be converted to the RLL channel sequence

$$1 \;\; -1 \;\; -1 \;\; -1 \;\; -1 \;\; 1 \;\; 1 \;\; 1 \;\; -1 \;\; -1 \;\; -1 \;\; -1 \;\; 1 \;\; -1 \;\; -1 \;\; 1 \ldots.$$

Waveforms that are transmitted without such an intermediate coding step are referred to as non-return-to-zero (NRZ). It can readily be verified that the minimum and maximum distance between consecutive transitions of the RLL sequence derived from a (dk) sequence is $d + 1$ and $k + 1$ symbols, respectively, or in other words, the RLL sequence has the virtue that at least $d + 1$ and at most $k + 1$ consecutive like symbols (runs) occur.

The outline of this chapter is as follows. We start with a discussion of the maximum rate of RLL sequences given the parameters d and k. Thereafter we will present various methods for constructing codes for generating RLL sequences.

3.2 Asymptotic Information Rate

3.2.1 Counting of Sequences

This section addresses the problem of counting the number of sequences of a certain length which comply with given dk constraints. We start for the sake of clerical convenience with the enumeration of (d) sequences. Let $N_d(n)$ denote the number of distinct (d) sequences of length n and define

$$N_d(n) = 0, \quad n < 0$$
$$N_d(0) = 1 \tag{3.1}$$

The number of (d) sequences of length $n > 0$ is found with the recursive relations [1]

$$(i) \quad N_d(n) = n + 1, \; 1 \leq n \leq d + 1$$
$$(ii) \quad N_d(n) = N_d(n - 1) + N_d(n - d - 1), \; n > d + 1 \tag{3.2}$$

The proof of Equation 3.2, taken from [1], is straightforward.

1. If $n \leq d + 1$, a (d) sequence can contain only a single 1 (and there are exactly n such sequences), or the sequence must be the all 0 sequence (and there is only one such sequence).
2. If $n > d + 1$, a (d) sequence can be built by one of the following procedures:
   i. To build any (d) sequence of length n starting with a 0, take the concatenation of a 0 and any (d) sequence of length $n - 1$. There are $N_d(n - 1)$ of such.
   ii. Any (d) sequence of length n starting with a 1 can be constructed by the concatenation of a 1 and d 0s followed by any (d) sequence of length $n - d - 1$. There are $N_d(n - d - 1)$ of such.

Table 3.1 lists the number of distinct (d) sequences as a function of the sequence length n with the minimum runlength d as a parameter.

When $d = 0$, we simply find that $N_0(n) = 2N_0(n - 1)$, or in other words, when there is no restriction at all, the number of combinations doubles when a bit is added, which is, of course, a well-known result.

TABLE 3.1 Number of Distinct (d) Sequences as a Function of the Sequence Length n and the Minimum Runlength d as a Parameter.

$d \setminus n$	2	3	4	5	6	7	8	9	10	11	12	13	14
1	3	5	8	13	21	34	55	89	144	233	377	610	987
2	3	4	6	9	13	19	28	41	60	88	129	189	277
3	3	4	5	7	10	14	19	26	36	50	69	95	131
4	3	4	5	6	8	11	15	20	26	34	45	60	80
5	3	4	5	6	7	9	12	16	21	27	34	43	55

The numbers $N_1(n)$ are

$$1, 2, 3, 5, 8, 13, \ldots,$$

where each number is the sum of its two predecessors. These numbers are called *Fibonacci numbers*.

The number of (dk) sequences of length n can be found in a similar fashion. Let $N(n)$ denote the number of (dk) sequences of length n. (For the sake of simplicity in notation no subscript is used in this case.) Define

$$N(n) = 0, \quad n < 0$$
$$N(0) = 1 \tag{3.3}$$

The number of (dk) sequences of length n is given by

$$
\begin{aligned}
N(n) &= n + 1, \ 1 \leq n \leq d+1 \\
N(n) &= N(n-1) + N(n-d-1), \quad d+1 \leq n \leq k \\
N(n) &= d + k + 1 - n + \sum_{i=d}^{k} N(n-i-1), \quad k < n \leq d+k \\
N(n) &= \sum_{i=d}^{k} N(n-i-1), \quad n > d+k
\end{aligned}
\tag{3.4}
$$

The proof of the above recursion relations is not interesting and therefore omitted (see [1]).

3.2.2 Capacity

An encoder translates arbitrary user (or source) information into, in this particular instance, a sequence that satisfies given dk constraints. On the average, m source symbols are translated into n channel symbols. What is the maximum value of $R = m/n$ that can be attained for some specified values of the minimum and maximum runlength d and k?

The maximum value of the rate, R, that can be achieved by any code is called the *capacity* of a (dk) code. The capacity, or asymptotic information rate, of (dk) sequences, denoted by $C(d,k)$, defined as the number of information bits per channel bit that can maximally be carried by the (dk) sequences, on average, is governed by the specified constraints and is given by

$$C(d,k) = \lim_{n \to \infty} \frac{1}{n} \log_2 N_{dk}(n) \tag{3.5}$$

We simply find

$$C(d,k) = \log_2 \lambda_{dk} \tag{3.6}$$

where λ_{dk} is the largest real root of the characteristic equation

$$z^{k+2} - z^{k+1} - z^{k-d+1} + 1 = 0 \tag{3.7}$$

Table 3.2 lists the capacity $C(d,k)$ versus the parameters d and k.

TABLE 3.2　Capacity $C(d, k)$ versus Runlength
Parameters d and k

k	$d = 0$	$d = 1$	$d = 2$	$d = 3$	$d = 4$
1	0.6942				
2	0.8791	0.4057			
3	0.9468	0.5515	0.2878		
4	0.9752	0.6174	0.4057	0.2232	
5	0.9881	0.6509	0.4650	0.3218	0.1823
6	0.9942	0.6690	0.4979	0.3746	0.2669
∞	1.000	0.6942	0.5515	0.4650	0.4057

3.3   Other Constraints

Besides sequences with simple runlength constraints as discussed above, there are a variety of channel constraints that have been reported in the literature.

3.3.1   MTR Constraints

The maximum transition run (MTR) codes, introduced by Moon and Brickner [2], $d = 0$, have different constraints on the maximum runs of 0s and 1s. The maximum 0 runlength constraint, k_0, is imposed, as in standard RLL constraints, for clock recovery, while the maximum runlength constraint on 1s, denote by k_1, is imposed to bound the maximum number of consecutive transitions (i.e., consecutive 1s). It has been shown by Moon and Brickner [2] that removing said vexatious sequences leads to improved robustness against additive noise. MTR (d, k) constraints, $d > 0$, have been advocated as they are said to improve the detection quality. The MTR constraint limits the number of consecutive strings of the form $0^d 1$, that is, repetitive occurrence of the minimum runlength are limited. In wireless infrared communications applications, the MTR constraint is imposed as otherwise catastrophic receiver failure under near-field may be induced [3, 4]. Implementations of these codes usually have $d = 1$ and rate equal to 2/3.

3.3.2 $(O, G/I)$ Sequences

Partial response signaling in conjunction with maximum likelihood detection [5–8] is a data detection technique commonly used in magnetic recording. Special runlength constraints are needed to avoid vexatious sequences which could foil the detection circuitry. These constraints are characterized by two parameters G and I. The parameter G stipulates the maximum number of allowed 0s between consecutive 1s, while the parameter I stipulates the maximum number of 0s between 1s in both the even and odd numbered positions of the sequence. The G constraint, as the k constraint in dk sequences, is imposed to improve the timing. The I constraint is used to limit the hardware requirements of the detection circuitry. Marcus et al. [9] showed that it is possible to represent $(O, G/I)$ constraints by state-transition diagrams.

To that end, we define three parameters. The quantity g denotes the number of 0s since the last 1, and a and b denote the number of 0s since the last 1 in the even and odd subsequence. It is immediate that

$$g(a, b) = \begin{cases} 2a + 1 & \text{if } a < b \\ 2b & \text{if } a \geq b \end{cases}$$

Each state in the state-transition diagram is labeled with 2-tuples (a, b), where by definition $0 \leq a, b \leq I$ and $g(a, b) \leq G$. A transition between the states numbered by (a, b) to $(b, a + 1)$ (emitting a 0) and (a, b) to $(b, 0)$ (emitting a 1) are easily attached.

By computing the maximum eigenvalue of the above state-transition matrix, we obtain the capacity of the $(O, G/I)$ sequences. Results of computations are listed in Table 3.3.

Examples of implementation of $(O, G/I)$ constrained codes were given by Marcus, Siegel and Patel [10], Eggenberger and Patel [11] and Fitzpatrick and Knudson [12].

TABLE 3.3 Capacity for Selected Values of G and I [9]

G	I	Capacity
4	4	0.9614
4	3	0.9395
3	6	0.9445
3	5	0.9415
3	4	0.9342
3	3	0.9157

3.3.3 Weakly Constrained Sequences

Weakly constrained codes do not follow the letter of the law, as they produce sequences that violate the channel constraints with probability p. It is argued that if the channel is not free of errors, it is pointless to feed the channel with perfectly constrained sequences. In the case of a dk-constrained channel, violation of the d-constraint will very often lead to errors at the receiving site, but a violation of the k-constraint is usually harmless. Clearly, the extra freedom offered by weak constraints will result in an increase of the channel capacity. An analytic expression between the capacity and violation probability of the k-constraint has been derived by Janssen and Immink [13]. Worked examples of weakly constrained codes have been given by Immink [14] and Jin et al. [15].

3.3.4 Two-Dimensional RLL Constraints

In conventional recording systems, information is organized along tracks. Interaction between neighboring tracks during writing and reading of the information cannot be neglected. During reading, in particular when tracking, either dynamic or static, is not optimal, both the information track itself plus part of the neighboring tracks are read, and a noisy phenomenon, called *crosstalk,* or *inter-track interference* (ITI) may disturb the reading process. Crosstalk is usually modeled as additive noise, and thus, essentially, the recording process is considered to be one-dimensional. Advanced coding systems that take into account inter-track interference, were developed by Soljanin and Georghiades [16].

It is expected that future mass data systems will show more of their two-dimensional character: the track pitch will become smaller and smaller relative to the reading-head dimensions, and, as a result, the recording process has to be modeled as a two-dimensional process. An example of a type of code, where the two-dimensional character of the medium is exploited to increase the code rate was introduced by Marcellin and Weber [17]. They introduced *multi-track (d,k)-constrained binary codes.* Such n-track codes are extensions of regular (d,k) codes for use in multi-track systems. In an n-track (d,k)-constrained binary code, the d constraint is required to be satisfied on each track, but the k constraint is required to be satisfied only by the bit-wise logical "or" of n consecutive tracks. For example, assume two parallel tracks, where the following sequences might be produced by a 2-track (d,k) code:

$$\text{track 1} \quad 000010100010100$$
$$\text{track 2} \quad 010000010000001$$

Note that the $d = 1$ constraint is satisfied in each track, but that the $k = 2$ constraint is satisfied only in a joint manner — there are never more than two consecutive occurrences of 0 on both tracks simultaneously. Although n-track codes can provide significant capacity increase over regular (d,k) codes, they suffer from the fact that a single faulty track (as caused by media defects, for example) may cause loss of synchronization and hence loss of the data on all tracks. To overcome this flaw Swanson and Wolf [18] introduced a class of codes, where a first track satisfies the regular (d,k) constraint, while the k-constraint of the second track is satisfied in the "joint" manner. Orcutt and Marcellin [19, 20] computed the capacity of *redundant* multi-track (d,k)-constrained binary codes, which allow only r tracks to be faulty at every time instant. Vasic computed capacity bounds and spectral properties [21–23]. Further improvements of n-track systems with faulty tracks were given by Ke and Marcellin [24].

In holographic recording, data is stored using optical means in the form of two-dimensional binary patterns. In order to safeguard the reliability of these patterns, certain channel constraints have been proposed. More information on holographic memories and channel constraints can be found in [25, 26].

Codes that take into account the two-dimensional character have been investigated by several authors. Talyansky, Etzion and Roth [27] studied efficient coding algorithms for two types of constraints on two-dimensional binary arrays. The first constraint considered is that of the t-conservative arrays, where each row and column of the array has at least t transitions. Blaum, Siegel, Sincerbox and Vardy [28–30] disclosed a code which eliminates long periodic stretches of contiguous light or dark regions in any of the dimensions of the holographic medium such that interference between adjacent images recorded in the same volume is effectively minimized.

Kato and Zeger [31] considered two-dimensional RLL constraints. A two-dimensional binary pattern of 1s and 0s arranged in an $m \times n$ rectangle is said to satisfy a two-dimensional (d, k) constraint if it satisfies a one-dimensional (d, k)-constraint both horizontally and vertically. In contrast to the one-dimensional capacity, there is little known about the two-dimensional capacity. It was shown by Calkin and Wilf that $C(d, k)$ is bounded as $0.587891 \leq C(d, k) \leq 0.588339$ [32]. Bounds on $C(d, k)$ have been derived by Kato and Zeger [31] and Siegel and Wolf [33].

3.4 Codes for the Noiseless Channel

In the present section, we take a look at the techniques that are available to produce constrained sequences in a practical manner. Encoders have the task of translating arbitrary source information onto a constrained sequence. It is most important that this be done as efficiently as possible within some practical considerations. Efficiency is usually measured in terms of the ratio of code rate R and capacity C of the constrained channel. A good encoder algorithm realizes a code rate close to the capacity of the constrained sequences, uses a simple implementation, and avoids the propagation of errors in the process of decoding.

In coding practice, the source sequence is partitioned into blocks of length p, and under the code rules such blocks are mapped onto words of q channel symbols. The rate of such an encoder is $R = p/q \leq C$. A code may be state dependent, in which case the codeword used to represent a given source block is a function of the channel or encoder state, or the code may be state independent. State independence implies that codewords can be freely concatenated without violating the sequence constraints. When the encoder is state dependent, it typically takes the form of a synchronous finite-state machine.

A decoder is preferably state independent. Due to errors made during transmission, a state-dependent decoder could easily lose track of the encoder state, and as a result the decoder could possibly make error after error with no guarantee of recovery. In order to avoid error propagation, a decoder should preferably use a finite observation interval of channel bits for decoding, thus limiting the span in which errors may occur. Such a decoder is called a *sliding block decoder*. A sliding block decoder makes a decision on a received word on the basis of the q-bit word itself, as well as a number of m preceding q-bit words and a upcoming q-bit words. Essentially, the decoder comprises a register of length $(m + a + 1)$ and a logic function $f(.)$ that translates the contents of the register into the retrieved q-bit source word. Since the constants m and a are finite, an error in the retrieved sequence can propagate in the decoded sequence only for a finite distance, at most the decoder window length $(m + a + 1)$. An important subclass of the sliding-block decoder is the *block decoder*, which uses only a single codeword for reproducing the source word, that is, $m = a = 0$ [34]. The above parameters define the playing field of the code designer. Early players are Tang and Bahl [1], Franaszek [35–38], and Cattermole [39]. In addition, important contributions were made by Jacoby [40, 41], Lempel [42], Patel [43], Cohen [44], and many others. Tutorial expositions can be found in [9, 45, 46].

In its simplest form, the set of encoder states, called *principal states*, is a subset of the channel states used to describe the constraints. From each of the principal states there are at least 2^p constrained words beginning at such a state and ending in a principal state. The set of principal states can be found by invoking Franaszek's procedure [35]. Flawless concatenation of the words is implied by the structure of the finite-state machine describing the constraints.

Concatenation of codewords can also be established by using *merging* bits between constrained words [40, 47, 48]. Merging bits are used, for example, in the EFM code employed in the Compact Disc [49]. Each source word has a unique q'-bit channel representation. We require one look-up table for translating source words into constrained words of length q' plus some logic circuitry for determining the $q - q'$ merging bits. Decoding is extremely simple: discard the merging bits and translate the q'-bit word into the p-bit source word. For (dk) codes the relation between Franaszek's principal state and the merging bit procedures was found by Gu and Fuja [50]. Immink [51] gave a constructive proof that (dk) codes with merging bits can be made for which $C - R < 1/(2q)$. As a result, (dk) codes with a rate only 0.1% less than Shannon's capacity can be constructed with codewords of length $q \approx 500$. The number of codewords grows exponentially with the codeword length, and the key obstacle to practically approaching capacity is the massive hardware required for the translation. The massiveness problem can be solved by using a technique called *enumeration* [52], which makes it possible to translate source words into codewords and vice versa by invoking an algorithmic procedure rather than performing the translation with a look-up table. Single channel bit errors could corrupt the entire data in the decoded word, and, of course, the longer the codeword the greater the number of data symbols affected. This difficulty can be solved by a special configuration of the error correcting code and the recording code [51, 53].

A breakthrough in code design occurred in the 1980s with the elegant construction method presented by Adler, Coppersmith, and Hassner (ACH) [54]. A generalized procedure was published by Ashley and Marcus [55]. The ACH algorithm, also called *state-splitting algorithm*, gives a step-by-step approach for designing constrained codes. The guarantee of a sliding-block decoder and the explicit bound on the decoder window length are the key strengths of the ACH algorithm. Roughly speaking, the state-splitting algorithm proceeds by iteratively modifying the FSTD. At each round of iteration, the maximum weight (greater than unity) is reduced, so that we eventually reach an FSTD whose approximate eigenvector has binary components. Complexity issues related to the number of encoder states and window length are an active field of research, which is exemplified by, for example, [56–58].

The *sequence replacement technique* [59] converts source words of length p into $(0, k)$-constrained words of length $q = p + 1$. The control bit is set to 1 and appended at the beginning of the p-bit source word. If this $(p + 1)$-bit sequence satisfies the prescribed constraint, it is transmitted. If, on the other hand, the constraint is violated, that is, a runlength of at least $k + 1$ 0s occur, we remove the trespassing $k + 1$ 0s. The position where the start of the violation was found is encoded in $k + 1$ bits, which are appended at the beginning of the $p + 1$-bit word. Such a modification is signaled to the receiver by setting the control bit to 0s. The codeword remains of length $p+1$. The above procedure is repeated until all forbidden subsequences have been removed. The receiver can reconstruct the source word as the position information is stored at a predefined position in the codeword. In certain situations the entire source word has to be modified which makes the procedure prone to error propagation. The class of rate $(q - 1)/q$, $(0, k)$-constrained codes, $k = 1 + \lfloor q/3 \rfloor$, $q \geq 9$, was constructed to minimize error propagation [60]. Error propagation is confined to one decoded byte irrespective of the codeword length q.

Recently, the publications by Fair et al. [61] and Immink and Patrovics [62] on *guided scrambling* brought new insights into high-rate code design. Guided scrambling is a member of a larger class of related coding schemes called *multi-mode* codes. In multi-mode codes, the p-bit source word is mapped into $(m + p)$-bit codewords. Each source word **x** can be represented by a member of a *selection set* consisting of $L = 2^m$ codewords. Examples of such mappings are the guided scrambling algorithm presented by Fair et al. [61], and the scrambling using a Reed-Solomon code by Kunisa et al. [63].

The encoder opts for transmitting that codeword that minimizes, according to a prescribed criterion, for example, the low-frequency spectral contents of the encoded sequence. There are two key elements which need to be chosen judiciously: (a) the mapping between the source words and their corresponding selection sets, and (b) the criterion used to select the "best" word. Provided that 2^m is large enough and the selection set contains sufficiently different codewords, multi-mode codes can also be used to satisfy almost any channel constraint with a suitably chosen selection method. A clear disadvantage is that the encoder needs to generate all 2^m possible codewords, compute the criterion, and make the decision.

References

[1] D.T. Tang and L.R. Bahl, Block codes for a class of constrained noiseless channels, *Info. Control,* vol. 17, pp. 436–461, 1970.

[2] J. Moon and B. Brickner, Design of a rate 6/7 maximum transition run code, *IEEE Trans. Magn.,* vol. 33, pp. 2749–2751, September 1997.

[3] M.A. Hassner, N. Heise, W. Hirt, B.M. Trager, Method and Means for Invertibly Mapping Binary Sequences into Rate 2/3, $(1, k)$ Run-Length-Limited Coded Sequences with Maximum Transition Density Constraints, U.S. Patent 6,195,025, February 2001.

[4] W. Hirt, M. Hassner, and N. Heise, IrDA-VFIr (16 Mb/s): Modulation code and system design, *IEEE Personal Commn.,* pp. 58–71, February 2001.

[5] H. Kobayashi, A survey of coding schemes for transmission or recording of digital data, *IEEE Trans. Commn.,* vol. COM-19, pp. 1087–1099, December 1971.

[6] R. Cideciyan, F. Dolivo, R. Hermann, W. Hirt, and W. Schott, A PRML system for digital magnetic recording, *IEEE J. Selected Areas in Commn.,* vol. 10, pp. 38–56, January 1992.

[7] H. Kobayashi and D.T. Tang, Application of partial response channel coding to magnetic recording systems, *IBM J. Res. Develop.,* vol. 14, pp. 368–375, July 1970.

[8] R.W. Wood and D.A. Petersen, Viterbi detection of class IV partial response on a magnetic recording channel, *IEEE Trans. Commn.,* vol. COM-34, pp. 454–461, May 1986.

[9] B.H. Marcus, P.H. Siegel, and J.K. Wolf, Finite-state modulation codes for data storage, *IEEE J. Selected Areas in Commn.,* vol. 10, no. 1, pp. 5–37, January 1992.

[10] B.H. Marcus, A.M. Patel, and P.H. Siegel, Method and Apparatus for Implementing a PRML Code, U.S. Patent 4,786,890, November 1988.

[11] J.S. Eggenberger and A.M. Patel, Method and Apparatus for Implementing Optimum PRML Codes, U.S. Patent 4,707,681, November 17, 1987.

[12] J. Fitzpatrick and K.J. Knudson, Rate 16/17, $(d = 0, G = 6/I = 7)$ modulation code for a magnetic recording channel, U.S. Patent 5,635,933, June 1997.

[13] A.J.E.M. Janssen and K.A.S. Immink, An entropy theorem for computing the capacity of weakly (d, k)-constrained sequences, *IEEE Trans. Inform. Theory,* vol. IT-46, no. 5, pp. 1034–1038, May 2000.

[14] K.A.S. Immink, Weakly constrained codes, *Electronics Letters,* vol. 33, no. 23, pp. 1943–1944, November 1997.

[15] Ming Jin, K.A.S. Immink, and B. Farhang-Boroujeny, Design techniques for weakly constrained codes, *Trans. Commn.,* vol. 51, no. 5, pp. 709–714, May 2003.

[16] E. Soljanin and C.N. Georghiades, Coding for two-head recording systems, *IEEE Trans. Inform. Theory,* vol. IT-41, no. 3, pp. 794–755, May 1995.

[17] M.W. Marcellin and H.J. Weber, Two-dimensional modulation codes, *IEEE J. Selected Areas Commn.,* vol. 10, no. 1, pp. 254–266, January 1992.

[18] R.D. Swanson and J.K. Wolf, A new class of two-dimensional RLL recording codes, *IEEE Trans. Magn.,* vol. 28, pp. 3407–3416, November 1992.

[19] E.K. Orcutt and M.W. Marcellin, Enumerable multi-track (d, k) block codes, *IEEE Trans. Inform. Theory,* vol. IT-39, pp. 1738–1743, September 1993.

[20] E.K. Orcutt and M.W. Marcellin, Redundant multi-track (d, k) codes, *IEEE Trans. Inform. Theory,* vol. IT-39, pp. 1744–1750, September 1993.

[21] B.V. Vasic, Capacity of channels with redundant multi-track (d, k) constraints: The $k < d$ Case, *IEEE Trans. Inform. Theory,* vol. IT-42, no. 5, pp. 1546–1548, September 1996.

[22] B.V. Vasic, Shannon capacity of M-ary redundant multi-track runlength limited codes, *IEEE Trans. Inform. Theory,* vol. IT-44, no. 2, pp. 766–774, March 1998.

[23] B.V. Vasic, Spectral analysis of maximum entropy multi-track modulation codes, *IEEE Trans. Inform. Theory,* vol. IT-44, no. 4, pp. 1574–1587, July 1998.

[24] L. Ke and M.W. Marcellin, A new construction for *n*-track (d, k) codes with redundancy, *IEEE Trans. Inform. Theory,* vol. IT-41, no. 4, pp. 1107–1115, July 1995.

[25] J.F. Heanue, M.C. Bashaw, and L. Hesselink, Volume holographic storage and retrieval of digital data, *Science,* pp. 749–752, 1994.

[26] J.F. Heanue, M.C. Bashaw, and L. Hesselink, Channel codes for digital holographic data storage, *J. Opt. Soc. Am.,* vol. 12, 1995.

[27] R. Talyansky, T. Etzion, and R.M. Roth, Efficient code construction for certain two-dimensional constraints, *IEEE Trans. Inform. Theory,* vol. IT-45, no. 2, pp. 794–799, March 1999.

[28] M. Blaum, P.H. Siegel, G.T. Sincerbox, and A. Vardy, Method and Apparatus for Modulation of Multi-Dimensional Data in Holographic Storage, U.S. Patent 5,510,912, April 1996.

[29] M. Blaum, P.H. Siegel, G.T. Sincerbox, and A. Vardy, Method and Apparatus for Modulation of Multi-Dimensional Data in Holographic Storage, U.S. Patent 5,727,226, March 1998.

[30] A. Vardy, M. Blaum, P.H. Siegel, and G.T. Sincerbox, Conservative arrays: multi-dimensional modulation codes for holographic recording, *IEEE Trans. Inform. Theory,* vol. IT-42, no. 1, pp. 227–230, January 1996.

[31] A. Kato and K. Zeger, On the capacity of two-dimensional run-length constrained channels, *IEEE Trans. Inform. Theory,* vol. IT-45, no. 5, pp. 1527–1540, July 1999.

[32] N.J. Calkin and H.S. Wilf, The number of independent sets in a grid graph, *SIAM J. Discr. Math.,* vol. 11, pp. 54–60, February 1998.

[33] P.H. Siegel and J.K. Wolf, Bit-stuffing bounds on the capacity of two-dimensional constrained arrays, *Proc. 1998 IEEE Int. Symp. Inform. Theory,* pp. 323, 1998.

[34] P. Chaichanavong and B. Marcus, Optimal block-type-decodable encoders for constrained systems, *IEEE Trans. Inform. Theory,* vol. IT-49, no. 5, pp. 1231–1250, May 2003.

[35] P.A. Franaszek, Sequence-state encoding for digital transmission, *Bell Syst. Tech. J.,* vol. 47, pp. 143–157, January 1968.

[36] P.A. Franaszek, Sequence-state methods for run-length-limited coding, *IBM J. Res. Develop.,* vol. 14, pp. 376–383, July 1970.

[37] P.A. Franaszek, Run-length-limited variable length coding with error propagation limitation, U.S. Patent 3,689,899, September 1972.

[38] P.A. Franaszek, On future-dependent block coding for input-restricted channels, *IBM J. Res. Develop.,* vol. 23, pp. 75–81, 1979.

[39] K.W. Cattermole, *Principles of Pulse Code Modulation,* Iliffe Books Ltd, London, 1969.

[40] G.V. Jacoby, A new look-ahead code for increasing data density, *IEEE Trans. Magn.,* vol. MAG-13, no. 5, pp. 1202–1204, September 1977.

[41] G.V. Jacoby and R. Kost, Binary two-thirds rate code with full word look-ahead, *IEEE Trans. Magn.,* vol. MAG-20, no. 5, pp. 709–714, September 1984.

[42] A. Lempel and M. Cohn, Look-ahead coding for input-restricted channels, *IEEE Trans. Inform. Theory,* vol. IT-28, no. 6, pp. 933–937, November 1982.

[43] A.M. Patel, Zero-modulation encoding in magnetic recording, *IBM J. Res. Develop.,* vol. 19, pp. 366–378, July 1975.

[44] M. Cohn and G.V. Jacoby, Run-length reduction of 3PM code via look-ahead technique, *IEEE Trans. Magn.,* vol. MAG-18, pp. 1253–1255, November 1982.

[45] B.H. Marcus, R.M. Roth, and P.H. Siegel, Constrained systems and coding for recording channels, in *Handbook of Coding Theory,* Brualdi R., Huffman C., and Pless V., Eds., Amsterdam, The Netherlands, Elsevier Press, 1996.

[46] K.A.S. Immink, Runlength-limited sequences, *Proc. IEEE,* vol. 78, no. 11, pp. 1745–1759, November 1990.

[47] G.F.M. Beenker and K.A.S. Immink, A generalized method for encoding and decoding runlength-limited binary sequences, *IEEE Trans. Inform. Theory,* vol. IT-29, no. 5, pp. 751–754, September 1983.

[48] K.A.S. Immink, Constructions of almost block-decodable runlength-limited codes, *IEEE Trans. Inform. Theory,* vol. IT-41, no. 1, pp. 284–287, January 1995.

[49] J.P.J. Heemskerk and K.A.S. Immink, Compact disc: system aspects and modulation, *Philips Techn. Review,* vol. 40, no. 6, pp. 157–164, 1982.

[50] J. Gu and T. Fuja, A new approach to constructing optimal block codes for runlength-limited channels, *IEEE Trans. Inform. Theory,* vol IT-40, no. 3, pp. 774–785, 1994.

[51] K.A.S. Immink, A practical method for approaching the channel capacity of constrained channels, *IEEE Trans. Inform. Theory,* vol. IT-43, no. 5, pp. 1389–1399, September 1997.

[52] T.M. Cover, Enumerative source coding, *IEEE Trans. Inform. Theory,* vol. IT-19, no. 1, pp. 73–77, January 1973.

[53] J.L. Fan and A.R. Calderbank, A modified concatenated coding scheme with applications to magnetic recording, *IEEE Trans. Inform. Theory,* vol. IT-44, pp. 1565–1574, July 1998.

[54] R.L. Adler, D. Coppersmith, and M. Hassner, Algorithms for sliding block codes: an application of symbolic dynamics to information theory, *IEEE Trans. Inform. Theory,* vol. IT-29, no. 1, pp. 5–22, January 1983.

[55] J.J. Ashley and B.H. Marcus, A generalized state-splitting algorithm, *IEEE Trans. Inform. Theory,* vol. IT-43, no. 4, pp. 1326–1338, July 1997.

[56] J.J. Ashley, R. Karabed, and P.H. Siegel, Complexity and sliding-block decodability, *IEEE Trans. Inform. Theory,* vol. IT-42, pp. 1925–1947, 1996.

[57] J.J. Ashley and B.H. Marcus, Canonical encoders for sliding-block decoders, *SIAM J. Discrete Math.,* vol. 8, pp. 555–605, 1995.

[58] B.H. Marcus and R.M. Roth, Bounds on the number of states in encoder graphs for input-constrained channels, *IEEE Trans. Inform. Theory,* vol. IT-37, no. 3, part 2, pp. 742–758, May 1991.

[59] A.J. de Lind van Wijngaarden and K.A.S. Immink, Construction of constrained codes using sequence replacement techniques, Submitted *IEEE Trans. Inform. Theory,* 1997.

[60] K.A.S. Immink and A.J. de Lind van Wijngaarden, Simple high-rate constrained codes, *Electronics Letters,* vol. 32, no. 20, pp. 1877, September 1996.

[61] I.J. Fair, W.D. Gover, W.A. Krzymien, and R.I. MacDonald, Guided scrambling: a new line coding technique for high bit rate fiber optic transmission systems, *IEEE Trans. Commn.,* vol. COM-39, no. 2, pp. 289–297, February 1991.

[62] K.A.S. Immink and L. Patrovics, Performance assessment of DC-free multimode codes, *IEEE Trans. Commn.,* vol. COM-45, no. 3, March 1997.

[63] A. Kunisa, S. Takahashi, and N. Itoh, Digital modulation method for recordable digital video disc, *IEEE Trans. Consumer Electr.,* vol. 42, pp. 820–825, August 1996.

4

Maximum Transition Run Coding

Barrett J. Brickner

Bermai, Inc.
Minnetonka, MN

4.1 Introduction

The written data and corresponding readback waveform in any recording system are subject to noise and distortions that limit reliability of the system. An error correction code allows a certain amount of data corruption to be corrected in the decoder by providing specific redundancy in the recorded data. If the system is more susceptible to errors in specific data patterns, a more direct approach is to employ a channel coding constraint that prevents these troublesome patterns so that the error simply does not occur.

In the text that follows, a specific class of codes designed to improve the performance of recording systems is explored. The discussion follows a methodology that is generally applicable to the development of code constraints to improve minimum distance properties of communications or recording systems. Initially, the system is characterized in terms of types of error events and the probability with which they may occur, a value expressed as a distance in a geometric context. Having identified the error types that are most likely to corrupt the system, pairs of code bit sequences that produce these errors are determined. An error occurs when noise, combined with intersymbol interference (ISI) causes the received signal produced by one code bit sequence to resemble that produced by another. This ambiguity is resolved by simply enforcing a constraint in the encoder, which prevents one or both of the code bit sequences. When only one of the two error-producing code bit sequences is removed, the detector and/or decoder must be modified to choose in favor of the valid sequence. The resulting encoder and detector/decoder work in concert to improve the system performance. However, the addition of a code constraint reduces the code rate and the amount of information conveyed in a sequence of code bits. An analysis or simulation is then used to verify that a net gain results from removing error events at the expense of a lower code rate.

A properly chosen maximum transition run (MTR) constraint, which limits the number of consecutive transitions, is shown to prevent minimum-distance errors for a variety of channel responses applicable to recording systems. This idea of using a code constraint to prevent problematic bit sequences is not new to recording. For years, RLL(d, k) codes which specify a minimum, d, and maximum, k, number of nontransitions between transitions have been used with $d > 0$ to help peak detector base read circuits by reducing the effects of ISI in adjacent transitions. A $d = 1$ code has also been used to improve the distance properties of a high order partial-response maximum likelihood (PRML) channel.[1] Moreover, while an MTR constraint is sufficient to remove certain, minimum-distance error events, it is not a unique solution to this problem. Other coding schemes based on forbidding one of the pairs of error generating code sequences have been shown to give similar distance gains.[2,3] For recording systems, RLL $d > 0$ and MTR constraints are of particular interest because many of the disturbances that fall outside the simple linear ISI and additive white Gaussian noise (AWGN) model occur when transitions are brought in close proximity to one another.

4.2   Error Event Characterization

For convenience and ease of analysis, the channel is assumed to be linear with additive white Gaussian noise such that the received signal is written as

$$r_k = s_k + n_k = \sum_{i=0}^{L-1} h_k a_{k-i} + n_k$$

where the discrete-time channel is represented as $h(D) = \sum_{k=0}^{L} h_k D^k$, and the input data are taken from the binary alphabet $\{0, 1\}$. Typically, $h(D)$ is formed by equalizing the received signal with a noise-whitened matched filter such that the noise statistics are preserved, and $h(D)$ is the combined response of the channel and equalizer. The noise is assumed to be additive white Gaussian noise with variance σ_n^2. The maximum likelihood sequence detection (MLSD) estimates an N-sample input sequence $\mathbf{a}_k = [a_k, a_{k-1}, \ldots, a_{k-N+1}]$ using

$$\hat{\mathbf{a}}_k = \arg\left\{ \min_{\mathbf{a}_k} \sum_{j=0}^{N+L-2} \left(r_{k-j} - \sum_{i=0}^{L-1} h_i a_{k-i-j} \right)^2 \right\}$$

where $a_{k-i} = 0$ for $i \geq N$. An error is produced whenever the estimated sequence does not match the input sequence in one or more locations, that is, $\hat{\mathbf{a}}_k \neq \mathbf{a}_k$ such that $e_k = \left\{ \begin{smallmatrix} a_k - \hat{a}_k, & k=0\ldots N-1 \\ 0, & \text{elsewhere} \end{smallmatrix} \right.$. Using Marcum's Q function

$$Q(x) = \Pr[X > x] = \frac{1}{\sqrt{2\pi}} \int_x^\infty e^{-z^2/2}\, dz$$

the probability of a particular error is $\Pr(\hat{\mathbf{a}} \neq \mathbf{a} \mid \mathbf{a}) \approx Q(d_{\hat{\mathbf{a}},\mathbf{a}}/2\sigma_n)$ where

$$d_{\hat{\mathbf{a}},\mathbf{a}} = \sqrt{\sum_{i=0}^{N+L-2} \left(\sum_{j=0}^{L-1} h_j e_{i-j} \right)^2}$$

This value is the Euclidean distance between two points whose coordinates correspond to the noiseless received signals generated by two valid input sequences. For low error rate situations, a small change in the distance d results in an exponential change in the error probability. Therefore, the performance of the system is dominated by those error events that produce the minimum distance, $d_{\min}$.

To compute the minimum distance as shown above, the corresponding error event must be known. In the general case, the length of the input sequence and, therefore, the possible error event sequences are unbounded. However, by considering all error events up to a particular length, the minimum distance can be bounded by

$$\min_{\mathbf{e}_k} \sum_{k=0}^{N-1} \left(\sum_{j=0}^{L-1} f_j e_{k-j} \right)^2 \leq d_{\min}^2 \leq \min_{\mathbf{e}_k} \sum_{k=0}^{N+L-2} \left(\sum_{j=0}^{L-1} f_j e_{k-j} \right)^2$$

where N is the length of the error event $\mathbf{e}_k$.[4] The lower-bound has particular relevance to the fixed delay tree search with decision feedback (FDTS/DF) detector as it gives the exact minimum distance when $N - 1$ is equal to the depth of the tree search.[5] It is often useful to compare the result of this calculation to the matched filter bound

$$d_{MF} = \sqrt{\sum_{k=0}^{L-1} f_k^2}$$

which indicates the distance for the case where a single bit is transmitted. For channels with little ISI, it is common for $d_{min} = d_{MF}$; however, for many of the responses seen in data storage channels, the ISI structure is more severe, and $d_{min} < d_{MF}$. In any case, the minimum distance for the uncoded channel is never larger than d_{MF}.

The error events of interest here are closed error events, or, in other words, those that are finite in duration. Open, or quasi-catastrophic, error events can extend indefinitely and are generally handled by additional code constraints in practical MLSD implementations. In this discussion, error events are shown to start and end with nonzero terms e_k. To be an independent error event, these sequences are both preceded and followed by $L - 1$ zero terms. Often, the errors of interest are short, or have a predictable form, so that they can be identified by a simple exhaustive search of all events up to a particular length N. However, to guarantee that all events have been identified, a more rigorous approach is followed. Altekar et al. describe a suitable method for finding error events in partial response channels up to a certain distance.[6]

With low-order partial response channels, such as PR1, PR4, and EPR4 characterized by the response polynomials $(1 + D), (1 - D)(1 + D)$, and $(1 - D)(1 + D)^2$, respectively, $d_{min} = d_{MF}$. However, as shown in Table 4.1, error events other than a single bit error are dominant for channels with a greater high frequency roll-off; that is, those with higher order $(1 + D)$ factors. In particular, these errors consist of groups of terms for which the signs of consecutive error bits alternate. An error event written as $\pm\{\mathbf{e}_1, \langle\mathbf{e}_2\rangle\}$ indicates that the same distance is obtained by concatenating error sequence $\mathbf{e}_1$ with a nonnegative integer number $(0, 1, 2, \ldots)$ of repetitions of sequence $\mathbf{e}_2$. Because these errors correspond to a signal difference with a significant high-frequency component, it is not too surprising that they are prevalent in channels that are inherently low-pass in nature.

The basic pattern that emerges in these error sequences for the high-order partial response channels is that they are defined by, or contain, the sequences $\pm\{+1, -1\}$ and $\pm\{+1, -1, +1\}$. Two particular cases of interest are PR2 with a response $(1 + D)^2$ and E2PR4 with a response $(1 - D)(1 + D)^3$. Error events with distances up through d_{MF} are given for these two channels in Table 4.2 and Table 4.3, respectively. In longitudinal magnetic disc recording, a Lorentzian pulse is often used to model the transition response

TABLE 4.1 Distance to Matched Filter Bound for Partial Response Channels

Response	Polynomial	d_{min}^2	d_{MF}^2	d_{min}/d_{MF} (dB)	Minimum Distance Errors
PR1	$(1 + D)$	2	2	0	$\pm\{+1, \langle-1, +1\rangle\}$
					$\pm\{+1, -1, \langle+1, -1\rangle\}$
PR2	$(1 + D)^2$	4	6	−1.76	$\pm\{+1, -1, \langle+1, -1\rangle\}$
					$\pm\{+1, -1, +1, \langle-1, +1\rangle\}$
EPR2	$(1 + D)^3$	10	20	−3.01	$\pm\{+1, -1\}$
Dicode	$(1 - D)$	2	2	0	$\pm\{+1, \langle+1\rangle\}$
PR4	$(1 - D)(1 + D)$	2	2	0	$\pm\{+1, \langle0, +1\rangle\}$
EPR4	$(1 - D)(1 + D)^2$	4	4	0	$\pm\{+1, \langle0, +1\rangle\}$
					$\pm\{+1, -1, +1, \langle-1, +1\rangle\}$
					$\pm\{+1, -1, +1, -1, \langle+1, -1\rangle\}$
E2PR4	$(1 - D)(1 + D)^3$	6	10	−2.22	$\pm\{+1, -1, +1\}$
E3PR4	$(1 - D)(1 + D)^4$	12	26	−3.36	$\pm\{+1, -1, +1\}$

TABLE 4.2   Dominant Closed Error Events for a PR2 $(1 + D)^2$ Channel

d^2	d^2/d_{MF}^2 (dB)	Error Events
4	−1.76	$\pm\{+1, -1, \langle +1, -1\rangle\}$
		$\pm\{+1, -1, +1, \langle -1, +1\rangle\}$
6	0.00	± 1
		$\pm\{+1, -1, 0, +1, -1, +1, \langle -1, +1\rangle\}$
		$\pm\{+1, -1, \langle +1, -1\rangle, 0, +1, -1, \langle +1, -1\rangle\}$
		$\pm\{+1, -1, +1, \langle -1, +1\rangle, 0, -1, +1, \langle -1, +1\rangle\}$

TABLE 4.3   Dominant Closed Error Events for an E2PR4 $(1 - D)(1 + D)^3$ Channel

d^2	d^2/d_{MF}^2 (dB)	Error Events
6	−2.22	$\pm\{+1, -1 +1, \langle -1, +1\rangle\}$
8	−0.97	$\pm\{+1, -1 +1, -1, \langle +1, -1\rangle\}$
		$\pm\{+1, -1, +1, -1, +1, \langle -1, +1\rangle\}$
		$\pm\{+1, -1, +1, 0, 0, +1, -1, +1\}$
10	0.00	± 1
		$\pm\{+1, 0, 0, +1, -1, +1\}$
		$\pm\{+1, -1, +1, 0, -1, +1, -1\}$
		$\pm\{+1, -1, +1, 0, 0, +1\}$
		$\pm\{+1, -1, +1, 0, 0, +1, -1, +1, -1, \langle +1, -1\rangle\}$
		$\pm\{+1, -1, +1, 0, 0, +1, -1, +1, -1, +1, \langle -1, +1\rangle\}$
		$\pm\{+1, -1, +1, 0, 0, 0 + 1, -1, +1\}$
		$\pm\{+1, -1, +1, 0, 0, +1, 0, 0, +1, -1, +1\}$
		$\pm\{+1, -1, +1, 0, 0, +1, -1, +1, 0, 0, +1, -1, +1\}$
		$\pm\{+1, -1, +1, -1, \langle +1, -1\rangle, 0, 0, -1, +1, -1\}$
		$\pm\{+1, -1, +1, -1, +1, \langle -1, +1\rangle, 0, 0, +1, -1, +1\}$

of the magnetic channel. This transition response is defined by

$$p(t) = \frac{A}{1 + \left(\frac{2t}{PW_{50}}\right)^2}$$

where PW_{50} parameterizes the width of the pulse. For a symbol clock period of T, the symbol density, defined as $D_s = PW_{50}/T$, gives a measure of the severity of the channel intersymbol interference (ISI). Figure 4.1 shows the relevant distance between several key error events as a function of density. As symbol density increases, the high-frequency roll-off of the channel response also increases, and just as with the partial response polynomials, the error events with alternating signs become dominant. Although the Lorentzian response with AWGN provides a reasonable approximation for the recording channel, it is not purported to be an exact model. As such, there will always be a place for analysis of errors in specific recording systems.[7]

4.3   Maximum Transition Run Codes

For channels where the minimum distance is not produced by a single bit error, a code can be constructed to prevent the error from occurring by prohibiting one or both of the pairs of sequences whose difference produces the error. Because the constraints used to prevent errors will require a greater density of code bits to maintain a particular data density, this approach does not always provide a net gain. Fortunately, there are a number of cases where the distance gain does exceed the code rate penalty.

First, consider the high density Lorentzian channels ($D_s \sim 3$) and E2PR4 response. For these, the dominant error events have the form $\pm\{+1, -1, +1\}$. The pairs of input sequences, which generate these

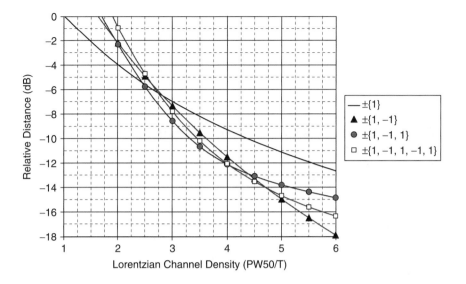

FIGURE 4.1 Euclidean distance for key error events in a Lorentzian channel.

errors, contain at least one pattern with three or more consecutive transitions as shown in Figure 4.2. If all input sequences that contain three or more consecutive transitions are eliminated, a detector can be constructed such that the original minimum distance error events are suppressed. A maximum transition run code with constraint parameter j is one that limits the number of consecutive transitions to j. In the case of the E2PR4 channel, a code with $j = 2$ is sufficient to prevent the $\pm\{+1, -1, +1\}$ error event. More generally, this constraint will prevent error events of the form $\pm\{+1, -1, +1, (-1, +1)\}$. Specifically, this constraint will allow the suppression of an error event where one or both of the sequences involved in the error contain $j + 1$ or more transitions. Because an MTR code usually includes a runlength limited k constraint to assist with timing recovery, the code parameters are encapsulated as MTR$(j; k)$.

The MTR code with $j = 1$ will also prevent any error events of the form $\pm\{+1, -1, \langle+1, -1\rangle\}$. This constraint is implied by RLL(d, k) codes with $d > 0$. Specifically, the MTR $j = 1$ constraint is the same as the more common RLL $d = 1$. Because $d = 1$ codes automatically prevent transition runs, they have the same distance-enhancing properties as MTR $j = 2$ codes, albeit at a lower code rate. This distance enhancing property of the $d = 1$ constraint has been exploited for both E2PR4 decoders and FDTS/DF.[1,5]

It is convenient to describe the MTR constraints in terms of an NRZI format where a 1 and 0 represent, respectively, the presence and absence of a transition. In this form, j indicates the maximum number of consecutive 1s, and k is the maximum number of consecutive 0s. A precoder of the form $P(D) = 1/(1 \oplus D)$ is used to generate the mapping from NRZI bits, denoted x_k, to the NRZ channel input symbols a_k. Thus, the precoder implements $a_k = a_{k-1} \oplus x_k$.

A simple rate 4/5 MTR$(2; 8)$ code can be constructed by removing from the list of 32 5-bit codewords, the all 0s codeword, all codewords with three or more consecutive transitions, and all codewords with more

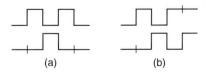

(a)   (b)

FIGURE 4.2 Pairs of write sequences which produce a $\pm\{+1, -1, +1\}$ error event.

than a single transition at the beginning or end. The set of 16 valid codewords is given by {00001, 00010, 00100, 00101, 00110, 01000, 01001, 01010, 01100, 01101, 10000, 10001, 10010, 10100, 10101, 10110}.

The MTR constraint is useful when the specific objective of the code designer is to limit the length of transition runs. With partial response channels such as PR4, an INRZI precoder given by $P(D) = 1/(1 \oplus D)^2$ is often used to convert $(0, G/I)$ codes into bit streams that avoid certain quasi-catastrophic error events (those errors that can occur in a Viterbi decoder truncated to have a finite decision delay). If an RLL$(0, k)$ coded sequence intended to be NRZI precoded is, instead, INRZI precoded, the output satisfies NRZI precoded MTR$(k + 1; k + 1)$ constraints. The MTR$(j; k)$ constraints remove all E2PR4 quasi-catastrophic error events except those that contain repetitions of $\pm\{+1, 0\}$. This particular event can be prevented by eliminating repetitions of the NRZI sequence $\{1, 1, 0, 0\}$ from the encoder output.[8] Cideciyan et al. define a formal constraint t to limit the number of consecutive $\{0, 0\}$ and $\{1, 1\}$ pairs.[9]

The use of the MTR constraint here is justified by the minimum distance gain it provides; however, an MTR constraint can provide additional benefits in the magnetic recording channel. In the presence of intertrack interference, the use of an MTR $j = 3$ constraint has been suggested as a means to reduce the $\pm(+1, -1, +1, -1, +1, \langle-1, +1\rangle)$ error event in EPR4 channels.[10] When transitions in a longitudinal recording system are written with a narrow spacing, the zigzag nature of the transitions allows portions of pairs of transitions to overlap and partially erase the transition pair. With an MTR $j = 2$ coded system, the dibit transition pairs can be recorded with the leading transition written early and the trailing transition written late to mitigate the effects of partial erasure.[11] However, if the increased separation is too large, it will be difficult to resolve the position of adjacent dibits as the two transitions separated by a nontransition are moved closer to one another. To avoid this difficulty, the code can be further constrained to require a minimum of two nontransitions between successive dibit patterns.[9]

Although the MTR $j = 2$ code will provide the desired distance gain, it is not a unique solution. In Figure 4.2, the error event produced by sequence pair (a) can be eliminated with MTR $j = 3$. Pair (b) is generated by a shifted tribit (three consecutive transitions). If these tribits are prevented from starting at either even or odd clock periods, then the decoder can uniquely resolve the correct sequence, and the error event will be prevented.[12,13] Although the time-varying nature of the constraints yields a more complicated detector, the available code rate is higher. These constraints are written as TMTR$(j - 1/j; k)$ to indicate that the MTR constraint alternates between $j - 1$ and j for sequences starting at either an even or odd time index. In fact, this code is also an MTR$(j; k)$ code, but the added constraint of $j - 1$ for transition runs starting on every other code bit period provides the same distance gain as the MTR$(j - 1; k)$ code.

For the strictly low-pass channels characterized by $(1 + D)^n$, the high-order response polynomials are also sensitive to error events with an even number of consecutive alternating signs. In particular, for PR2$(n = 2)$, minimum distance error event is $\{+1, -1\}$. Although there are several pairs of sequences which can produce this error, if an MTR $j = 2$ constraint is employed, this type of error is only produced by a shifted dibit (two consecutive transitions). Just as a time-varying TMTR$(2/3; k)$ constraint can be used to prevent a shifted tribit, so too can a TMTR$(1/2; k)$ constraint be used to prevent a shifted dibit error, thus yielding a 1.76 dB distance gain.[14] A set of codewords suitable for implementing a rate 3/4 TMTR$(1/2; 6)$ code is given by {0001, 0010, 0100, 0101, 0110, 1000, 1001, 1010}. Because the MTR constraint is time-varying for even and odd time indices, choosing codewords with an even length simplifies the design by making the constraints uniquely position dependent within the codeword. Although the use of parity codes is beyond the scope of this discussion, it is worth noting that a parity code combined with an MTR$(1/2; k)$ code can prevent single occurrences of error events with distances up to 3.98 dB from d_{min}.[15]

TMTR codes prevent errors due to shifted transition runs by allowing these patterns to begin only on alternating sample indices. A sufficient condition for avoiding shifted transition run errors is to enforce a constraint that uniquely specifies a valid starting position for the transition run. The even/odd timing requirements of the TMTR code provide this constraint. Another approach, proposed as an alternative to the TMTR$(1/2; k)$ code, is to combine an MTR$(2; k)$ code with a constraint that forces dibits to be preceded by one of either an even or odd number of nontransitions.[16] Herein, this type of code is denoted

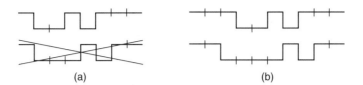

FIGURE 4.3 Pairs of write sequences which produce an error containing a shifted tribit.

MTR($j; k$, odd) if the number of nontransitions preceding a run of j transitions is odd. The same idea can be applied to shifted tribits in an E2PR4 channel with an MTR(3; k, odd) code. An error event with a shifted tribit must be preceded by a single bit error during the period of nontransitions for the error to occur. This condition is illustrated in Figure 4.3; the bottom sequence for the pair labeled (a) is disallowed because it contains an even number of nontransitions before the tribit. In the same figure, the sequence pair labeled (b) shows two valid sequences that can produce an error containing a shifted tribit. Note that the squared distance for this error, $\pm\{+1, 0, 0, +1, -1, +1\}$, in an E2PR4 channel is 10, which is the same as d_{MF}. For isolated transitions and dibits (runs of transitions up to $j - 1$), no constraint is placed on the number of 0s preceding the transition run. A set of codewords which give a rate 3/4 MTR(3; 7, odd) code are given by {00010, 00100, 00101, 00110, 01000, 01010, 01100, 01101, 10000, 10001, 10010, 10100, 10101, 10110, 11000, 11010}.

All the MTR variants discussed to this point will serve to eliminate the targeted error sequences. What varies from one to another is the detector/decoder complexity and available code capacity. The capacity for a set of constraints is an upper bound on the achievable code rate, R. Recording channels, which suffer from ISI are particularly sensitive to code rate because the SNR loss is greater than the $10 \cdot \log_{10} R$ that would be expected if the only penalty were increased noise bandwidth. Assuming that the rate loss is incurred to remove error events with distances less than d_{MF}, the penalty can be computed from Figure 4.1 as a function of density. For convenience, define user density as the number of data bits per PW_{50}, that is, the symbol density is $D_s = D_u/R$. The rate loss penalty for a code with rate $R = 1/2$ increases from 4 dB at user density $Du = 1$ to 5.31 dB at $Du = 2$, and 5.67 dB at $Du = 3$. Ultimately, the penalty will be higher in a real system; the computation here is for a linear ISI channel, and additional rate dependent loss mechanisms such as partial erasure and transition noise are neglected.

To compute capacity, a finite state transition diagram (FSTD) representing the code constraints is constructed. All valid coded bit streams can be obtained by traversing states in the FSTD and concatenating the corresponding edge labels. In addition to computing capacity, the FSTD can be used as the basis for certain code construction techniques.[17] The FSTD for a MTR($j; k$) code is shown in Figure 4.4 where an NRZI format is assumed for the code bits. Note that the dotted lines indicate additional states, like those previous, may be added as necessary to give the proper number of states.

For an FSTD with N states, an adjacency matrix describing the edges is constructed. This is an N-by-N matrix **A**, where each entry a_{ij} is the number of edges from state i to state j. As an example, consider an

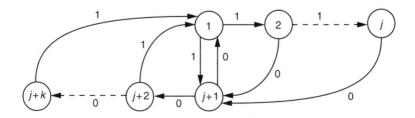

FIGURE 4.4 Finite state transition diagram for an MTR($j; k$) code.

TABLE 4.4   Capacities for MTR($j; k$) Codes

	$j = 1$	$j = 2$	$j = 3$	$j = 4$	$j = 5$	$j = 6$
$k = 1$	0.0000	0.4057	0.5515	0.6174	0.6509	0.6690
$k = 2$	0.4057	0.6942	0.7947	0.8376	0.8579	0.8680
$k = 3$	0.5515	0.7947	0.8791	0.9146	0.9309	0.9388
$k = 4$	0.6174	0.8376	0.9146	0.9468	0.9614	0.9684
$k = 5$	0.6509	0.8579	0.9309	0.9614	0.9752	0.9818
$k = 6$	0.6690	0.8680	0.9388	0.9684	0.9818	0.9881
$k = 7$	0.6793	0.8732	0.9427	0.9718	0.9850	0.9912
$k = 8$	0.6853	0.8760	0.9447	0.9735	0.9865	0.9927
$k = 9$	0.6888	0.8774	0.9457	0.9744	0.9873	0.9934
$k = 10$	0.6909	0.8782	0.9462	0.9748	0.9877	0.9938
$k = 11$	0.6922	0.8786	0.9465	0.9750	0.9879	0.9940
$k = 12$	0.6930	0.8789	0.9466	0.9751	0.9880	0.9941
$k = \infty$	0.6942	0.8791	0.9468	0.9752	0.9881	0.9942

MTR(2; 4) constraint. The adjacency matrix is then

$$
A = \begin{bmatrix}
0 & 1 & 1 & 0 & 0 & 0 \\
0 & 0 & 1 & 0 & 0 & 0 \\
1 & 0 & 0 & 1 & 0 & 0 \\
1 & 0 & 0 & 0 & 1 & 0 \\
1 & 0 & 0 & 0 & 0 & 1 \\
1 & 0 & 0 & 0 & 0 & 0
\end{bmatrix}
$$

Given A, capacity is then computed as

$$ C = \log_2 \lambda(A) $$

where $\lambda(A)$ is the largest real eigenvalue of the matrix A.[18] For the MTR(2; 4) example, the capacity is $C = \log_2(1.7871) = 0.8376$. Capacities for different values of j and k are provided in Table 4.4. For MTR constraints of $j = 4$, the impact to capacity is minimal, and codes with rates of 24/25 can be constructed. Typically, MTR codes are specified with $j < k$, but because j and k represent the maximum number of consecutive 1s and 0s, respectively, the capacity for MTR($j; k$) is the same as for MTR($k; j$).

The TMTR codes were proposed as a means of obtaining the same distance gain, but at a higher capacity. The FSTD for a TMTR($j - 1/j; k$) code is shown in Figure 4.5. In this figure, the states are shown with

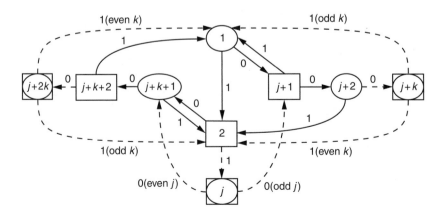

FIGURE 4.5   Finite state transition diagram for a TMTR($j - 1/j; k$) code.

TABLE 4.5 Capacities for MTR($j - 1/j; k$) Codes

	$j = 1/2$	$j = 2/3$	$j = 3/4$	$j = 4/5$
$k = 1$	0.0000	0.5000	0.5840	0.6358
$k = 2$	0.5706	0.7507	0.8170	0.8482
$k = 3$	0.6804	0.8423	0.8974	0.9231
$k = 4$	0.7381	0.8802	0.9312	0.9543
$k = 5$	0.7619	0.8983	0.9466	0.9685
$k = 6$	0.7764	0.9070	0.9540	0.9753
$k = 7$	0.7831	0.9115	0.9576	0.9786
$k = 8$	0.7874	0.9137	0.9594	0.9802
$k = 9$	0.7894	0.9149	0.9604	0.9810
$k = 10$	0.7908	0.9156	0.9608	0.9814
$k = 11$	0.7915	0.9159	0.9611	0.9816
$k = 12$	0.7919	0.9161	0.9612	0.9817
$k = \infty$	0.7925	0.9163	0.9613	0.9818

circles or squares to indicate alternating clock periods, that is, the squares represent an odd time index and the circles an even time index, or vice versa. The end states for runs of 0s and ones are determined by whether k and j are even or odd. To be valid, edges can only connect from squares to circles and from circles to squares. Table 4.5 lists the capacities for different values of j and k. Of interest to E2PR4 channels are the TMTR($2/3; k$) codes. For large values of k, the capacity of these codes approaches 0.9163, which is significantly better than the 0.8791 attained by MTR($2; k$) codes. For PR2 channels, the MTR $j = 1$ (or RLL $d = 1$) codes are limited to a capacity of 0.6942, while TMTR($1/2; k$) codes are available with capacities up to 0.7925.

Even greater capacities are available from the MTR($j; k$, odd) codes where an odd number of 0s must precede a run of j 1s. The state diagram for this code is shown in Figure 4.6. If the RLL k constraint is an

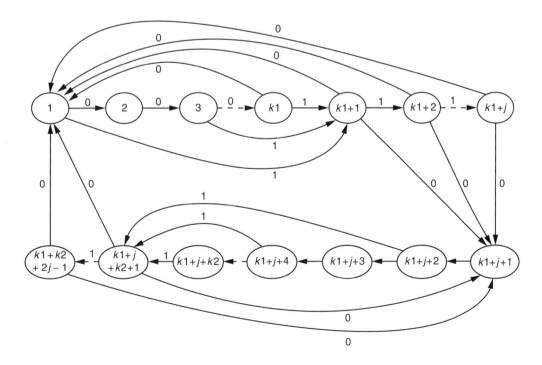

FIGURE 4.6 Finite state transition diagram for an MTR($j; k$, odd) code.

TABLE 4.6   Capacities for MTR($j; k$, odd) Codes

	$j = 2$	$j = 3$	$j = 4$	$j = 5$
$k = 1$	0.0000	0.0000	0.0000	0.0000
$k = 2$	0.6054	0.7618	0.8232	0.8510
$k = 3$	0.7366	0.8569	0.9049	0.9263
$k = 4$	0.7720	0.8903	0.9363	0.9566
$k = 5$	0.7996	0.9089	0.9519	0.9709
$k = 6$	0.8086	0.9165	0.9588	0.9774
$k = 7$	0.8161	0.9211	0.9625	0.9807
$k = 8$	0.8187	0.9231	0.9642	0.9823
$k = 9$	0.8210	0.9243	0.9651	0.9831
$k = 10$	0.8218	0.9248	0.9656	0.9835
$k = 11$	0.8225	0.9252	0.9658	0.9837
$k = 12$	0.8227	0.9253	0.9659	0.9838
$k = \infty$	0.8232	0.9255	0.9661	0.9839

odd number, then $k_1 = k$, and $k_2 = k - 1$; otherwise, $k_1 = k - 1$, and $k_2 = k$. In any case, $k_1 + k_2 = 2k - 1$, and the total number of states is $2(k + j - 1)$. Capacities for different parameter values are listed in Table 4.6. For E2PR4 channels, an MTR(3; k, odd) code provides capacities up to 0.9255 compared with 0.9163 for the TMTR(2/3; k) code. The increase in capacity is even more dramatic for PR2 channels where an MTR(2; k, odd) code has capacities up to 0.8232 compared with 0.7925 for TMTR(1/2; k).

4.4 Detector Design for MTR Constraints

The code constraints considered to this point provide a distance gain by allowing sequences that contain, at most, one type of bit sequence involved in generating an error event. However, it is the detector which must determine the correct sequence from the receive signal waveform. In order to resolve the ambiguity about which pair of error-generating sequences was actually written, the detector must be modified to prevent detection of the disallowed sequences.

A common detector choice is MLSD, implemented with a Viterbi detector.[19] If the channel response has length L time-samples (e.g., $L = 5$ for E2PR4), each detector state may correspond to a sequence with up to $L - 2$ transitions. Therefore, if $L > j + 2$, at least two states in the detector will correspond to forbidden sequences. Considering edges from previous states, one additional transition can be implied, so for $L > j + 1$, at least two edges in the trellis will correspond to illegal sequences. In the case of a TMTR code, these illegal states and edges will also be time-varying to match the code constraints.

As an example, consider the $N = 2^{L-1} = 16$ state trellis section shown in Figure 4.7. For the TMTR(2/3; k) code, the states with the NRZ sequence labels 0101 and 1010, which correspond to three consecutive transitions, are illegal during time periods for which only sequences of two transitions are allowed. At all time periods, transitions between 0101 and 1010 are removed because they correspond to four consecutive transitions. By removing these states and transitions, the Viterbi path memory can only contain sequences that are permitted by the TMTR constraints. For a static MTR(2; k) constraint, the states 0101 and 1010 are always illegal. The additional states and edges removed by $j = 2$ are shown with dotted lines in the figure. The MTR($j; k$, odd) codes require a slightly different approach. Accommodating the constraint within the trellis structure by pruning states and edges would require a 2^{j+k+1} state trellis such that a sequence with a transition, followed by k nontransitions and then j transitions is represented. Clearly, this is unreasonable for large values of k. Alternatively, for states corresponding to j consecutive transitions, the detector can look back through the path memory associated with the edges leading into the state in question. If the number of nontransitions preceding the j-transition run is even, the edge is disallowed.

Fixed delay tree search with decision feedback (FDTS/DF) is another detector structure that has been suggested for use with MTR coded system. This detector uses a decision feedback equalizer (DFE) to remove ISI due to symbols beyond a chosen truncation point in the response. A distance metric, such as is

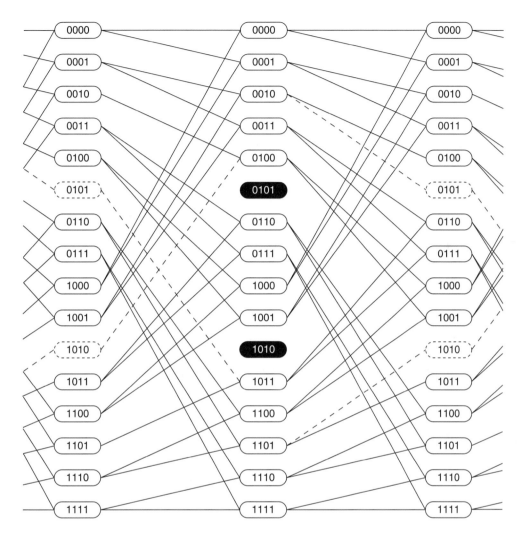

FIGURE 4.7 E2PR4 Viterbi trellis shown with states and edges removed for a TMTR$(2/3; k)$ code and in dashed lines for an MTR$(2; k)$ code.

used with MLSD, is computed for possible sequences constructed from the remaining ISI terms. Decisions are made at a fixed time-delay determined by the tree depth or number of ISI terms considered. For an MTR $j = 2$ code and FDTS/DF with $t = 2$ (sequence estimation using the $f_0 + f_1 D + f_2 D$ portion of the channel response), Figure 4.8 shows the relevant sequences in the tree search eliminated by the code. As shown, the previous decision is used to dynamically deselect one of two detector paths, thereby preventing the illegal tribit patterns.

4.5   Simulation Results

The net SNR gain for a particular MTR coding scheme can be estimated by subtracting the rate loss penalty from the distance gain. Because this value is only an approximation, simulations are used to provide a more accurate result. Here, the coding gain on a Lorentzian channel equalized to an E2PR4 target is examined. Although it is the difference in SNR that is of interest here, the definition used in the plot is the squared peak of the isolated transition response to the integrated noise power in the $1/PW_{50}$ frequency band.

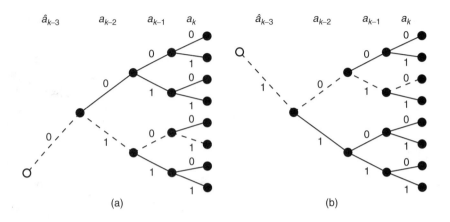

FIGURE 4.8   Fixed delay tree search detector shown with sequences disallowed for an MTR $j = 2$ code.

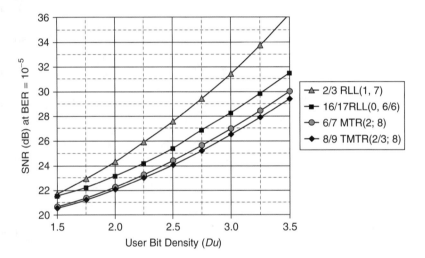

FIGURE 4.9   Simulation results for an E2PR4 equalized Lorentzian channel.

For several types of codes, the SNR required to obtain a bit error rate (BER) of 10^{-5} is shown as a function of user density in Figure 4.9. To isolate the distance gain of the code, ideal clock synchronization and gain control are assumed. The receive signal is equalized to the E2PR4 response with a relatively long FIR filter designed to minimize the mean squared error at the equalizer output.

A rate 16/17 RLL(0, 6/6) code is used in lieu of an uncoded reference. From the figure, it is clear that although the RLL(1,7) code provides a significant distance gain, its code rate loss exceeds the distance gain, giving a net loss. Of course, such a code is often chosen to mitigate transition noise and nonlinearities, which are not reflected in this model. At a user density of $D_u = 2.5$, the rate 6/7 MTR codes provides a net gain of about 1 dB while, as expected, the rate 8/9 TMTR provides a slightly larger gain.

4.6 Summary

Performance in recording channels, subject to severe intersymbol interference, is limited by error event sequences that produce a minimum distance less than the matched filter bound. High density Lorentzian, PR2, and E2PR4 channels are all subject to error events that can be avoided by properly constraining the maximum number of consecutive transitions. A variety of code constraints with varying capacities can be formed around this basic idea. For a simple linear channel with additive Gaussian noise, these codes provide a net performance gain, despite the fact that the MTR constraint reduces the code rate.

References

[1] Behrens, R. and Armstrong, A., An advanced read/write channel for magnetic disk storage, *Proc. 26th Asilomar Conf. on Signals, Systems, and Computers*, 956, 1992.

[2] Karabed, R. and Siegel, P. H., Coding for higher order partial response channels, *Proc. SPIE Int. Symp. on Voice, Video, and Data Commn.*, 2605, 115, 1995.

[3] Karabed, R., Siegel, P. H., and Soljanin, E., Constrained coding for binary channels with high intersymbol interference, *IEEE Trans. Inform. Th.*, 45, 1777, 1999.

[4] Messerschmitt, D. G., A geometric theory of intersymbol interference, Part II: Performance of the maximum likelihood detector, *Bell Syst. J.*, 52, 1973.

[5] Moon, J. and Carley, L. R., Efficient sequence detection for intersymbol interference channels with run-length constraint, *IEEE Trans. Commn.*, 42, 2654, 1994.

[6] Altekar, S. A. et al., Error-event characterization on partial-response channels, *IEEE Trans. Info. Th.*, 45, 241, 1999.

[7] Xu, C. and Keirn, Z., Error event analysis of EPR4 and ME2PR4 channels: Captured head signals versus Lorentzian signal models, *IEEE Trans. Magn.*, 36, 2200, 2000.

[8] Brickner, B. *Maximum Transition Run Coding and Pragmatic Signal Space Detection for Digital Magnetic Recording*, Ph.D. Thesis, University of Minnesota, 1998.

[9] Cideciyan, R. D., et al., Maximum transition run codes for generalized partial response channels, *IEEE J. Sel. Areas Commn.*, 19, 619, 2001.

[10] Soljanin, E., On-track and off-track distance properties of class 4 partial response channels, *Proc. SPIE Int. Symp. on Voice, Video, and Data Commn.*, 2605, 92, 1995.

[11] Brickner, B. and Moon, J., Combatting partial erasure and transition jitter in magnetic recording, *IEEE Trans. Magn.*, 36, 532, 2000.

[12] Bliss, W. G., An 8/9 rate time-varying trellis code for high density magnetic recording, *IEEE Trans. Magn.*, 33, 2746, 1997.

[13] Knudson Fitzpatrick, K. and Modlin, C. S., Time-varying MTR codes for high density magnetic recording, *Proc. IEEE Global Telecom. Conf.*, 1997.

[14] Moision, B. E., Siegel, P. H., and Soljanin, E., Distance-enhancing codes for digital recording, *IEEE Trans. Magn.*, 34, 69, 1998.

[15] Brickner, B. and Padukone, P., Partial Response Channel Having Combined MTR and Parity Constraints, U.S. Patent 6,388,587, 2002.

[16] Liu, Pi-Hai, Distance-Enhancing Coding Method, U. S. Patent 6,538,585, 2003.

[17] Marcus, B., Siegel, P. H., and Wolf, J. K., Finite-state modulation codes for data storage, *IEEE J. Sel. Areas Commun.*, 10, 5, 1992.

[18] Shannon, C. E., A mathematical theory of communication, *Bell Syst. Tech. J.*, 27, 379, 1948.

[19] Forney, D. G., Maximum-likelihood sequence estimation of digital sequences in the presence of intersymbol interference, *IEEE Trans. Inform. Th.*, 18, 363, 1972.

5

Spectrum Shaping Codes

Stojan Denic
University of Ottawa
Ottawa, ON

Bane Vasic
University of Arizona
Tucson, AZ

5.1 Introduction

The first application of spectrum shaping codes was related to digital communication systems that used transformers to connect two communication lines. Because transformers do not convey dc-component, and suppress low frequency components, direct transmission of source signals whose power spectral densities contain these frequency components were not possible without significant distortion. That is why dc-free or dc-balanced codes were devised [1–3, 10–13]. Their role is to transform a source sequence into a channel sequence whose spectral characteristic corresponds to spectral characteristic of communication channel. At the end of communication line, the sequence is received by the decoder that generates original sequence without errors in the case of noiseless channel. In recording systems that can be modeled as any communication system, this kind of codes have been widely used. For instance in digital audio tape systems, they prevent write signal distortion that can occur due to transformer-coupling in write electronics [2]. In optical recording systems they are used to circumvent the interference between recorded signal and servo tracking system. Further development of spectrum shaping codes for recording systems was driven by requirements for better codes in the sense of larger rejection of low frequency components. The codes providing this feature are codes with higher order spectral zero at $f = 0$, and can be found in [14, 16, 17]. Although the width of suppressed frequencies of these codes is smaller than in the case of dc-balanced codes, the rejection in the vicinity of $f = 0$ is significantly larger. Another class of spectrum shaping codes were invented in order to support the use of frequency multiplexing technique for track following [1], and partial response technique for high density data storage [15]. Both techniques require the spectral nulls of the recorded signal at frequencies that can be different than $f = 0$ in order to enable reliable data storage. The typical example of such codes are codes that have spectral zeros at submultiple of channel symbol frequency.

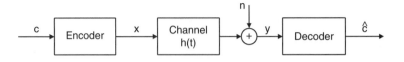

FIGURE 5.1 Recording system.

For this type of codes see, for example, [18–20]. The fourth characteristic group of spectrum shaping codes are those that give rise to spectral lines. Their purpose is to give the reference information to the head positioning servo system that positions and maintains the head accurately over a track in digital recorders [1].

Besides spectral constraint, recorded sequences have to comply with certain time constraints. That is why it is interesting to say something about compound codes, which generate sequences that in the same time satisfy more than one constraint. Typical representatives are RLL (runlength limited) dc-free codes. RLL dc-free sequences have confined minimal, and maximal consecutive like channel symbols, and in the same time, their spectrum has zero at $f = 0$, [21, 22]. Important classes of spectrum shaping codes are dc-free error correcting codes. It was mentioned earlier that the decoder of dc-free code will decode channel sequence without errors if the channel is noiseless. Because a recording channel is not noiseless, a lot of effort was put into design of dc-free error correcting codes, both block and convolutional codes. The examples of these codes can be found in [27–30].

The goal of this chapter is to give a survey of spectrum shaping codes for digital recording systems from the theoretical, and practical point of view. The organization of the article is as follows. Section 5.2 contains short description of recording system with respect to the role of spectrum shaping codes. In Section 5.3, dc-free codes are considered. Some theoretical basis for studying of dc-free codes are given, which will be important for studying of all types of spectrum shaping codes [4–8]. Section 5.4 discusses codes with higher order spectral zeros at $f = 0$, and codes with zeros at frequencies that are submultiple of channel symbol frequency. In Section 5.5, certain compound constraint codes are mentioned. The specific example of RLL-dc code is given. All codes are considered from the point of view of maxentropic sequences. The channel capacity of maxentropic sequences are computed, and corresponding power spectral densities are given. Also, the basic encoding techniques are described.

5.2   Recording System and Spectrum Shaping Codes

The encoder of spectrum shaping code is the last in the chain of encoders preceding the recording channel, and the decoder of spectrum shaping code is the first in the chain of the decoders that follows the recording channel. As shown in Figure 5.1, the encoder receives a symbol stream $c = \{c_k\}_{k=0}^{\infty}$ from an error correcting code encoder, and transforms it into the stream of channel symbols $x = \{x_k\}_{k=0}^{\infty}$ that matches the spectral characteristics of the recording channel with impulse response $h(t)$. $n(t)$ is additive noise. The decoder accepts data stream from the channel $y = \{y_k\}_{k=0}^{\infty}$, and transforms it to the symbol stream $\hat{c} = \{\hat{c}_k\}_{k=0}^{\infty}$ that is the noisy version of c, and fed to the error correcting code decoder. From the point of view of error correcting code encoder, and decoder, the spectrum shaping code encoder, and decoder are merely part of the recording channel. One can also define the code rate R of the spectrum shaping codes. In the case of block codes, the input stream c is divided into the sourcewords of length k, which are encoded in codewords of length n forming the output stream x. The coding rate is defined as the ration of the word length at the input k, and the codeword length n, $R = k/n$.

5.3   Dc-free Codes

5.3.1   Introduction

Dc-free codes belong to a class of spectrum shaping codes that transform the spectrum of the input sequence into an encoded sequence whose spectrum has zero at the zero frequency. Dc-free codes emerged first in digital communications. The intrinsic part of communication lines were coupling devices whose characteristic as well as the characteristic of digital recording channels is that they suppress the low frequency components of the transmitted signal or recorded data respectively. This implies the necessity of signal processing techniques, which will reshape the spectrum of the original source sequence, and in that way match the characteristics of the communication or recording channels. Not only that dc-free codes give zero at dc, but also the low frequency components of the encoded sequence are suppressed depending on the choice of code parameters. But this is not the only reason for employing dc-free codes. The other reason for using dc-free codes is that the optical data storage systems use high-pass filters to diminish the effect of dirt on system performance caused for example by fingerprints on an optical medium. Another reason is to prevent the mutual interference between recorded data, and servomechanism for track tracking that operates at a low frequency [1]. All those arguments indicate the importance of dc-free codes for data recording systems.

Dc-free codes attracted considerable attention of the data storage community, and a number of papers, and patents have been published till now, and a large number of references can be found in [2].

This part of the chapter is organized in the following manner. In the Section 5.3.2, theoretical background of dc-free constraint sequences is given. Section 5.3.3 considers the finite state transition diagram (FSTD) description of dc-free constraints and method for computing the noiseless channel capacity of the constrained channel. Section 5.3.4 presents power spectral density (PSD) characteristics of dc-free constraints, and discusses the most important parameters that determine PSD, and channel capacity, and their mutual dependence. Section 5.3.5 contains survey of simple and well-known coding techniques for generating dc-free sequences.

5.3.2   Dc-free Constraint Sequences

The starting point for analysis and design of dc-free codes is the result by Pierobon [3]. Before stating the result of Pierobon the notion of running digital sum is defined. The running digital sum (RDS) at moment n, z_n, of sequence of symbols $x = \{x_k\}_{k=0}^{\infty}$ is defined as

$$z_n = \sum_{k=0}^{n} x_k \tag{5.1}$$

Pierobon proved that the power spectral density $S_x(f)$ of a sequence x vanishes at zero frequency, regardless of the sequence distribution, if and only if the running digital sum $z = \{z_k\}_{k=0}^{\infty}$ of the encoded sequence is bounded, that is, $|z_n| \leq N$, for each n, when $N < \infty$. This condition describes the constraint posed on the sequences that are eligible for recording on a recording medium. In other words, any sequence that violates the condition on RDS will not be permitted in the recording channel, and sequences that satisfy this constraint are called dc-free sequences. It follows that recording channels are channels with the constraint that is called dc-free constraint. Further, the effect of RDS is twofold; first, it affects the shape of power spectral density of constrained sequences, and second the upper bound N on RDS determines the capacity of a constrained recording channel [1]. In general case, the capacity of noiseless constrained channels was defined by Shannon [4] as

$$C = \lim_{T \to \infty} \frac{\log_2 n(T)}{T} \tag{5.2}$$

$n(T)$ is total number of admissible sequences, and T is sequence duration. The channel capacity determines the maximal theoretical code rate R of an constrained code. It will be explained later that the upper bound of RDS, N, has contradictory effects on the channel capacity, and desired power spectral density of recording sequences so that code design is a compromise between two opposite requirements.

5.3.3   Capacity of Dc-free Constraint

In order to compute the channel capacity of dc-free constrained channels, the notion of digital sum variation (DSV) is introduced. If $N_{\max}$ is the maximum value of z, and $N_{\min}$ is its minimum value then digital sum variation is defined as $N = N_{\max} - N_{\min} + 1$. Actually, DSV is equal to the number of different values that z can take. Intuitively, it can be seen that the total number of admissible sequences $n(T)$ depends on DSV, that is, the larger N, the larger $n(T)$, and the larger channel capacity C.

The common tool for description of constrained sequences is a finite state transition diagram (FSTD). It was shown by Shannon [4] that the FSTD description of a constrained channel can be used for computation of its channel capacity. In this section that result is used to determine the channel capacity of M-ary dc-free constraint, where M is the number of levels that are permitted in the recording channel [5]. One example of FSTD representing dc-free constraint is given in Figure 5.2. It depicts M-ary dc-free constraint when the channel symbol alphabet is $\{-(M-1)/2, \ldots, -1, 0, 1, \ldots, (M-1)/2\}$, M odd. If M is even, then the channel symbol alphabet can be chosen as $\{-(M-1), -(M-3) \ldots, -1, 0, 1, \ldots, (M-3), (M-1)\}$. The state of the FSTD represents the value of RDS, z_k, at time instant k. An edge between two states represents the transition between states, while a label above the edge denotes the input symbol generated during the transition between states. The number of states is equal to DSV, N. To compute the channel capacity, the connection matrix of FSTD is introduced. The connection matrix D of FSTD is a square matrix of dimension N, where the entry d_{ij} of matrix D represents number of edges emanating from state i, and ending at state j. The connection matrix for FSTD shown in Figure 5.2 is

$$D = \begin{bmatrix} 1 & 1 & 0 \\ 1 & 1 & 1 \\ 0 & 1 & 1 \end{bmatrix} \tag{5.3}$$

On the other hand, because the transition from one state to another depends only on the current state, each FSTD may be assigned corresponding Markov chain $s = \{s_k\}_{k=0}^{\infty}$, where each state of FSTD is related to a value that Markov chain can take. The finite number of values that Markov chain takes is equal to the number of states of FSTD. Each edge is assigned a transition probability p_{ij}, from state i to state j, where $i, j \in \sum$, where $\sum$ is the set of all states of Markov chain. The measure of uncertainty or measure of information generated by a Markov chain is equal to the entropy of Markov chain. It is calculated as

$$H(X) = \sum_{i=1}^{N} p_i H_i \tag{5.4}$$

where p_i is probability of state $i (i = 1, \ldots, N)$, and H_i is the entropy of the state i, that is, uncertainty of being in state $i (i = 1, \ldots, N)$ at any time instant. It is defined as the entropy of the transition probabilities p_{ij}, $H_i = \sum_{j=1}^{N} p_{ij} \log \frac{1}{p_{ij}}$.

It was proven by Shannon [4] that channel capacity of constrained noiseless channel is

$$C = \max H(X) = \log_2 \lambda_{\max} \tag{5.5}$$

λ_{max} is the maximum eigenvalue of connection matrix D of FSTD representing channel constraint. The Equation (5.5) connects the channel capacity of the constrained channel to maximum entropy notion.

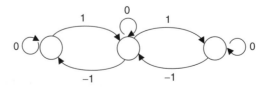

FIGURE 5.2   FSTD for $M = 3$, $N = 3$ dc-free constraint.

TABLE 5.1 Capacity of Dc-free Constraint

N	M=2	M=3	M=4	M=5	M=6	M=7	M=9
3	0.5	1.2716	—	1.5850	—	—	—
4	0.6492	1.3885	1	1.8325	—	2	—
5	0.7925	1.4500	1.2925	1.9765	—	2.2159	2.3219
6	0.8495	1.4864	1.4500	2.0642	1.5850	2.3559	2.5114
7	0.8858	1.5098	1.5665	2.1223	1.7952	2.4529	2.6423
8	0.9103	1.5258	1.6508	2.1626	1.9227	2.5211	2.7388
9	0.9276	1.5371	1.7111	2.1919	2.0283	2.5713	2.8117
10	0.9403	1.5455	1.7573	2.2137	2.1085	2.6094	2.8672

It shows that maximum entropy Markov chain generates maximal number of sequences of certain length. The constrained sequences generated by a Markov chain having maximum entropy are called maxentropic sequences. Table 5.1 contains the channel capacities for some M-ary dc-free constraints. Table 5.1 shows that channel capacity increases when two degrees of freedom DSV, and M increase. For $M = 2$, the closed form expression for channel capacity was derived [1]. The derivation is based on the recurrent relation that exists between characteristic polynomials of connection matrices D_N, D_{N-1}, D_{N-2}, where the subscripts denote the DSV of the corresponding channels described by connection matrices. The formula for channel capacity is

$$C(N) = \log_2 2 \cos \frac{\pi}{N+1} \tag{5.6}$$

The results obtained by Equation 5.6 agree with those found in Table 5.1.

5.3.4   Spectral Characteristics of Dc-free Constraint

In Section 5.3.2, a necessary and sufficient condition for the null of a power spectral density at $f = 0$ is introduced. In this section the importance of variance of RDS will be considered, and the power spectral densities of maxentropic dc-free sequences will be shown.

In [6], Justesen derived interesting relation between the sum variance (variance of RDS) $E[z_k^2] = \sigma_z^2(N)$ where N is a digital sum variation, and so-called cut-off frequency ω_0

$$2\sigma_z^2(N)\omega_0 \approx 1 \tag{5.7}$$

The cut-off frequency is defined as the value of the frequency $\omega_0 = 2\pi f_0$, such that $S_x(f_0) = 0.5$. The cut-off frequency determines the bandwidth, called notch width, from $f = 0$ to $f = f_0$, within which the power spectral density of recorded sequence $S_x(f)$ is low. The performance is better if the cut-off frequency f_0 is larger giving the larger notch width. On the other hand, from Equation 5.7, it is clear that the larger cut-off frequency f_0, the smaller sum variance $\sigma_z^2(N)$. Having in mind that the sum variance $\sigma_z^2(N)$, and the channel capacity $C(N)$ both decrease as N decreases, it can be assumed that the capacity of the constrained channel $C(N)$ is smaller as cut-off frequency f_0 grows, implying that the redundancy $1 - C(N)$ is bigger [1]. The previous discussion also points out the equal importance of redundancy $1 - C(N)$, and sum variance $\sigma_z^2(N)$ as performance measures for spectrum shaping codes. That is why in [1], the new measure of performance for spectrum shaping codes was introduced as the product of redundancy, and sum variance $(1 - C(N))\sigma_z^2(N)$. It was shown that this product is tightly bounded, from below, and above for $N > 9$ in the case of maxentropic dc-free sequences. This shows the significance of the relation Equation 5.7, that reveals the conflicting demands posed on the constrained sequences. If one wants to get better suppression of low frequency components of power spectral density, one must pay with bigger redundancy, that is, code inefficiency. Although the previous result is derived for binary recording channels, the conclusions are true for $M > 2$.

In what follows, the power spectral densities for some maxentropic M-ary dc-free sequences will be computed. In order to compute the power spectral density of the constrained channel, FSTD of Moore type, and corresponding Markov chain representation of the constrained channel are used. For example,

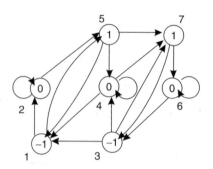

FIGURE 5.3   Moore type FSTD for $M = 3$, $N = 3$ dc-free constraint.

in Figure 5.3, the FSTD of Moore type for $M = 3$, $N = 3$, dc-free constrained channel is depicted. The characteristic of the Moore type FSTD, as opposed to the FSTD found in Figure 5.2, is that all edges entering the same state have the same label, that is, the same channel symbol is generated by visiting particular state. In this example the alphabet of channel symbols is $\{-1, 0, 1\}$. If the initial state is $s_0 = 4$, and $z_0 = 0$, then $z_k \in \{-1, 0, 1\}$ meaning $N = 3$. It can be seen that states 1 and 2 correspond to $z_k = -1$, states 3, 4, and 5 to $z_k = 0$, and 6 and 7 to $z_k = 1$.

Associated with FSTD, there exists a Markov chain generating maxentropic M-ary dc-free sequences with transition probabilities given by [4]

$$p_{ij} = \frac{1}{\lambda_{\max}} d_{ij} \frac{v_j}{v_i} \tag{5.8}$$

The quantities v_i, and v_j are the entries of the right eigenvector corresponding to the maximal eigenvalue $\lambda_{\max}$ of the connection matrix D. The methods for computing power spectral density $S_x(f)$ of Markov sources were proposed in [7, 8], and are used here in order to get the spectra of maxentropic M-ary dc-free sequences. The equation for continuous part of power spectral density of Markov source is given as

$$S_x(f) = C_x(0) + 2 \sum_{k=1}^{\infty} C_x(k) \cos 2\pi k f \tag{5.9}$$

where $C_x(k), k = 0, 1, 2, \ldots$, is autocovariance function of generated sequence x, and can be written as

$$C_x(k) = \xi^T \Pi (P^{|k|} - P_\infty) \xi \tag{5.10}$$

In Equation 5.10, each entree of vector ξ, which has as many entrees as there are states of Markov chain, represents symbol emitted when particular state of Markov chain is visited. Further, matrix Π is a diagonal matrix whose diagonal elements are steady state probabilities of Markov chain, P is a transition probability matrix, and each row of P_∞ is equal to the steady state probability vector of Markov chain. The importance of maxentropic sequences is that their power spectral density corresponds to the power spectral density of the codes whose code rate R is near the channel capacity C, and consequently can be used as a good approximation of the spectrum of those codes. The power spectral density of maxentropic sequences does not depend on the specific encoder and decoder realization, and therefore it is easier to compute than the spectrum of the specific code.

Figure 5.4 shows PSD of maxentropic sequences for $M = 3$, and $N = 3, 4, 5$ versus normalized frequency f where normalization is done by $f_s = 1/T_s$, where T_s is a signaling interval. It is interesting to check the validity of the formula Equation 5.7 for previously computed PSD. For instance, the normalized cut-off frequency for $M = 3, N = 4$, is $f_0 = 0.1$, $\omega_0 = 0.6283$, and by formula Equation 5.7, approximate sum variance is 0.7958. The formula for exact value of the sum variance is given in [6], according to which $\sigma_z^2 = -\sum_{k=1}^{\infty} k R_x(k) = 0.8028$, meaning that Equation 5.7 provides good approximation for relation between sum variance and cut-off frequency.

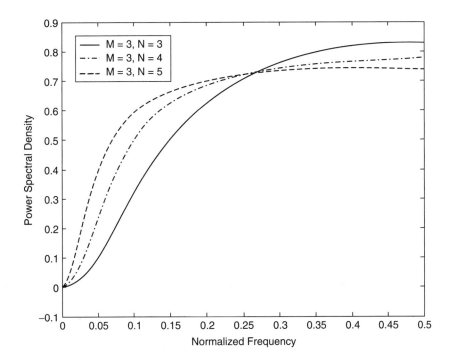

FIGURE 5.4 Continuous part of the power spectral density of $M = 3, N = 3, 4, 5$ maxentropic sequences.

5.3.5 Encoding and Decoding of Dc-free Constraints

Finite state transition diagrams of dc-free constraints, and their maxentropic Markov chains are of great importance. They are analytical tools that provide a number of parameters, such as channel capacity, cut-off frequency, sum variance, representing theoretical bounds of dc-free constraints. Those parameters have a practical meaning for the designers of encoders and decoders of dc-free codes. But these analytical tools do not give a recipe how to build practical encoders, and decoders. This section considers the standard solutions for realization of encoders and decoders of dc-free constraints. To determine the quality of particular encoder for constraint channels, two parameters are introduced. The first, called code rate efficiency, is defined as the ratio of code rate R, and channel capacity C, $\eta = R/C$. While the first one is commonly used for any channel coding scheme, the second was specially defined for dc-free codes [1]. The encoder efficiency E is

$$E = \frac{(1 - C(N))\sigma_z^2(N)}{(1 - R)s_z^2} \tag{5.11}$$

The encoder efficiency compares the product of redundancy, and sum variance of maxentropic sequence for particular channel, and product of redundancy, and sum variance of specific encoder for the same channel. The maxentropic sequence was taken as a reference because it achieves the channel capacity of dc-free constraint channel. The importance of sum variance was explained in the previous section.

In general, encoders, and decoders can be divided into two groups, state independent and state dependent encoders and decoders. State independent encoders are usually realized by look-up tables where there exists one-to-one correspondence between sourcewords and codewords. As opposed to state independent encoders, state dependent encoders are designed as a synchronous finite state machines [9] where the next codeword is the function of a current internal state of the encoder, and a current sourceword. The advantage of state dependent encoding is that sometimes more efficient codes can be constructed (in terms of code rate) [1] with shorter codewords.

Similarly, the output of state dependent decoder depends on the decoder current state, the current input codeword, as well as finitely many upcoming codewords [9]. The weakness of this type of decoders is that if an error occurs due to the noisy channel the decoder could lose track of states. In turn, this can lead to the series of errors, that is, error propagation. In order to prevent this kind of event, state independent encoders are introduced. The decoder's output depends only on the certain number of preceding codewords, current codeword, and certain number of upcoming codewords. The name for such type of decoders is sliding window decoders [9].

Here, both types of encoders will be presented. It is assumed that recording channel is binary channel with channel alphabet $\{-1, 1\}$. The main idea behind dc-free encoder realization is the fact that the RDS of channel sequence has to be bounded. Having that in mind, the notion of codeword disparity is introduced. The disparity d of codeword $\mathbf{x} = (x_1 x_2 \ldots x_n)$ of length n is defined as

$$d = \sum_{i=1}^{n} x_i \tag{5.12}$$

The simplest idea is to use only codewords that have zero disparity. Each sourceword is assigned a unique codeword of zero disparity. It means that RDS at the end of each codeword will be equal to zero, if $z_0 = 0$. This type of encoding is called zero disparity encoding, and it is obviously state independent. It is simple, but the main shortcoming is inefficiency in terms of code rate R for a fixed codeword length n, because the number of possible codewords with zero disparity of length n is finite, $\binom{n}{n/2}$. The larger efficiency η can be achieved by increasing the codeword length n, which in the same time makes the look-up table realization of encoder more complex.

Another way for achieving better efficiency is to use so-called low disparity codes. The drawback of this technique is the increase of the power of low frequency components as compared to zero disparity coding. The codes that belong to this class of encoding can be found in [10]. Again the goal is to keep the value of RDS within some prescribed bounds. In addition to zero disparity codewords, it is allowed to use the codewords with low disparity. As explained in [1], let S_+ denote all codewords with positive disparity, and let S_- denote all codewords with negative disparity. S_+ is the union, $S_+ = \cup_{j=0}^{K} S_j$, $K \leq n/2$, where S_j is the set of all codewords with disparity $d = 2j$. The set S_- is obtained by inverting codewords of set S_+. $K = 0$ represents the case of zero disparity encoding. If $K = 1$, each sourceword is assigned a pair of codewords of opposite disparity or a codeword of zero disparity. In process of encoding, the value of RDS is tracked. Because the encoder has a choice of two codewords to send to the channel, it chooses one whose disparity minimizes absolute value of RDS. Actually, the encoder has two codebooks, and chooses the codeword from one codebook such that an instantaneous value of RDS is within some bounds.

The polarity bit coding is yet another simple method that generates dc-free sequences [11, 12]. One extra bit, called polarity bit, is added to $n - 1$ source bits comprising codeword of length n. The polarity bit is set to 1. If the disparity of the codeword has the same sign as RDS at the moment of sending the codeword, the inverted codeword is recorded. Otherwise original codeword is recorded. The decoder, based on the polarity bit, recognizes if the codeword was inverted or not.

In [1], the encoder efficiency E versus codeword length n of above coding techniques is shown. The conclusion can be drawn that low disparity codes outperform zero disparity codes, and zero disparity codes outperform polarity bit codes. Also, for small codeword lengths, low, and zero disparity codes have unit efficiency, while the efficiency drops as the codeword length increases.

It is worth to mention two more coding techniques. Their realizations do not rely on look-up tables so that they are convenient if better efficiency is necessary. One is based on enumerating the set of sequences denoted by $T = T(z_n, n, N, z_0)$ [1], where n is a sequence length, N is the maximal value of RDS within the sequence, z_n and z_0 are final and initial value of RDS. It is understood that the minimal value of RDS is 1. The enumeration algorithm establishes one to one correspondence between set T and the set of integers $0, 1, \ldots, |T| - 1$. The enumeration represents the foundation of decoding algorithm. Also algorithm that performs reverse operation was derived, that enables mapping from the set of integers to the set of constrained sequences. The second technique uses the fact that each k-bit binary sourceword can be divided in

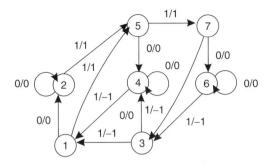

FIGURE 5.5   Encoder for $M = 3$, $N = 3$ dc-free constraint.

TABLE 5.2   Decoding Table
for M=3, N=3, Dc-free Code

Window	Decoded Bit
1	1
0	0
-1	1

two segments, each having equal disparity [13]. The zero disparity codeword is obtained by inverting one of two segments. Additional l bits are used to mark the position splitting these two segments. Those additional l bits are usually coded with zero disparity word. It follows that code rate is $R = k/(k + l)$. The encoding is possible because there always exists one to one correspondence between sourcewords and codewords.

5.3.5.1   State Dependent Encoding

Here, the example of state dependent encoder, and sliding window decoder is presented, for the following parameters $M = 3$, $N = 3$. It is assumed that the source generates binary sequences with alphabet $\{0, 1\}$. The set of channel symbols is $\{-1, 0, 1\}$. The capacity of this constraint is $C = 1.2716$ bits/symbol. The code rate of $R = 1/1$ bits/symbol is chosen. It means that both sourceword length, and codeword length is 1. The code rate efficiency is $\eta = 0.77$. The starting point for encoder construction is FSTD presented at Figure 5.3, that represents Moore type FSTD of $M = 3$, $N = 3$ dc-free constraint. This type of FSTD belongs to almost finite type according to Proposition 7 in [9]. According to Theorem 3 in [9], there exists a noncatastrophic finite state encoder accompanied with sliding window decoder. Approximate eigenvector is $v = [1\,1\,1\,1\,1\,1\,1]$ that guarantees the existence of enough number of edges from each state, and there is no need for state splitting. The encoder is shown in Figure 5.5. Every label above an edge consists of pair, sourceword/codeword. From every state emanate as many edges as there are sourcewords, and in this case, it is $2^1 = 2$. The input bits are assigned in such a way to minimize the size of the decoder window. Table 5.2 gives sliding window decoder table. It can be seen that only one symbol is enough to decode a bit.

5.4   Codes with Higher Order Spectral Zeros

5.4.1   Introduction

The first representatives of spectrum shaping codes were dc-free codes discussed in the previous section. Further development of spectrum shaping codes went into two directions. One direction is the improvement in suppressing of low frequencies, and the other is the introduction of spectral zeros at frequencies different than zero frequency.

As it was pointed out at the very beginning, the role of spectrum shaping codes is to match the spectral characteristics of source sequences to the channel characteristics, and in the case of dc-free codes to cancel out dc, and to reduce the low frequency components of the recorded signal. The goodness of some code is measured by notch width determined by cut-off frequency. This parameter determines the frequency bandwidth within which the power spectral density is less than some value, but it does not tell anything about how well these spectral components are rejected. Further improvement in terms of better suppression of low frequency components is achieved by generalizing of basic concepts given in previous sections. The better reduction of low frequency components is obtained by putting constraint on so-called kth order RDS [14]. That results in higher order zeros of power spectrum density.

The importance of spectrum shaping codes was rediscovered with introducing of partial response signaling techniques in high density recording devices [15]. The spectrum of such kind of recording channels can have zero not only at $f = 0$ but also at Nyquist frequency $f = f_s/2$, where f_s is a channel symbol frequency. The spectrum of codes that are designed to match partial response recording channels, have to have zeros not only at $f = 0$, but at some other frequencies as well, as required by specific channel. It was revealed that spectrum shaping codes on partial response channels have an additional virtue. It turned out that they exhibit enhanced error correcting capabilities on partial response channels.

In this section codes that give rise to higher order spectral nulls at frequencies $f = 0$, and $f = pf_s/q$, where p, and q are relatively prime, are presented. The results discussed here are due to Eleftheriou, and Cideciyan [14], and Immink [1]. Among the authors that considered these problems are Immink [16], Immink, and Beenker [17], Marcus, and Siegel [18], and Karabed, and Siegel [19].

5.4.2   Sequences with Higher Order Zeros at $f = 0$

First time the notion of sequences whose power spectral densities have higher order spectral nulls was introduced by Immink in [16]. He defined running digital sum sum (RDSS) of sequence $\{x_k\}_{k=0}^n$ as

$$y_n = \sum_{i=0}^n z_i = \sum_{i=0}^n \sum_{j=0}^i x_j \tag{5.13}$$

where z_k is the value of RDS at moment k. It can be proven that power spectral density of sequence $\{x_k\}_{k=0}^\infty$ has second order zero at $f = 0$, that is, $S_x(0) = 0$, and $S_x^{(2)}(0) = 0$, if and only if RDSS is bounded. The sequences whose RDSS is bounded are called dc^2-constrained sequences, and corresponding codes are dc^2-constrained codes. The odd derivatives of power spectral density $S_x(f)$ are zero because $S_x(f)$ is an even function of frequency f.

As in the case of dc-free constraint, similar encoding techniques are applied for higher order constraints, for example, zero disparity encoding, enumerative encoding, state dependent encoding. Zero disparity encoding employs the codewords that have zero disparity with respect to both RDS, and RDSS. The enumerative encoding counts the number of codewords of fixed length n, $x = (x_1 x_2 \ldots x_n)$, denoted by $A_n(d_z, d_y)$, that satisfy RDS disparity $d_x = \sum_{i=1}^n x_i$, and RDSS disparity $d_y = \sum_{j=1}^n \sum_{i=1}^j x_i$. The state dependent encoding are organized in the following way. All codewords are of zero d_z disparity, and are divided in two groups; one consists of codewords having zero, and positive d_y disparity, and the other consists of codewords having zero, and negative d_y disparity. The encoder chooses the next codeword such that at the end of that codeword the RDSS is close to zero.

The larger rejection of low frequencies is accomplished if the higher derivatives of power spectral density $S_x^{(k)}(f)$, $k > 2$ are zero at $f = 0$. The condition for vanishing of higher order derivatives is defined in [17], and it uses the concept of codeword moment. The kth moment of codeword x is defined as

$$\mu_k^0(x) = \sum_{i=1}^n i^k x_i \tag{5.14}$$

where $k \in \{0, 1, 2, \ldots\}$. The superscript 0 means that the spectral null is at $f = 0$. The first $2K + 1$ derivatives of power spectral density $S_x(f)$ vanish at $f = 0$, if the first $K + 1$ codeword moments $\mu_k^0(x)$

are zero. The concept of kth codeword moment is very useful in computing the error correcting capabilities of codes with higher order spectral zeros, and it is related to kth order running digital sum that is introduced in the following section [14].

5.4.3 K-RDS$_f$ Sequences

In [14], more general approach was taken. The concepts of kth order running digital sum at frequency f, and K-RSD$_f$ FSTD are introduced. They are used to describe necessary, and sufficient conditions that an FSTD has to satisfy in order to be able to generate sequences whose power spectral densities have desired characteristics. Two main theorems of [14], Theorem 2, and Theorem 4, completely describe FSTDs that generate sequences with spectral nulls of order K at $f = 0$, and $f = pf_s/q$, respectively. The frequency f_s is a channel symbol frequency, and p, and q are relatively prime numbers. In this section just Theorem 4 will be presented.

The kth order RDS$_f$ at frequency $f = pf_s/q$ of sequence $x = \{x_k\}_{k=0}^n$ is defined as

$$\sigma_k^f(x) = \sum_{i_1=0}^n \sum_{i_2=0}^{i_1} \cdots \sum_{i_k=0}^{i_{k-1}} w^{i_k} x_{i_k} \tag{5.15}$$

where $w = e^{-j2\pi p/q}$. In order to be able to state the main result, another notion has to be defined. For an FSTD is said that it is K-RDS$_f$ FSTD if there is a mapping ψ from the set of states $\sum$ onto a finite set of complex numbers ς such that $x(D) = wz(D)(1 - w^{-1}D)^K$, where $x = \{x_k\}_{k=0}^n$ is a channel sequence, $z_n = \psi(s_{n+1})$, $s = \{s_k\}_{k=0}^n$ is a state Markov process ($s_k \in \sum$), and $a(D) = \sum_{k=0}^\infty a_k$ defines D transform of sequence $a = \{a_k\}_{k=0}^\infty$. The spectrum of sequence x generated by state process s assigned to an FSTD has a spectral null of order K at $f = pf_s/q$, if and only if FSTD is an K-RDS$_f$ FSTD. The main result of [14], Theorem 4, that completely describes K-RDS$_f$ FSTD, says that the following statements are equivalent:

1. There are K functions, $\psi_k, 1 \leq k \leq K$, that map the set of states $\sum$ onto a finite set of complex numbers ς such that

$$x_n = w\psi_1(s_{n+1}) - \psi_1(s_n)$$
$$\psi_{k-1}(s_n) = \psi_k(s_n) - w^{-1}\psi_k(s_{n-1}), \quad 2 \leq k \leq K \tag{5.16}$$

2. There are K functions, $\psi_k, 1 \leq k \leq K$, that map the set of states $\sum$ onto a finite set of complex numbers ς such that the kth order RDS$_f$, $1 \leq k \leq K$, of any channel sequence is given by

$$\sigma_1^f(x) = w^{n+1}\psi_1(s_{n+1}) - \psi_1(s_0)$$
$$\sigma_k^f(x) = w^{n+1}\psi_k(s_{n+1}) - \psi_k(s_0) - \sum_{i=1}^{k-1} \psi_{k-i}(s_0)\binom{n+i}{i}, \quad 2 \leq k \leq K \tag{5.17}$$

3. There are $K - 1$ functions, $\psi_k, 1 \leq k \leq K - 1$, that map the set of states Σ onto a finite set of complex numbers ς such that for every cycle of states $s_0, s_1, \ldots, s_{n+1} = s_0$ of length that is multiple of q, the following equations are satisfied

$$\sigma_1^f(x) = 0$$
$$\sigma_k^f(x) = -\sum_{i=1}^{k-1} \psi_{k-i}(s_0)\binom{n+i}{i}, \quad 2 \leq k \leq K \tag{5.18}$$

4. K-RDS$_f$ FSTD that goes through the state sequence $s = \{s_k\}_{k=0}^n$ generates channel sequence $x = \{x_k\}_{k=0}^n$, that has a spectral null of order K at $f = pf_s/q$.

It should be noted that spectral null of order K guarantees that first $2K - 1$ derivatives of power spectral density $S_x(f)$ vanish at frequency f. If, in formula (Equation 5.15), and expressions in Theorem 4, w is replaced by 1, and superscript f with 0, Theorem 4 becomes Theorem 2 from [14] that completely characterizes FSTD that generates sequences with K-th order spectral null at $f = 0$. A usefulness of previous results can be seen from manipulation of two formulas in (Equation 5.16), which gives

$$\sigma_{n+1}^f = A\sigma_n^f + w^n 1 x_n \tag{5.19}$$

where $\sigma_n^f = w^n \psi_n$ is K-dimensional column vector, A is a lower triangular all one matrix, and $\mathbf{1}$ is K-dimensional all one vector. Again if $w = 1$ then this formula is valid for the case of spectral null at $f = 0$. If $\sigma_0^f = 0$, the kth element of vector σ_n^f represents kth-order RDS$_f$ of sequence x at moment n. The Equation (5.19) describes dynamics of K-RDS$_f$ FSTD with respect to time. Also, the Equation (5.19) is related to the concept of canonical state transition diagrams [18], denoted by D_K^f. These diagrams have a countably infinite number of states, and there exists a finite state subdiagram of D_K that generates sequences with Kth order spectral null at frequency f. One such finite state subdiagram, D_2^0, is shown in Figure 5.6, for the following values of parameters: cardinality of channel symbol alphabet $M = 2$, 1-RDS$_0$, and 2-RDS$_0$ assume values from $\{-1, 0, 1\}$, and $K = 2$. One can notice that 1-RDS$_0$ is obtained by setting $w = 1$, and $k = 1$ in Equation 5.15, which is classical RDS as defined by Equation 5.1, while 2-RDS$_0$ is obtained by setting $w = 1$, and $k = 2$ in Equation 5.15, which is classical RDSS as defined by Equation 5.15. In this case vector σ_n^f is two-dimensional, where first entree is z_n, and the second is y_n, if the notation from previous section is used. The graph can be drawn such that abscissa represents z_n, and ordinate represents y_n. If V_k denotes the number of different values that kth-order RDS$_f$ can assume then for $K = 2$, the following relation holds between V_1, the number of values that 1-RDS$_0$ can assume, V_2, the number of values that 2-RDS$_0$ can assume

$$V_1 = 2\left\lfloor \sqrt{V_2 - 1} \right\rfloor + 1 \tag{5.20}$$

where $\lfloor x \rfloor$ defines the largest integer smaller than x. In general, it is proven that finite bound on V_k implies the finite bound on all V_j, $j < k$. These FSTD enables the computation of capacities, and power spectral densities of the constraints, and encoder/decoder design. Table 5.3 gives capacities for some values of constraint parameters. In Figure 5.7, the power spectral densities of two maxentropic sequences, (1) with first order spectral null, and parameters $M = 2$, $N = 3$, and (2) with second-order spectral null, and parameters $M = 2$, $V_1 = V_2 = 3$, 2-RDS$_0$, at frequency $f = 0$ are shown. It can be seen that sequences with second-order spectral null has better rejection of low frequency components although the notch width is wider in the case of sequences with first-order spectral null. The relation between cut-off frequency, and sum variance, Equation 5.7, is no longer valid. Figure 5.8 depicts, power spectrum of two binary memoryless codes with spectral zero at $f = 0.5 f_s$, whose codeword length are (1) $n = 4$, and (2) $n = 8$.

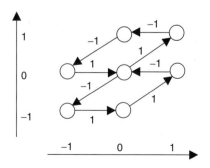

FIGURE 5.6 Canonical state diagram for $M = 2$, $V_1 = V_2 = 3$, 2-RDS$_0$ constraint.

TABLE 5.3 Capacity of Higher Order Zero Constraint for M=2, f=0

K	V_k	C
2	$V_2 = 3$	0.2500
2	$V_2 = 4$	0.3471
2	$V_2 = 5$	0.4428
2	$V_2 = 6$	0.5155
2	$V_2 = 7$	0.5615
3	$V_3 = 5$	0.1250
3	$V_3 = 6$	0.1250
3	$V_3 = 7$	0.1735
3	$V_3 = 8$	0.1946
3	$V_3 = 9$	0.2500

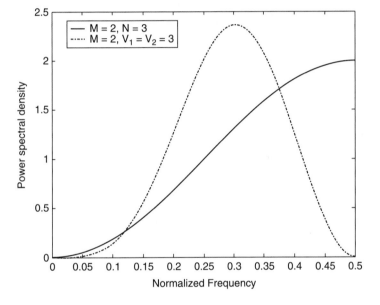

FIGURE 5.7   Power spectral density: (i) $M = 2$, $N = 3$ dc-free constraint; (ii) $M = 2$, $V_1 = V_2 = 3$, 2-RDS$_0$ constraint.

These codes satisfy condition that $\sigma_1^{f_s/2}(x) = 0$, and closed form expression for their power spectral densities is derived in [1]. The larger codeword length, the smaller notch width, saying that better rejection of low frequency components has to be paid by redundancy.

A number of constructions of codes with higher order spectral zeros, based on FSTDs, can be found in [14].

At the end, we give a useful relation between kth codeword moment at frequency f, $\mu_k^f(x)$, and kth-order RDS$_f$. Lemma 1 from [14] says that the following two statements are equivalent

$$\sigma_f^k(x) = 0, \quad 1 \le k \le K$$

$$\mu_f^k(x) = \sum_{i=0}^{n} i^k w^i x_i = 0, \quad 0 \le k \le K - 1$$

for codeword $x = (x_0 x_1 \ldots x_n)$.

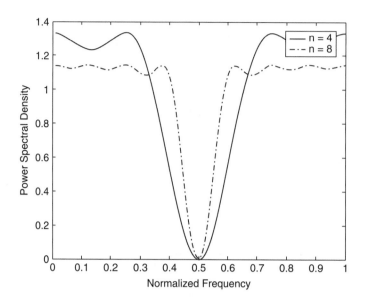

FIGURE 5.8 Power spectral density of codes with spectral zero at $f = f_s/2$: (i) Codeword length $n = 4$; (ii) Codeword length $n = 8$.

5.4.4 K-RDS$_f$ Sequences on Partial Response Channels

As it was mentioned in introduction, codes that generates K-RDS$_f$ constrained sequences improve the performance of recording systems utilizing partial response channels. The most interesting cases are of channels whose transfer functions have $(1 - D)^P$, $(1 + D)^P$ or both as factors. Here, two results will be given that concerns partial response channels. One is related to Hamming distance of binary sequences having one Kth order spectral null [17], and the other considers Euclidean distance of encoded sequences that was conveyed through partial response channel [14]. It is assumed that used codes have spectral zeros identical to spectral zeros of the channels. This notion of matched spectral zeros of codes, and partial response channels was introduced in [20].

Hamming distance. According to [17], the lower bound of minimum Hamming distance $d_{\min}^H$ between two binary sequences that have Kth order spectral null at $f = 0$ or $f = f_s/2$ is given as

$$d_{\min}^H \geq 2K \tag{5.21}$$

Euclidean distance. Before stating the result, the Euclidean distance for this kind of codes and channels will be discussed. Let's consider two sequences of channel symbols $x = \{x_i\}_{i=0}^n$, and $\hat{x} = \{\hat{x}_i\}_{i=0}^n$, generated by the following sequences of states, $\psi = \{\psi_i\}_{i=0}^{n+1}$, and $\hat{\psi} = \{\hat{\psi}_i\}_{i=0}^{n+1}$ respectively, such that $\psi_0 = \hat{\psi}_0$, $\psi_{n+1} = \hat{\psi}_{n+1}$. The error sequence at the input of the channel is defined as $e_i = x_i - \hat{x}_i, 0 \leq i \leq n$. If the error sequence at the output of the channel with memory H is denoted as $\varepsilon = \{\varepsilon_i\}_{i=0}^{n+H}$, then Euclidean distance is defined as

$$d_{\min}^2 = \min_{\varepsilon} \sum_{i=0}^{n+H} |\varepsilon_i|^2 \tag{5.22}$$

where minimum is taken over all allowable output error sequences and all n.

The lower bound on Euclidean distance $d_{\min}^2$ for sequences with Kth order spectral null at $f = 0$ or $f = f_s/2$ at the output of a partial response channel with a spectral null of order P at $f = 0$ or $f = f_s/2$ is given by

$$d_{\min}^2 \geq 8d^2(K + P) \tag{5.23}$$

where $2d$ is the minimum distance between two amplitude levels. The lower bound in Equation 5.23 can be reached if $M + P \leq 10$.

5.5 Composite Constrained and Combined Encoding

Previous two sections considered the codes that main purpose is to shape the spectrum of the channel stream in order to match spectral characteristics of the channel, and in that way make recording reliable. This type of encoding represents just one of many types of encoding used in digital recording systems. Like any communication system, a recording system employs different encoding techniques such as source encoding, channel encoding, and modulation encoding. It means that a channel sequence has to satisfy different kinds of constraints to be reliable recorded. That's why there is a need for codes that generate sequences satisfying composite constraints. One example of such codes are codes that are in the same time RLL (run length limited), and dc-free codes [21, 22]. RLL codes are widely used in digital recording systems. They confine minimal, and maximal number of consecutive like symbols in a recording channel to fight intersymbol interference, and to enable clock recovery. In wider sense, to this group of codes belongs the combination of dc-free codes and error correcting codes that improves the performance of dc-free codes on noisy recording channels. For instance those codes can be found in [27–30].

The construction of composite constrained codes can be based on so-called composite graphs generating composite constraint. A composite graph represents the composition of two or more graphs that generate different constraints, and the sequence satisfying composite constraint is the one that in the same time satisfies the constraints of composition constituents. Here, the formal approach of graph composition will be given. The basis for spectral analysis of composite constraints can be found in [23–25].

A sofic or constrained system S is the set of all biinfinite sequences generated by walks on a directed graph $G = G(S)$ whose edges are labeled by symbols in a finite alphabet A. The graph $G = (V, E, \pi)$ is given by a finite set of vertices (or states) V, a finite set of directed edges E, and a labeling $\pi : E \to A$. So, for a given sequence of edges $\{e^{(k)}\}(e^{(k)} \in E)$, we have the output sequence $\{a^{(k)} = \pi(e^{(k)})\}$. A graph G is strongly connected if for every two vertices $u, v \in V$ there exists a path (sequence of edges) from u to v. A graph, G, is deterministic if for each state $v \in V$, the outgoing edges from v, $E(v)$, are distinctly labeled.

The connection matrix (or vertex transition matrix) $\mathbf{D}(G) = \mathbf{D} = [D(u, v)]_{u,v \in V}$ of graph G is $|V| \times |V|$ matrix where entry $D(u, v)$ is the number of edges from vertex u to vertex v, and $|V|$ is the number of vertices of the graph G.

One type of composition of two graphs is given by Kronecker's product. The Kronecker's product of the graphs $G_0 = (V_0, E_0, \pi_0)$ and $G_1 = (V_1, E_1, \pi_1)$, $G = G_0 \otimes G_1$, is the graph $G = (V, E, \pi)$ for which $V = V_0 \times V_1$ ($\times$ denotes Cartesian product of the sets) and for every edge e_0 from u_0 to v_0 in G_0 and every edge e_1 from u_1 to v_1 in G_1, there exists an edge $e(e = (e_0, e_1))$ in G, emanating from vertex $u = (u_0, u_1) \in V$ and terminating at $v = (v_0, v_1) \in V$ with the vector label $\pi(e) = \pi(e_0, e_1) = [\pi_0(e_0)\pi_1(e_1)]$.

So, the graph $G = G_0 \otimes G_1$ generates vector sequences $\{a^{(k)}\} = \{[a_0^{(k)}a_1^{(k)}]\}$. If connection matrices of the graphs G_0 and G_1 are $\mathbf{D}_0$ and $\mathbf{D}_1$, then the adjacency matrix of the graph $G = G_0 \otimes G_1$, $\mathbf{D}$, is $\mathbf{D} = \mathbf{D}_0 \otimes \mathbf{D}_1$, wherein now $\otimes$ denotes the Kronecker's product of matrices [26]

$$\mathbf{D} = [D(u, v)]_{u,v \in V} = [D(u = (u_0, u_1), v = (v_0, v_1))]_{u_0,v_0 \in V_0; u_1,v_1 \in V_1}$$
$$= \mathbf{D}_0(u_0, u_1) \otimes \mathbf{D}_1(u_1, v_1) = [D_0(u_0, v_0) \cdot D_1(u_1, v_1)]_{u_0,v_0 \in V_0; u_1,v_1 \in V} \tag{5.24}$$

As an example, Figure 5.9 shows two deterministic graphs G_0 and G_1 and their Kronecker's product, G.

The Kronecker's product of the graphs represents the vector constrained system in which the component subsequences of the constrained systems, are generated independently. So the vector sequence carries an average amount of information equal to the sum of average amounts of information of the constituent sequences. Consequently the capacity of the Kronecker's product $G = \otimes_{0 \leq i \leq N-1} G_i = G_0 \otimes G_1 \otimes \cdots \otimes G_{N-1}$ is $C(G) = \sum_{0 \leq i \leq N-1} C(G_i)$. The proof of this statement follows directly from the fact that the set of eigenvalues of $\mathbf{D}(G)$ is the set of all products of eigenvalues of the factor graph connection matrices [26].

As an example we show the composition of two graphs, G_1 representing $M = 3$, $(d = 1, k = 2)$ RLL constraint (Figure 5.10), and G_2 representing $M = 3$, $N = 3$ dc-free constraint (Figure 5.3). Resulting

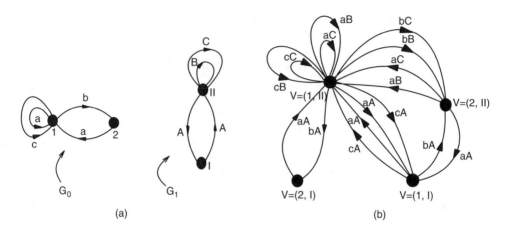

(a)   (b)

FIGURE 5.9 (a) An example of two graphs G_0 and G_1 ($V_0 = \{1,2\}$, $V_1 = \{I, II\}$, $A_0 = \{a,b,c\}$, $A_1 = \{A,B,C\}$), and (b) Their Kronecker's product.

graph G_3 generates sequences that are in the same time RLL, and dc-free (Figure 5.11), with parameters $M = 3$, ($d = 1, k = 2$), $N = 3$. Both constrained systems use the same alphabet $A = \{-1, 0, 1\}$. This composition is slightly modified as compared to previously defined Kronecker's product. Namely, there exists transition between the states in composite graph, $u = (u_0, u_1) \in V$, and $v = (v_0, v_1) \in V$, if $\pi_0(e_0) = \pi_1(e_1)$. Then, the label of corresponding edge will be $\pi(e) = \pi_0(e_0) = \pi_1(e_1)$. It should be noted that resulting graph is not always strongly connected. That's why, the irreducible component of G_3 should be found that has the same Shannon capacity C as reducible graph. The final form of the composite graph is obtained by finding the Shannon cover of G_3 [9]. The graph in Figure 5.11 has ten states and it is Moore's type FSTD. The Shannon capacity is $C = 0.4650$ bits/sym, and the chosen code rate is $R = 1/3$ bits/sym, which is less than C. Table with capacities of this constraint is given in [22]. In order to get the encoder, the algorithm from [9] was applied, and it is shown in Figure 5.12. The sliding window decoder for this code can be found in [22].

At the end, in Figure 5.13, the spectrum of M-ary RLL dc-free maxentropic sequences are show, for fixed DSV, $N = 7$, and different values of parameter M.

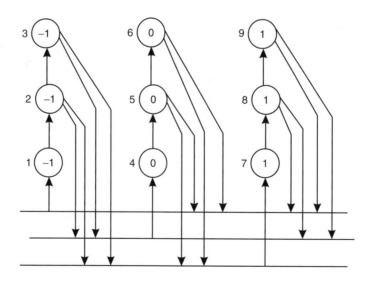

FIGURE 5.10   FSTD of M-ary RLL constraint, $M = 3$, ($d = 1, k = 2$).

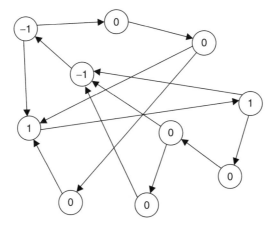

FIGURE 5.11 FSTD of M-ary RLL dc-free constraint, $M = 3$, $(d = 1, k = 2)$, $N = 3$.

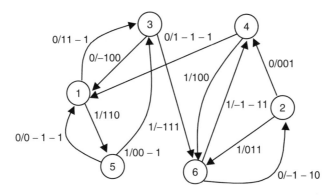

FIGURE 5.12 Encoder of $M = 3$, $(d = 1, k = 2)$, $N = 3$, RLL dc-free codes.

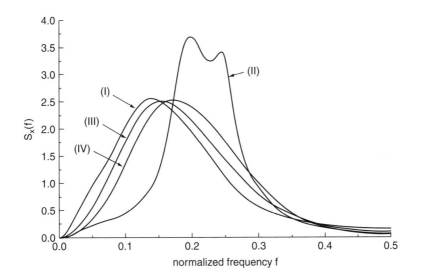

FIGURE 5.13 Spectrum of M-ary RLL dc-free maxentropic sequences: (i) $M = 3$, $(d = 1, k = 3)$, $N = 7$; (ii) $M = 4$, $(d = 1, k = 3)$, $N = 7$; (iii) $M = 5$, $(d = 1, k = 3)$, $N = 7$; (iv) $M = 7$, $(d = 1, k = 3)$, $N = 7$.

5.6   Conclusion

This chapter gives a survey of basic concepts and ideas of spectrum shaping codes for digital recording systems. We considered theoretical, and practical aspects of four groups of spectrum shaping codes, dc-free codes, codes with higher order spectral null at $f = 0$, codes with spectral nulls at the submultiples of channel symbol frequency, and codes with composite constraints. We provided constrained channel capacities, power spectral densities, and some practical solutions for encoding and decoding schemes.

Acknowledgment

This work is supported in part by the NSF under grant CCR-0208597.

References

[1] Immink, K.A.S., *Coding Techniques for Digital Recorders,* 1st ed., Prentice Hall International, New York, 1991.

[2] Immink, K.A.S., Siegel, P.H, and Wolf, J.K., Codes for digital recorders, *IEEE Trans. Inform. Theory,* 44, 2260, 1998.

[3] Pierobon, G.L., Codes for zero spectral density at zero frequency, *IEEE Trans. Inform. Theory,* 30, 435, 1984.

[4] Shannon, C.E., A mathematical theory of communications, *Bell Syst. Tech. J.,* 27, 379, 1948.

[5] Spielman, S. et al., Using pit-depth modulation to increase the capacity and data transfer rate in optical discs, invited paper, *1997 Optical Data Storage Society Annual Meeting.*

[6] Justesen, J., Information rate and power spectra of digital codes, *IEEE Trans. Inform. Theory,* 28, 457, 1982.

[7] Cariolaro, G.L. and Tronca, G.P., Spectra of block coded digital signals, *IEEE Trans. Commun.,* 22, 1555, 1974.

[8] Bilardi, G., Padovani, R., and Pierobon, G.L., Spectral analysis of functions of Markov chains with applications, *IEEE Trans. Commun.,* 31, 853, 1983.

[9] Marcus, B.H., Siegel, P.H., and Wolf, J.K., Finite state modulation codes for data storage, *IEEE J. Selec. Area Commun.,* 10, 5, 1992.

[10] Cattermole, K.W., Principles of digital line coding, *Int. J. Electron.,* 55, 3, 1983.

[11] Bowers, F.K., U.S. Patent 2957947, 1960.

[12] Carter, R.O., Low disparity binary coding system, *Elect. Lett.,* 1, 65, 1965.

[13] Knuth, D.E., Efficient balanced codes, *IEEE Trans. Inform. Theory,* 32, 51, 1986.

[14] Eleftheriou, E. and Cideciyan, R.D., On codes satisfying Mth order running digital sum constraint, *IEEE Trans. Inform. Theory,* 37, 1294, 1991.

[15] Kobayashi, H., Correlative level coding and maximum likelihood decoding, *IEEE Trans. Inform. Theory,* 17, 586, 1971.

[16] Immink, K.A.S., Spectrum shaping with binary dc^2-constrained channel codes, *Philips J. Res.,* 40, 40, 1985.

[17] Immink, K.A.S. and Beenker, G.F.M, Binary transmission codes with higher order spectral zeros at zero frequency, *IEEE Trans. Inform. Theory,* 33, 452, 1987.

[18] Marcus, B.H. and Siegel, P.H., On codes with spectral nulls at rational submultiples of the symbol frequency, *IEEE Trans. Inform. Theory,* 33, 557, 1987.

[19] Karabed, R. and Siegel, P.H., Matched spectral-null codes for partial-response channels, *IBM Res. Rep.* RJ 7092, April 1990.

[20] Karabed, R. and Siegel, P.H., Matched spectral-null trellis codes for partial-response channels, part II: High rate codes with simplified Viterbi detectors, *IEEE Int. Symp. Inform. Theory,* Kobe, Japan, p. 143, 1988.

[21] Norris, K. and Bloomberg, D.S., Channel capacity of charge-constrained run-length limited codes, *IEEE Trans. Magn.*, 17, 3452, 1981.

[22] Denic, S.Z., Vasic, B., and Stefanovic, M.C., M-ary RLL dc-free codes for optical recording channels, *Elect. Lett.*, 36, 1214, 2000.

[23] Vasic, B., Spectral analysis of multitrack codes, *IEEE Transactions on Information Theory*, Vol. 44, no. 4, pp. 1574–1587, July 1998.

[24] Vasic, B. and Stefanovic, M., Spectral analysis of coded digital signals by means of difference equation systems, *Elect. Lett.*, 27, 2272, 1991.

[25] Vasic, B., Spectral analysis of codes for magnetic and optical recording, Ph.D. dissertation, University of Nis, May 1993 (in Serbian).

[26] Cvetkovic, D., *Combinatorial Theory of Matrices*, Naucna knjiga, Belgrade, 1985 (in Serbian).

[27] Popplewell, A. and O'Reilly, J., A simple strategy for constructing a class of DC-free error-correcting codes with minimum distance 4, *IEEE Trans. Inform. Theory*, 41, 1134, 1995.

[28] Deng, R.H., Li, Y.X., Herro, and M.A., DC-free error-correcting convolutional codes, *Elect. Lett.*, 29, 1910, 1993.

[29] Lee, S.I., DC- and Nyquist-free error correcting convolutional codes, *Elect. Lett.*, 32, 2196, 1996.

[30] Etzion, T., Constructions of error-correcting DC-free block codes, *IEEE Trans. Inform. Theory*, 36, 899, 1990.

[31] Immink, K.A.S. and Patrovics, L., Performance assessment of dc-free multimode codes, *IEEE Trans. Commun.*, 45, 293, 1997.

Introduction to Constrained Binary Codes with Error Correction Capability

Hendrik C. Ferreira
Rand Afrikaans University
Auckland Park, South Africa

Willem A. Clarke
Rand Afrikaans University
Auckland Park, South Africa

6.1 Introduction

Constrained codes (alternatively called modulation codes, line codes or transmission codes) impose runlength or disparity constraints on the coded sequences, in order to either comply with the input restrictions of some communications channels, as determined by intersymbol interference or bandwidth limitations, or to aid in receiver synchronization and detection processes. Usually, these codes are not designed for error correction although they sometimes have limited error detection capabilities.

Binary runlength constrained codes, that is, (d, k) or (d, k, C) codes, find application on digital magnetic and optical recorders [54, 56]. Here d is the minimum number and k the maximum number of code zeros between consecutive code ones in the nonreturn to zero inverse (NRZI) representation and C the upper bound on the running disparity between ones and zeros in the nonreturn to zero (NRZ) representation.

Binary dc free codes have a bounded running disparity of ones and zeros in the coded sequences. These codes have been employed in early disc systems and later on in tape drives, and they also find widespread

(a) Traditional approach to coding for error control and channel with input constraints.

(b) Combined coding scheme

FIGURE 6.1   Traditional and combined coding schemes.

application on metallic and optical cable systems. Since the minimum Hamming distance d_{min} is at least 2, dc free codes can detect at least one error. Balanced binary codes are dc free codes with equal numbers of ones and zeros in every codeword. This class of codes can also be considered as a subset of the class of constant weight codes.

In Figure 6.1(a), we show the traditional concatenated coding scheme, used to achieve both the goals of conforming to the channel input constraints and of providing error correction. To date, this approach is still used in many recording standards and products. One disadvantage of this scheme is the error propagation at the output of the constraint code's decoder: a single channel error may trigger multiple decoding errors. In general, the closer the coding rate R of the constrained code approaches the capacity C of the input restricted channel, the higher the complexity of such a constrained coding scheme, and the more the errors propagated. The propagated errors furthermore tend to be bursty in nature, which poses an additional load on the error-correcting scheme, and hence even more redundancy may be required. Consequently, since the early 1980s, several researchers have investigated the coding scheme in Figure 6.1(b). In the literature, codes for this scheme have been called combined codes, and on occasion also combi-codes or transcontrol codes.

During the late 1970s and early 1980s, the question was also posed whether some soft decision coding gain could be obtained from the constrained codes employed at that time, by using Viterbi decoding — see for example, [95]. This was partly inspired by the benefits of trellis-coded modulation for bandlimited channels, which was introduced at that time and which furthermore contributed to the impetus to develop combined codes.

It should also be noted that during the period under review here, several combined code constructions were furthermore aimed at partial response channels and later also on channels with two dimensional runlength constraints (e.g., [21]), as well as on channels with multilevel (M-ary instead of binary) symbols (e.g., [76]). However, a complete overview of code constructions for all these channels, is beyond the scope of this chapter.

In this chapter, we thus emphasize codes for the channel with one dimensional (d, k) constraints. The development of these codes went hand in hand with the development of error correcting dc free codes during the same time period, often by the same researchers and often using the same construction techniques, hence we also include some results and references on the latter class of codes.

6.2   Bounds

At first, the construction of error-correcting constrained codes, that is, constrained codes with minimum Hamming distance $d_{min} \geq 3$, appeared to be an elusive goal. Consequently, to find an existence proof, some early lower bounds on the Hamming distance achievable with runlength constrained block codes or balanced block codes were set up, using Gilbert type arguments, and published in [30]. To find an existence proof, the freedom of having infinitely long codewords was assumed.

Briefly, a constrained code with Hamming distance d_{min} can be formed by selecting a word from the set of all constrained words of length n bits as first code word, and by purging all other words at $d < d_{min}$ from the set. Subsequently, a second word at $d = d_{min}$ from the first word can be selected, and all words remaining at $d < d_{min}$ from this word can also be purged. This purging process can be continued until only the desired set of code words remains.

Thus, since $\binom{n}{i}$ is the maximum number of binary words which may be at distance $i < d_{min}$ from a code word, we can arrive at the following lower bound on the minimum Hamming distance achievable with an (n, k) constrained block code with $n \to \infty$:

$$2^{nC} \geq 2^k \sum_{i=0}^{d_{min}-1} \binom{n}{i} \tag{6.1}$$

where C is the capacity of the noiseless input restricted channel. Stated differently, if $C > k/n$, we can obtain an (n, k) block code with desirable d_{min} by making n large enough.

This bound is also a rather loose lower bound on the minimum k achievable with a specified n and d_{min}, since firstly not all binary words at distance i from a retained codeword satisfy the constraints, and secondly the number of words which has to be purged as the purging process continues, may grow smaller, as some of these words have been purged earlier.

As shown in [30], the bound can be tightened for balanced dc free codes, for which n will be even. Note that the distance between two words with the same weight and hence also the parameter i in Equation 6.1, can only be even for words of the same weight. Setting $i = 2a$, and making use of the balance between ones and zeros in both the constrained words retained and those purged, we can form a bound for balanced (d, k, C) sequences as

$$2^{nC} \geq 2^k \sum_{a=0}^{(d_{min}-2)/2} \binom{n/2}{a}^2 \tag{6.2}$$

which simplifies for a balanced code with constant weight $w = n/2$ (and minimum runlength $d = 0$) to

$$\binom{n}{n/2} \geq 2^k \sum^{(d_{min}-2)/2} \binom{n/2}{a}^2 \tag{6.3}$$

For further work and improvements on the topics of both lower and upper bounds, refer to [2, 43, 60, 73, 94, 97, 98]. A few results from these references will be briefly discussed next.

Ytrehus [98] derived recursive upper bounds for several choices of (d, k). Kolesnik and Krachkovsky [62] generalised the Gilbert-Varshamov bound to constrained systems by estimating the average volume of such constrained spheres. They then used a generating function for the distribution of pair wise distances of words in the constrained system, together with the Perron-Frobenius theorem, in order to obtain asymptotic existence results relating the attainable code rate R to a prescribed relative minimum distance d_{min} when $n \to \infty$. Gu and Fuja [43] improved on this Gilbert-Varshamov bound even further, still using average volume sphere arguments. Marcus [73], use labelled graphs to improve on the bound of Kolesnik and Krachkovsky.

Abdel-Ghaffar and Weber in [2] derive explicit sharp lower and upper bounds on the size of optimal codes that avoid computer search techniques. Using sphere packing arguments, they derive general upper bounds on the sizes of error-correcting codes and apply these bounds to bit-shift error correcting (d, k) codes. These bounds improve on the bounds given by Ytrehus [98].

In [43], Gu and Fuja provide a generalised Gilbert-Varshamov bound, derived via analysis of a code-search algorithm. This bound is applicable to block codes whose codewords must be drawn from irregular sets. It is demonstrated that the average volume of a sphere of a given radius approaches the maximum such volume and so a bound previously expressed in terms of the maximum volume can in fact be expressed in terms of the average volume. This bound is then applied specifically to error-correcting (d, k)-constrained codes.

6.3   Example: A Trellis Code Construction

Some early work on the actual construction of error-correcting constrained codes, can be traced back to [29–34], and interestingly, as later discovered, also to [4]. Subsequently, many widely different and sometimes ad hoc approaches to constructing such codes, evolved. We next present as simple example, suitable for an introductory and tutorial presentation, an approach from the earlier work on trellis codes. In the following section, we shall give an overview and classification of other block and trellis code constructions in the literature.

Trellis code constructions have the advantage that a general decoding algorithm, namely the Viterbi algorithm is immediately available. Constrained codes are usually nonlinear and suitable decoding algorithms may thus be difficult to find or complex to implement. Furthermore, soft decisions may be utilized when doing Viterbi decoding, in order to obtain additional coding gain.

The construction procedure in this section can be used to obtain trellis codes having various coding rates, constraint lengths and free distances, and with complexity commensurate with that of the class of linear binary convolutional codes widely used in practice. It uses the distance preserving mapping technique first described in [34, 35]. For related work, refer to [14, 38, 40, 41, 90].

Refer to Figure 6.2. The mapping table in Figure 6.2 maps the output binary n-tuple code symbols from a $R = k/n$ convolutional code (henceforth called the base code) into constrained binary m-tuples, which in this application are code words from a (d, k) code. The key idea is to find an ordered subset of 2^n m-tuples, out of the set of all possible constrained m-tuples with cardinality $N(m)$, such that the Hamming distance between any two constrained m-tuples is at least as large as the distance between the corresponding convolutional code's output n-tuples which are mapped onto them. This property may be called *distance preserving*, since the Hamming distance of the base code will be at least be conserved, and may sometimes even be increased in the resulting trellis code.

To illustrate this idea, we first present an example. We use the simple, "generic text book example" four state, binary $R = 1/2, v = 2, d_{\text{free}} = 5$ convolutional code in Figure 6.3(a) as base code. At the output of the encoder, we can map the set of binary 2-tuple code symbols, $\{00, 01, 10, 11\}$ onto constrained 4-tuples with $d = 1$, specifically using the set $\{0100, 0010, 1000, 1010\}$. Note that the last bit of each constrained 4-tuple used here is a merging bit, initially set to 0, to allow concatenation of the constrained symbols, without violating the d constraint. The state system of the resulting code appears in Figure 6.3(b).It should be stressed that the intention is to decode this resulting code in one step with the Viterbi algorithm.

In general, the property of *distance preserving* can be verified by setting up the matrices $D = [d_{ij}]$ and $E = [e_{ij}]$. Briefly, let d_{ij} be the Hamming distance between the binary code symbols i and j, where

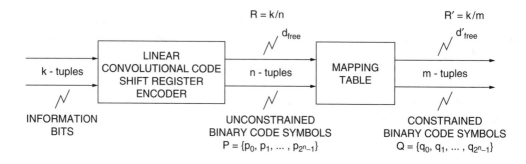

FIGURE 6.2   Distance preserving trellis code: encoding process.

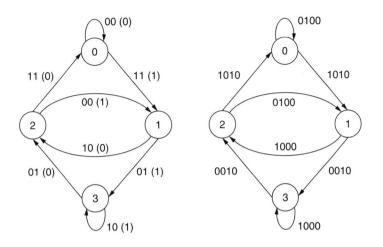

FIGURE 6.3 State systems: (a) Convolutional base code, (b) Constrained trellis code.

$0 \le i, j \le 2^n - 1$. The key to this code construction technique is thus to find an ordered subset of 2^n constrained m-tuples such that $e_{ij} \ge d_{ij}$, for all i and j, and where e_{ij} is the Hamming distance between the i'th and j'th m-tuples in the subset. Thus D and E can be set up to verify the mapping for our example code in Figure 6.3:

$$D = \begin{bmatrix} 0 & 1 & 1 & 2 \\ 1 & 0 & 2 & 1 \\ 1 & 2 & 0 & 1 \\ 2 & 1 & 1 & 0 \end{bmatrix} \quad \text{and} \quad E = \begin{bmatrix} 0 & 2 & 2 & 3 \\ 2 & 0 & 2 & 1 \\ 2 & 2 & 0 & 1 \\ 3 & 1 & 1 & 0 \end{bmatrix} \quad (6.4)$$

In this example the base code has $d_{\text{free}} = 5$, consequently the resulting $(d = 1, k)$ constrained code will also have $d_{\text{free}} \ge 5$. It can be shown by inspection that the maximum runlength k can be reduced to $k = 3$, by inverting the merging bit whenever possible.

Furthermore, note that the same mapping can now be applied to any $R = 1/2$ convolutional base code. In this way, we can thus easily construct more powerful trellis codes achieving larger free distances.

The procedure in the above example can be formalized as follows. Let the unconstrained binary n-tuples form a set $U = \{u_i\}$, with cardinality 2^n. The n-bit code symbols of a linear $R = k/n$ convolutional code will always be contained in U. The set of all constrained binary m-tuples may be represented by set $C = \{c_i\}$, with cardinality $N(m)$. The m-bit code symbols of the desired $R' = k/m$ constrained trellis code will be a subset of C.

In general, we want to transform a linear rate $R = k/n$ convolutional code with free distance d_{free} into rate $R' = k/m$, d'_{free} trellis code with $d'_{\text{free}} \ge d_{\text{free}}$. Usually, $m > n$, and to maximize the rate R' of the constrained trellis code derived from a given $R = k/n$ base code, we thus choose to only investigate mappings such that

$$n = \lfloor \log_2 N(m) \rfloor \quad (6.5)$$

The transformation will be per invariable mapping table as shown in Figure 6.3, that is, the same unconstrained binary code symbol will always be mapped onto the same constrained code symbol, irrespective of the base code.

To formalize the above: we want to map the ordered subset containing all unconstrained binary n-tuples, that is, $P = \{p_0, p_1, \dots, p_{2^n-1}\}$, with elements $p_i \in U$, $p_i < p_{i+1}$, onto an ordered subset of constrained m-tuples, that is, $Q = \{q_0, q_1, \dots, q_{2^n-1}\}$, with $Q \subset C$. To devise the mapping table, we first compute the Hamming distance matrix D with elements d_{ij}, being the Hamming distance between unconstrained binary n-tuples p_i and p_j, or,

$$d_{ij} = w(p_i + p_j) \quad 0 \leq i, j \leq 2^n - 1 \tag{6.6}$$

where $w(x)$ denotes the Hamming weight of x and addition is modulo 2. The task is now to find a suitable ordered subset Q, with 2^n elements q_i from the set C, such that the Hamming distance matrix E with elements

$$e_{ij} = w(q_i + q_j) \quad 0 \leq i, j \leq 2^n - 1 \tag{6.7}$$

has

$$e_{ij} \geq d_{ij}, \quad \text{for all } i, j \tag{6.8}$$

Note that the matrices D and E are square symmetric matrices with all-zero diagonal elements. Furthermore, the total number of permutations of the $N(m)$ constrained m-tuples, taken 2^n at a time, grows very rapidly with the tuple length m. For this reason in [35], we modeled the search as a tree search, similar to the Fano algorithm, but with the number of branches per node in the tree decreasing by one at each new depth. Note also that the main diagonal divides D into an upper and a lower triangular array with equal valued entries, and due to the symmetry only one of these need to be used in the search. The search may furthermore be sped up by reordering P such that the maximum valued elements in D are encountered earlier. See [35] for more details.

A few tests can also be performed before the tree search to establish the nonexistence of a suitable subset of constrained m-tuples or to prove that a specific m-tuple cannot be a member of the ordered subset of m-tuples — refer to [35].

Using prefix constructions, a mapping for some m may be used as kernel and extended to find a mapping for $m + 1$. The principle can be explained as follows. The set of binary $(n + 1)$-tuples, can be ordered following normal lexicography, that is, setting up the standard table of $(n + 1)$-bit binary numbers. It is easy to see that this set is partitioned into two subsets each containing 2^n elements. The first subset of $(n + 1)$-bit binary numbers are obtained by prefixing the set of n-bit binary numbers with a most significant bit 0, and the second subset of $(n + 1)$-bit binary numbers by prefixing the set of binary n-bit binary numbers with a most significant bit 1. Within each subset, the intradistance between elements is determined by the $n \times n$ D matrix, and stays the same. However, the binary prefixes of 0 and 1 account for an additional one unit of distance between two elements from the two different subsets. In a similar way, the ordered subset Q, containing 2^n constrained binary m-tuples can be extended to an ordered subset containing 2^{n+1} constrained $(m + 2)$-tuples by using two prefixes with Hamming distance at least one unit. For the $d = 1$ constraint, we can use the prefixes 00 and 10 and still satisfy the minimum runlength requirement. When using balanced symbols to construct dc free codes, we can use the prefixes 01 and 10.

Explicit mappings for m-tuples with $d = 1$, where it was attempted to maximize the combined code's rate $R' = k/m$, and to minimize k, are reported in [35], as well as mappings using balanced m-tuples from which dc free codes can be obtained. The highest achievable code rates with the mappings published, were $R' = 4/9$ and $R' = 3/6$ respectively. The maximum achievable free distances, will be determined by the underlying $R = 4/5$ and $R = 3/4$ base codes — in the literature on convolutional codes there are many codes of these rates available with different free distances and constraint lengths. Song and Shwedyk [90] later investigated a graph theoretic procedure to enumerate all the distance preserving mappings as in [35].

Finally, as discussed in [38], note that we can expand and generalize the concept of a distance preserving mapping to include a distance conserving mapping (DCM), as well as a distance increasing mapping (DIM) and a controlled distance reducing mapping (DRM).

In this section, we have thus presented a code construction procedure, capable of constructing powerful constrained trellis codes. By using this construction procedure, advantage can be taken of the many results on, and vast literature covering good convolutional codes. Furthermore an important reason for presenting this procedure is that it can also be applied to other constraints and channels as in [38, 14].

6.4 An Overview of Some Other Code Constructions

A literature search, which was not exhaustive, revealed more than one hundred papers on the topic of combining error correction with constrained codes. We attempted to include a representative selection of papers in our bibliography. As can be seen, several disjoint, and sometimes ad hoc procedures evolved. We next present a short overview, attempting to indicate some of the most important trends and directions.

6.4.1 Channel Models and Error Types

Much of the work on error correcting constrained codes focused on the binary symmetric channel, since these codes dominate the theory of linear error correcting codes, and also exhibit certain robustness. Refer to [13, 18, 67] for a few examples of constructions aimed at correcting additive or reversal errors on the binary symmetric channel. It should however be noted that these codes cannot be directly interleaved to correct burst errors — usually the channel's input constraints will be violated.

Some experimental work (see e.g., [51]) showed that peak shift errors, that is, errors represented in NRZI as $010 \rightarrow 001$ or $010 \rightarrow 100$, dominate on many recording channels, and this inspired a body of work on suitable code constructions. In one test of the IBM 3380 disk, 85% of the observed errors were shift errors [47]. For examples of such code constructions, refer to [11, 47, 63, 64, 70, 87].

During the late 1980s, as recording densities increased, it was observed that the electronic circuits for bit synchronization might fail more often; hence the topic of correcting bit insertion/deletion errors also received some attention. However, these errors, although they may have very destructive consequences, have a much lower probability of occurrence than the above error types and hence not many papers were published on this topic — see [11, 37, 46, 59] for a few examples of code constructions.

6.4.2 Input Constraints

Simultaneously with the interest in combined codes for magnetic recording, several researchers investigated combined codes for cable systems with somewhat different input constraints, such as a dc free power spectral density, or maximum runlength constraint, with no restriction on the minimum runlength — see for example, [5, 60–61, 79, 82–86].

In this regard, it is interesting to note that the balanced codes, a subset of the family of constant weight codes, are dc free, and bounds on the cardinalities of balanced codes as a function of d_{min}, have been tabulated in a few text books on coding for error correction and papers in the literature even before the new interest in combined codes during the 1980s.

In terms of (d, k) constraints, the $(1, 7)$ and $(2, 7)$ constraints dominated the magnetic recording industry for a long time, hence most researchers of combined codes attempted to conform with these constraints, or at least with $d = 1$ or 2 — see for example, [1, 6].

So far, the success of investigations into the construction of new combined codes appears to be proportional to the capacity of the input constrained channel. For (d, k, C) parameters of practical interest, the channel capacity increases in the same order if the input constraint is relaxed from (d, k, C) to (d, k) to $(0, k, C)$ to $(0, k)$. Consequently, very few results have been published on error correcting (d, k, C) codes. On the other hand, when it was proposed to relax the d constraint for magnetic recording systems, results on $(0, k)$ combined codes followed readily — for example, [93].

6.4.3 Block and Trellis Code Constructions

Most constructions in the vast field of linear error correcting codes can be classified as either a block code or a trellis code, and the same holds for combined codes. For a few results on block codes, refer to [2, 30, 67, 72, 87, 91], and for some results on trellis codes (or convolutional codes), refer to [15, 19, 45, 48–50, 100].

In the highly competitive recording industry, an increase in storage density of a few percentage points can be an important advantage. Although trellis codes have the advantages of Viterbi decoding and soft decision coding gain, some of the combined code constructions using block codes achieve higher code rates and hence the best exploitation of the Shannon capacity of the input-constrained noiseless channel. Furthermore, interleaving of the sequences of combined trellis codes in order to correct burst errors, can violate the (d,k) constraints. On the other hand, block-coding schemes are sometimes based on linear error correcting codes, burst error correcting in nature, such as Reed-Solomon codes.

Several authors (see e.g., [10, 36, 55]) have considered a block code construction which is systematic over a (d,k) constrained sequence. In its simplest form, a finite state machine or lookup table encoder may firstly map the bits from the information source onto a (d,k) sequence of length k bits. Next the parity bits of an (n,k) error correcting codeword can be computed. Finally, these parity bits are appended to the information (d,k) sequence in such a way that the parity sequence also complies with the channel's (d,k) constraints. This constrained parity sequence can be obtained by using means such as a lookup table, buffer bits between parity bits, etc.

Some authors have also parsed the (d,k) sequence into substrings, starting with a 1 and followed by between d and k zeros, or alternatively in reverse order. In [36, 17] the authors went a step further and showed that a unique integer composition can be associated with each (d,k) sequence. By imposing compositional restrictions on the (d,k) sequences, some error detection becomes possible. Error correction can be done by further appending parity bits. An advantage of applying the theory of integer compositions here, is that generating functions and channel capacities followed naturally. However, due to the compositional restrictions, code rates were too low for practical implementations, except when $d > 4$, while historically recording systems employed codes with $d = 1$ or 2, as dictated by physical factors such as detection window width. The parity bits in [36] typically keep track of the number of parts in the composition, or of the sum of the indexes of the positions in which a part occurs, both expressed modulo a small integer, and hence the number of parity bits could stay fixed, irrespective of the codeword length.

6.4.4   Combined Codes Directly Derived from Linear Error Correcting Codes

A natural question posed early in the development of combined codes, was whether a subcode of a linear error correcting code, or a coset code, having the required constraints, might be used as a combined code. This may have the advantage of making the powerful theory of linear codes applicable to the input constrained channel and perhaps using off-the-shelf decoders.

Examples considering subcodes of linear block codes can be found for example in [80], while subcodes of linear convolutional codes can be found in [78].

Pataputian and Kumar in [80] presented a (d,k) subcode of a linear block code. The modulation code is treated as a subcode of the error correcting code (a Hamming code), and they find a subcode of an error correcting code satisfying the additional (d,k) constraints required. They do this by selecting the coset in the standard array of Hamming codes which has the maximum number of (d,k) constrained sequences. This approach requires modulation codes that have very large block sizes when compared to conventional modulation codes in practice. One advantage of this scheme is that off-the-shelf decoders can be used. Systematic (but suboptimum) subcodes are also presented. In a similar approach, Liu and Lin [72] describes a class of (d,k) block codes with minimum Hamming distance of 3 based on $dklr$-sequences (the l and r represent the maximum number of consecutive zeros at the beginning and end of a (d,k)-sequence respectively) and cyclic Hamming codes. A codeword in the constructed code is formed by two subblocks satisfying the $dklr$ constraint. One is the message subblock and the other is the parity check subblock.

Similarly, employing cosets in new code constructions, has been considered in for example [18, 48–50]. Hole in [48] presented cosets of convolutional codes with short maximum zero-run lengths, that is, the k parameter. He achieved this by using cosets of (n,k) convolutional codes to generate the channel inputs. For $k \leq n - 2$ it is shown that there exist cosets with short maximum zero-run length for any constraint length. Any coset of an $(n, n-1)$ code with high rate and/or large constraint length is shown to have a large maximum zero-run length. A systematic procedure for obtaining cosets with short maximum zero-run length (n, k) codes is also given.

6.4.5 Constrained Codes Carefully Matched to Error Correcting Codes

Another natural approach was to carefully match the constrained codewords to the symbols of the error correcting code. In this way error correcting performance may be optimized. See for example [6]. The distance preserving mappings in [35] also fall into this category.

6.4.6 Constructions Employing Ideas from Contemporary Developments in Coding Techniques

One important contemporary idea was the principle of set partitioning applied to the channel signals, borrowed from the field of trellis coded modulation (TCM) where it was applied very successfully to amplitude/phase signals to develop combined coding and modulation schemes for the bandlimited channel. Attempts to apply it to the binary input constrained channel, met with limited success, due to the signal set lacking the same degree of symmetry.

Another important development was the ACH or state splitting algorithm for constructing finite state machine modulation codes for the input restricted channel — see for example, [21, 77] for application to combined codes. Application, again, does not follow directly. Nasiri-Kenari and Rushforth [77] show how the state-splitting and merging procedure can be adapted and applied to the problem of finding efficient (d, k) codes with guaranteed minimum Hamming distance. A second procedure in [77] partitions the encoder for a state-dependent (d, k) code into two subsections, one with memory and one without, and then combine the subsystem having memory with a matched convolutional code.

6.4.7 Restrictions on the (d, k) Sequence

By imposing compositional restrictions on (d, k) sequences, or by alternating between sets of (d, k) sequences with either only odd or even parity, error detecting and ultimately error correcting codes could be constructed — see, [36, 67].

Lee and Wolf [67] first derived a codebook Q consisting of all codewords adhering to the (d, k) constraint. However, concatenation of codewords from this codebook usually violates the (d, k) constraints. Therefore, the codebook Q is divided into maximal concatenatable subsets. Using a finite state code construction algorithm, they then generate a code with minimum free distance of three. Ferreira and Lin [36] presented combinatorial and algebraic techniques for systematically constructing different (d, k) block codes capable of detecting and correcting single bit errors, single peak-shift errors, double adjacent errors and multiple adjacent erasures. Their constructions are based on representing constrained sequences using integer compositions. Codes are obtained by imposing restrictions on such compositions and by appending parity bits to the constrained sequences.

Lee and Madisetti [69] proposes a general construction scheme for error correcting (d, k) codes having a minimum Hamming distance of 4. The proposed method uses a codeword set with a minimum Hamming distance of 2, instead of using two sets obtained by partitioning a concatenatable codeword set as in Lee and Wolf [67]. This means the code rate is always higher than that of Lee and Wolf. This proposed coding scheme is especially beneficial when the code lengths are short.

6.4.8 Multilevel Constructions

Several researchers have considered algorithms with multilevel partitioning of the symbol set, which is especially suitable for the constructing of dc free codes — see for example [13, 57].

On the other hand, M-ary multilevel (d, k) codes with error correction abilities have also received attention [76]. M-ary (d, k) codes are used for recording media that support unsaturated M-ary $(M \geq 3)$ signalling. This is different than the normal binary case $(M = 2)$ where the media is saturated. In [76], McLaughlin presented codes that achieve high coding densities, with improved minimum distance over an ordinary Adler-Coppersmith-Hassner code designed with the state-splitting algorithm. It is also shown that these codes have comparable minimum distance to Ungerboeck style amplitude modulation trellis codes.

6.4.9 Constructions Using the Lee or Levenshtein Metrics

Most of the work under review, employed the Hamming distance metric, since it is widely known and the best understood. Consequently most constructions are directed towards the correction of additive (reversal) errors. After it was realized that codes over the Lee metric could be used to correct shift errors and to some extent insertions/deletion errors, some publications followed [11, 47, 63, 87].

Bours [11] suggested the use of the Lee distance when considering peak-shift errors. Roth and Siegel [87] showed that some of the Lee-metric BCH codes could be used to provide efficient protections against bit-shift or synchronisation errors. For bit-shift correction, these codes possess a smaller redundancy than the codes in Hamming metric. In [63], Krachkovsky et al. proposed another class of fixed length, t-error-correcting codes in the Lee metric. In their codes, the Galois field characteristics may be chosen independently of t and metric parameter q (where q is the alphabet size).

Similarly, a few papers employed the Levenshtein metric to correct insertions/deletions – see for example, [7, 37]. These errors cause catastrophic failures due to loss of synchronisation. Although these errors are not as common as the other types of errors, it is important to have a method to detect them and differentiate them from regular errors. Insertions/deletions, also known as synchronisation errors, are different from peak-shift or bit-shift errors in that all the 1s are shifted after an insertion/deletion. In a peak-shift error, only one 1 is shifted. Often, the Levenshtein distance is used in this environment rather than the Hamming distance.

Blaum et al. [7] used the $(1, 7)$ modulation code to present two methods to recover from insertions or deletions. The first method (based on variable length codes) allows for the identification of up to three insertions and/or deletions in a given block, permitting quick synchronisation recovery. The second method, based on block codes, allows for the detection of large numbers of insertions and deletions. Kuznetsov and Vinck [64] presented codes for the correction of one of the following: a peak-shift of type $(k - d)/2$ or less, a deletion of $(k - d)/2$ or less zeros between adjacent ones and an insertion of $(k - d)/2$ or less zeros between adjacent ones.

Klove [59] presented codes correcting a single insertion/deletion of a zero or a single peak-shift. Particularly, he considers variable length (d, k) codes of constant Hamming weight.

6.4.10 Spectral Shaping Constraints and Error Correction

Codes with higher order spectral nulls [20, 52] were initially constructed for shaping the spectrum in the frequency domain, without consideration for error correction. Later, it was realized that these codes may also have good additive error correction properties, and in fact also insertion/deletion error correction properties [37]. Thus there seems to be a link to be further investigated.

6.4.11 Maximum Likelihood Decoding of Standard Constrained Codes

Some researchers revisited the approach as in [95], and investigated maximum likelihood or Viterbi decoding of standard constrained codes as employed in products on the market, reporting some gain for the digital compact cassette [12] and the DVD [44].

6.5 Post Combined Coding System Architectures

The movement to construct combined codes, or error correcting constrained codes as defined in Figure 6.1(b), experienced a peak during the 1990s, although a few new papers still appear every year. The emphasis of later work in order to achieve the same goals, has shifted to some extent to reversed concatenation (or post modulation coding) [10, 24, 42, 55] and to iterative decoding techniques – see for example, [25–27].

With these alternative approaches, higher coding rates and hence storage densities may be achieved. In the traditional concatenated coding scheme, the efficiency (R/C) of the constrained code, was more of a limiting factor than the efficiency of the error correcting code, and this influenced the post combined coding architectures later proposed. Many of the proposed combined coding schemes described in this chapter, also suffered the limitation of being too narrowly focused on one type of error.

Immink describes "a practical method for approaching the channel capacity of constrained channels" in [55], expanding a scheme previously proposed by Bliss [10]. Immink's scheme employs very long constrained code words while avoiding the possibly massive error propagation which may be triggered by a single channel error. The technique can be used in conjunction with any constrained code and it reverses the normal hierarchy of the error correction and channel codes. Block diagrams of Immink's scheme are shown in Figure 6.4.

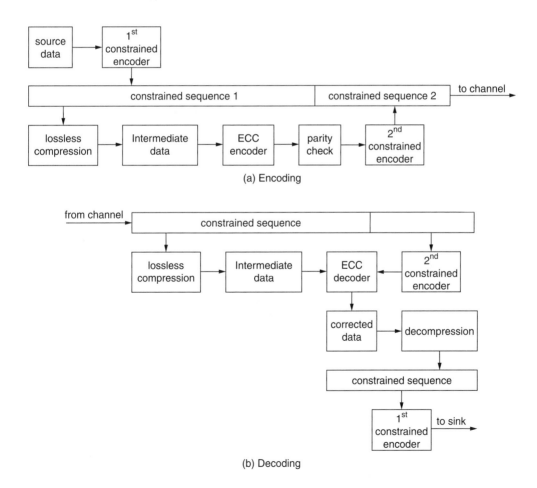

FIGURE 6.4 Immink's post modulation scheme.

Essential to Immink's scheme is the lossless compression step with limited error propagation, used to create the intermediate coding layer. A symbol error-correcting code, such as a byte orientated Reed-Solomon code is used to encode the intermediate layer. Look up tables, which are carefully matched, are used between the layers. Finally, the two sequences generated by the first and second constrained codes are cascaded and transmitted.

At the receiving end, the received sequence is decoded by firstly retrieving the parity symbols under the decoding rules of the second channel decoder. Next the constrained sequence is blocked into q-tuples and by using a lookup table are translated into the symbols of the intermediate sequence. Transmission errors in the intermediate sequence or parity symbols are corrected with the aid of the ECC code. The corrected intermediate sequence is decompressed and a constrained sequence which is essentially error-free, is obtained. This sequence is decoded and the original data retrieved.

In [55] the systematic design of the long block-decodable (d, k) constrained block codes, essential for the new coding method, is also considered. To this effect, Immink employs enumerative encoding and concatenatable (d, k, l, r) sequences, that is, (d, k) sequences with at most l consecutive leading zeros preceding the first one, and at most r consecutive trailing zeros succeeding the last one.

Examples of explicit results in [55] include a rate $R = 256/371$ $(d, k) = (1, 12)$ code which achieves 99.6% of the capacity of the $(1, 12)$ constrained channel, and a $R = 256/466$ $(d, k) = (2, 15)$ code achieving 99.86% of the capacity of the $(2, 15)$ constrained channel.

In conclusion, Immink's scheme made possible the use of long constrained codewords while avoiding error propagation, in order to approach the capacity of the input constrained channel. Important, it also offers the capability of correcting random and burst channel errors with powerful state of the art error correction codes such as Reed-Solomon codes.

6.6   Conclusion

In this chapter, we have attempted to give the reader an introduction to, and overview of the topic of combined codes. It currently appears that, in magnetic recording applications, this field and approach may have been overtaken by other systems architectures. However, it has stimulated much research and debate and has thus contributed to the development of newer architectures. Also, combined codes have potentially other applications in digital communications and the transmission of information. Hopefully, this presentation will thus help to stimulate further research.

References

[1] Abdel-Ghaffar, K.A.S., Blaum M., and Weber J.H., Analysis of coding schemes for modulation and error control, *IEEE Transactions on Information Theory*, vol. 41, no. 6, pp. 1955–1968, November 1995.

[2] Abdel-Ghaffar Khaled A.S. and Weber J.H., Bounds and constructions for runlength-limited error–control block codes, *IEEE Transactions on Information Theory*, vol. 37, no. 3, pp. 789–800, May 1991.

[3] Barg A.M. and Litsyn S.N., DC-constrained codes from Hadamard matrices, *IEEE Transactions on Information Theory*, vol. 37, no. 3, pp. 801–807, May 1991.

[4] Bassalygo A., Correcting codes with an additional property, *Problems of Information Transmission*, vol. 4, no1, pp. 1–5, Spring 1968.

[5] Bergmann E.E., Odlyzko A.M., and Sangani S.H., Half weight block codes for optical communications, *AT&T Technical Journal*, vol. 65, no. 3, pp. 85–93, May–June 1986.

[6] Blaum M., Combining ECC with modulation: performance comparisons, *Proceedings of the IEEE Globecom '90 Conference*, San Diego, California, USA, vol. 3, pp. 1778–1781, December 2–5, 1990.

[7] Blaum M., Bruck J., Melas C.M., and van Tilborg H.C.A., Resynchronising (d, k)-constrained sequences in the presence of insertions and deletions, *Proceedings IEEE International Symposium on Information Theory*, p. 126, 17–22 January 1993.

[8] Blaum M., Litsyn S., Buskens V., and van Tilborg H.C.A., Error-correcting codes with bounded running digital sum, *IEEE Transactions on Information Theory,* vol. 39, no. 1, pp. 216–227, January 1993.

[9] Blaum M., Bruck J., Melas C.M., and van Tilborg H.C.A., Methods for Synchronising (d,k)-constrained sequences, *Proceedings of the 1994 IEEE International Conference on Communications,* vol. 3, pp. 1800–1808, 1–5 May 1994.

[10] Bliss W.G., Circuitry for performing error correction calculations on baseband encoded data to eliminate error propagation, *IBM Technological Disclosure Bulletin.* vol. 23, pp. 4633–4634, 1981.

[11] Bours P.A.H., Construction of fixed-length insertion/deletion correcting runlength-limited codes, *IEEE Transactions on Information Theory,* vol. 40, no. 6, pp. 1841–1856, November 1994.

[12] Braun V., Schouhamer Immink K.A., Ribeiro M.A., and van den Enden G.J., On the application of sequence estimation algorithms in the digital compact cassette (DCC), *IEEE Transactions on Consumer Electronics,* vol. 40, no. 4, pp. 992–998, November 1994.

[13] Calderbank A.R., Herro M.A., and Telang V., A multilevel approach to the design of dc-free line codes, *IEEE Transactions on Information Theory,* vol. 35, no. 3, pp. 579–583, May 1989.

[14] Chang J.C., Chen R.J., Klove T, and Tsai S.C., Distance preserving mappings from binary vectors to permutations, *IEEE Transactions on Information Theory,* vol. 49, no. 4, pp. 1054–1059, April 2003.

[15] Chiu M.C., DC-free error-correcting codes based on convolutional codes, *IEEE Transactions on Communications,* vol. 49, no. 4, pp. 609–619, April 2001.

[16] Coene W., Pozidis H., and Bergmans J., Run-length limited parity-check coding for transition-shift errors in optical recording, *Proceedings IEEE GLOBECOM, 2001,* pp. 2982–2986, November 2001.

[17] Coetzee C.S., Ferreira H.C., and van Rooyen P.G.W., On the performance and implementation of a class of error and erasure control (d,k) block codes, *IEEE Transactions on Magnetics,* vol. MAG-26, no. 5, pp. 2312–2314, September 1990.

[18] Deng R.H. and Herro M.A., DC-free coset codes, *IEEE Transactions on Information Theory,* vol. 34, no. 4, pp. 786–792, July 1988.

[19] Deng R.H., Li Y.X., and Herro M.A., DC-free error-correcting convolutional codes, *Electronics Letters,* vol. 29, no. 22, pp. 1910–1911, 28th October 1993.

[20] Eleftheriou E. and Cideciyan R.D., On codes satisfying Mth-order running digital sum constraints, *IEEE Transactions on Information Theory,* vol. 37, no. 5, pp. 1294–1313, September 1991.

[21] Erxleben W.H. and Marcellin M.W., Error-correcting two-dimensional modulation codes, *IEEE Transactions on Information Theory,* vol. 41, no. 4, pp. 1116–1126, July 1995.

[22] Etzion T., Cascading methods for runlength-limited arrays, *IEEE Transactions on Information Theory* vol. 43, no. 1, pp. 319–324, January 1997.

[23] Fair I.J. and Xin Y., A method of integrating error control and constrained sequence codes, *2000 Canadian Conference on Electrical and Computer Engineering,* vol. 1, pp. 63–67, 7–10 March 2000.

[24] Fan J.L., Calderbank A.R., A modified concatenated coding scheme with applications to magnetic data storage, *IEEE Trans. Inform. Theory,* vol. 44, pp. 1565–1574, July 1998.

[25] Fan J.L., Constrained coding and soft iterative decoding, *Proceedings IEEE Information Theory Workshop,* 2001, pp. 18–20, 2–7 September 2001.

[26] Farkas P., Pusch W., Taferner M., and Weinrichter H., Turbo-codes with run length constraints, *International Journal of Electronics and Communications,* vol. 53, no. 3, pp. 161–166, January 1999.

[27] Farkas P., Turbo-codes with RLL properties, IEE Colloquium on Turbo Codes in Digital Broadcasting — Could It Double Capacity? (Ref. No. 1999/165), pp. 13/1–13/6, 22 November 1999.

[28] Fernandez E.M.G. and Baldini F.R., A method to find runlength limited block error control codes, *Proceedings of the 1997 IEEE International Symposium on Information Theory,* p. 220, 29 June–4 July 1997.

[29] Ferreira H.C., On dc free magnetic recording codes generated by finite state machines, *IEEE Transactions on Magnetics,* vol. 19, no. 6, pp. 2691–2693, November 1983.

[30] Ferreira H.C., Lower bounds on the minimum Hamming distance achievable with runlength constrained or dc free block codes and the synthesis of a (16, 8) dmin = 4 dc free block code, *IEEE Transactions on Magnetics*, vol. 20, no. 5, pp. 881–883, September 1984.

[31] Ferreira H.C., The synthesis of magnetic recording trellis codes with good Hamming distance properties, *IEEE Transactions on Magnetics*, vol. 21, no. 5, pp. 1356–1358, September 1985.

[32] Ferreira H.C., Hope J.F., and Nel A.L., Binary rate four eighths, runlength constrained, error correcting magnetic recording modulation code, *IEEE Transactions on Magnetics*, vol. 22, no. 5, pp. 1197–1199, September 1986.

[33] Ferreira H.C., Hope J.F., Nel A.L., and van Wyk M.A., Viterbi decoding and the power spectral densities of some rate one half binary dc free modulation codes, *IEEE Transactions on Magnetics*, vol. 23, no. 3, pp. 1928–1934, May 1987.

[34] Ferreira H.C., Wright D.A., and Nel A.L., On generalized error correcting trellis codes with balanced binary symbols, *Proceedings of the 25th Annual Allerton Conference on Communication, Control and Computing*, Monticello, Illinois, USA, pp. 596–597, September 30–October 2, 1987.

[35] Ferreira H.C., Wright D.A., and Nel A.L., Hamming distance preserving mappings and trellis codes with constrained binary symbols, *IEEE Transactions on Information Theory*, vol. 35, no. 5, pp. 1098–1101, September 1989.

[36] Ferreira H.C. and Lin S., Error and erasure control (d, k) block codes, *IEEE Transactions on Information Theory*, vol. 37, no. 5, pp. 1399–1408, September 1991.

[37] Ferreira H.C., Clarke W.A., Helberg A.S.J., Abdel-Ghaffar K.A.S., and Vinck A.J., Insertion/deletion correction with spectral nulls, *IEEE Transactions on Information Theory*, vol. 43, no. 2, pp. 722–732, March 1997.

[38] Ferreira H.C., and Vinck A.J., Interference cancellation with permutation trellis codes, *Proceedings of the IEEE Vehicular Technology Conference Fall 2000*, Boston, MA, USA, pp. 2401–2407, September 24–28, 2000.

[39] Fredrickson L.J. and Wolf J.K., Error detecting multiple block (d, k) codes, *IEEE Transactions on Magnetics*, vol. 25, no. 5, pp. 4096–4098, September 1989.

[40] French C.A., Distance preserving run-length limited codes, *IEEE Transactions on Magnetics*, vol. 25, no. 5, pp. 4093–4095, September 1989.

[41] French C.A. and Lin Y., Performance comparison of combined ECC/RLL codes, *Proceedings of the 1990 IEEE International Conference on Communications*, Atlanta, USA, pp. 1717–1722, April 1990.

[42] Fitingof B. and Mansuripur M., Method and apparatus for implementing post-modulation error correction coding scheme, U.S. Patent 5,311,521, May 1994.

[43] Gu J. and Fuja T., A generalized Gilbert-Varshamov bound derived via analysis of a code-search algorithm, *IEEE Transactions on Information Theory*, vol. 39, no. 3, pp. 1089–1093, May 1993.

[44] Hayashi H., Kobayashi H., Umezawa M., Hosaka S., and Hirano H., DVD players using a Viterbi decoding circuit, *IEEE Transactions on Consumer Electronics*, vol. 44, no. 2, pp. 268–272, May 1998.

[45] Helberg A.S.J. and Ferreira H.C., Some new runlength constrained binary modulation codes with error-correcting capabilities, *Electronics Letters*, vol. 28, no. 2, pp. 137–139, 16th January 1992.

[46] Helberg A.S.J., Clarke W.A., Ferreira H.C., and Vinck A.J.H., A class of dc free, synchronization error correcting codes, *IEEE Transactions on Magnetics*, vol. 29, no. 6, pp. 4048–4049, November 1993.

[47] Hilden H.M., Howa D.G., and Weldon E.J. Jr., Shift error correcting modulation codes, *IEEE Transactions on Magnetics*, vol. 27, no. 6, pp. 4600–4605, November 1991.

[48] Hole K.J., Cosets of convolutional codes with short maximum zero-run lengths, *IEEE Transactions on Information Theory*, vol. 41, no. 4, pp. 1145–1150, July 1995.

[49] Hole K.J. and Ytrehus O., Further results on cosets of convolutional codes with short maximum zero-run lengths, *Proceedings IEEE International Symposium on Information Theory*, pp.146, 17–22 September 1995.

[50] Hole K.J. and Ytrehus O., Cosets of convolutional codes with least possible maximum zero- and one-run lengths, *IEEE Transactions on Information Theory*, vol. 44, no. 1, pp. 423–431, January 1998.

[51] Howell T.D., Analysis of correctable errors in the IBM 3380 disk file, *IBM Journal of Research and Development*, vol. 28, no. 2, pp. 206–211, March 1984.

[52] Immink K.A.S. and Beenker G.F.M., Binary transmission codes with higher order spectral zeros at zero frequency, *IEEE Transactions on Information Theory*, vol. 33, no. 3, pp. 452–454, May 1987.

[53] Immink K.A.S., Coding techniques for the noisy magnetic recording channel: a state-of-the art report, *IEEE Transactions on Communications*, vol. COM-37, no. 5, pp. 413–419, May 1989.

[54] Immink K.A.S., *Coding Techniques for Digital Recorders*, Prentice-Hall, Englewood Cliffs, NJ, 1991.

[55] Immink K.A.S., A practical method for approaching the channel capacity of constrained channels, *IEEE Transactions on Information Theory*, vol. 43, no. 5, pp. 1389–1399, September 1997.

[56] Immink K.A.S., Coding for mass data storage systems, Shannon Foundation, The Netherlands, 1999.

[57] Jeong C.K. and Joo E.K., Generalized algorithm for design of DC-free codes based on multilevel partition chain, *IEEE Communications Letters*, vol. 2, pp. 232–234, August 1998.

[58] Kamabe H., Combinations of finite state line codes and error correcting codes, *Proceedings of the 1999 IEEE Information Theory and Communications Workshop*, p. 126, 20–25 June 1999.

[59] Klove T., Codes correcting a single insertion/deletion of a zero or a single peak-shift, *IEEE Transactions on Information Theory*, vol. 41, no. 1, pp. 279–283, January 1995.

[60] Kokkos A., Popplewell A., and O'Reilly J.J., A power efficient coding scheme for low frequency spectral suppression, *IEEE Transactions on Communications*, vol. 41, no.11, pp. 1598–1601, November 1993.

[61] Kokkos A., O'Reilly J.J., Popplewell A., and Williams S., Evaluation of class of error control line codes: an error performance perspective, *IEE Proceedings-I*, vol. 139, no. 3, pp. 128–132, April 1992.

[62] Kolesnik V.D. and Krachkovsky V.Y., Generating functions and lower bounds on rates for limited error-correcting codes, *IEEE Transactions on Information Theory*, vol. 37, no. 3, pp. 778–788, May 1991.

[63] Krachkovsky V.Y., Yuan Xing Lee and Davydov V.A. A new class of codes in Lee metric and their application to error-correcting modulation codes, *IEEE Transactions on Magnetics*, vol. 32, no. 5, pp. 3935–3937, September 1996.

[64] Kuznetsov A. and Vinck A.J.H., A coding scheme for single peak-shift correction in (d, k)-constrained channels, *IEEE Transactions on Information Theory*, vol. 39, no. 4, pp. 1444–1449, July 1993.

[65] Laih S. and Yang C.N., Design of efficient balanced codes with minimum distance 4, *IEE Proceedings-I*, vol. 143, no. 4, pp. 177–181, August 1996.

[66] Lee J. and Lee J., Error correcting RLL codes using high rate RSC or turbo code, *Electronics Letters*, vol. 37, no. 17, pp. 1074–1075, 16th August 2001.

[67] Lee P., and Wolf J.K., A general error-correcting code construction for run-length limited binary channels, *IEEE Transactions on Information Theory*, vol. 35, no. 6, pp. 1330–1335, November 1989.

[68] Lee J. and Madisetti V.K., Error correcting run-length limited codes for magnetic recording, *IEEE Transactions on Magnetics*, vol. 31, no. 6, pp. 3084–3086, November 1995.

[69] Lee J. and Madisetti V.K., Combined modulation and error correction codes for storage channels, *IEEE Transactions on Magnetics*, vol. 32, no. 2, pp. 509–514, March 1996.

[70] Levenshtein V.I. and Han Vinck A.J., Perfect (d, k)-codes capable of correcting single peak-shifts, *IEEE Transactions on Information Theory*, vol. 39, no. 2, pp. 656–662, May 1993.

[71] Lin Y. and Wolf J.K., Combined ECC/RLL codes, *IEEE Transactions on Magnetics*, vol. 24, no. 6, pp. 2527–2529, November 1988.

[72] Pi-Hai Liu and Yinyi Lin, A class of (d, k) block codes with single error correcting capability, *IEEE Transactions on Magnetics*, vol. 33, no. 5, pp. 2758–2760, September 1997.

[73] Marcus B.H. and Roth R.M., Improved Gilbert-Varshamov bound for constrained systems, *IEEE Transactions on Information Theory*, vol. 38, no. 4, pp. 1213–1221, July 1992.

[74] Markarian G. and Honary B., Trellis decoding technique for block RLL/ECC, *IEE Proceedings-I,* vol. 141, no. 5, pp. 297–302, October 1994.

[75] Markarian G., Honary B., and Blaum M., Maximum-likelihood trellis decoding technique for balanced codes, *Electronics Letters,* vol. 31, no. 6, pp. 447–448, 23rd March 1995.

[76] McLaughlin S.W., Improved distance M-ary (d, k) codes for high density recording, *IEEE Transactions on Magnetics,* vol. 31, no. 2, pp. 1155–1160, March 1995.

[77] Nasiri-Kenari M. and Rushforth C.K., Some construction methods for error-correcting (d, k) codes, *IEEE Transactions on Communications,* vol. 42, no. 2/3/4, pp. 958–965, February/March/April 1994.

[78] Nasiri-Kenari M. and Rushforth C.K., A class of DC-free subcodes of convolutional codes, *IEEE Transactions on Communications,* vol. 44, no. 11, pp. 1389–1391, November 1996.

[79] O'Reilly J.J. and Popplewell A., Class of disparity reducing transmission codes with embedded error protection, *IEE Proceedings-I,* vol. 137, no. 2, pp. 73–77, April 1990.

[80] Patapoutian A. and Kumar P.V., The (d, k) subcode of a linear block code, *IEEE Transactions on Information Theory,* vol. 38, no. 4, pp. 1375–1382, July 1992.

[81] Perry P.N., Runlength-limited codes for single error detection in the magnetic recording channel, *IEEE Transactions on Information Theory,* vol. 41, no. 3, pp. 809–815, May 1995.

[82] Popplewell A. and O'Reilly J.J., Spectral characterisation and performance evaluation for a new class of error control line codes, *IEE Proceedings-I,* vol. 137, no. 4, pp. 242–246, August 1990.

[83] Popplewell A. and O'Reilly J.J., Runlength limited codes for random and burst error correction, *Electronics Letters,* vol. 28, no. 10, pp. 970–971, 7th May 1992.

[84] Popplewell A. and O'Reilly J.J., Runlength limited binary error control codes, *IEE Proceedings-I,* vol. 139, no. 3, pp. 349–355, June 1992.

[85] Popplewell A. and O'Reilly J.J., Manchester-like coding with single error correction and double error detection, *Electronics Letters,* vol. 29, no. 6, pp. 524–525, 18th March 1993.

[86] Popplewell A. and O'Reilly J.J., A simple strategy for constructing a class of DC-free error-correcting codes with minimum distance 4, *IEEE Transactions on Information Theory,* vol. 41, no. 4, pp. 1134–1137, July 1995.

[87] Roth R.M. and Siegel P.H., Lee-metric BCH codes and their application to constrained and partial-response channels, *IEEE Transactions on Information Theory,* vol. 40, no. 4, pp. 1083–1096, July 1994.

[88] Saeki K. and Keirn Z., Optimal combination of detection and error correction coding for magnetic recording, *IEEE Transactions on Magnetics,* vol. 37, no. 2, pp. 708–713, March 2001.

[89] Sechny M. and Farkas P., Some new runlength-limited convolutional codes, *IEEE Transactions Communications,* vol. 47 no. 7, pp. 962–966, July 1999.

[90] Song S. and Shwedyk E., Graph theoretic approach for constrained error control codes, *Proceedings of the 1990 International Symposium on Information Theory and Its Applications,* Honolulu, HI, USA, pp. 17–18, November 27–30, 1990.

[91] van Tilborg H. and Blaum M., On error-correcting balanced codes, *IEEE Transactions on Information Theory,* vol. 35, no. 5, pp. 1091–1095, September 1989.

[92] van Wijngaarden A.J. and Soljanin E., A combinatorial technique for constructing high-rate MTR-RLL codes, *IEEE Journal Selected Areas in Communication,* vol. 19, pp. 582–588, April 2001.

[93] van Wijngaarden A.J. and Immink K.A.S., Maximum runlength-limited codes with error control capabilities, *IEEE Journal Selected Areas in Communication,* vol. 19, pp. 602–611, April 2001.

[94] Waldman H. and Nisenbaum E., Upper bounds and Hamming spheres under the DC constraint, *IEEE Transactions on Information Theory,* vol. 41, no. 4, pp. 1138–1145, July 1995.

[95] Wood R.W., Viterbi reception of Miller-squared code on a tape channel, *Proceedings of the IERE International Conference on Video and Data Recording,* Southampton, England, pp. 333–344, 26–23 April 1982.

[96] Wood R.W., Further comments on the characteristics of the Hedeman H-1, H-2 and H-3 codes, *IEEE Transactions on Communications,* vol. 31, no. 1, pp. 105–110, January 1983.

[97] Yang S.H. and Winick K.A., Asymptotic bounds on the size of error-correcting recording codes, *IEE Proceedings-I,* vol. 141, no. 6, pp. 365–370, December 1994.

[98] Ytrehus O., Upper bounds on error-correcting runlength-limited block codes, *IEEE Transactions on Information Theory,* vol. 37, no. 3, pp. 941–945, May 1991.

[99] Ytrehus O., Runlength-limited codes for mixed-error channels, *IEEE Transactions on Information Theory,* vol. 37 no. 6, pp. 1577–1585, November 1991.

[100] Ytrehus O. and Hole K., Convolutional codes and magnetic recording, *In Proceedings of the URSI International Symposium on Signals, Systems, and Electronics,* Paris, pp. 803–804, September 1992.

7

Constrained Coding and Error-Control Coding

John L. Fan
Flarion Technologies
Bedminster, NJ

7.1   Introduction

A *constraint* imposes a restriction on the set of sequences that are allowed to be transmitted on a channel. *Constrained coding* is a process by which a user data sequence is encoded in a lossless manner into a sequence that satisfies the constraint. The most common use of constrained coding is for runlength limitations, as discussed in Chapter 3. Constrained codes (also referred to as modulation codes, line codes, or runlength codes) are used in various magnetic and optical data storage systems. They are used for various reasons, such as improving timing recovery and handling physical limitations of the recording media.

In *error-control coding* (ECC), the transmission is restricted to a subset of sequences in such a way that the receiver can decode the original transmitted sequence in the presence of errors introduced by a noisy channel. In the case of linear codes, it is possible to put the ECC into systematic form, in which the parity bits are appended to the user bits. Recently, there have been breakthroughs in the development of ECCs such as Turbo codes and low-density parity-check (LDPC) codes that make use of soft information for near-optimal decoding performance.

Both error-control codes and constrained codes are widely used in digital storage systems, where the situation corresponds to a noisy channel whose input is required to satisfy a modulation constraint. Although constrained coding and error-control coding share many similarities, they are typically designed and implemented separately.

This chapter considers several configurations for integration of the constrained code and the error-control code:

- The most commonly used configuration is called *standard concatenation*, in which the constrained code is the inner code and the error-control code is the outer code in a serial concatenation.
- A promising alternate configuration is *reverse concatenation*, in which a constrained code forms the outer code and a systematic error-control code forms the inner code, where the parity bits from the

ECC are passed through another constrained code. Reverse concatenation possesses the advantages of reducing error-propagation and facilitating soft decoding.

- The *bit insertion method* is a variation of reverse concatenation in which the parity is inserted into the sequence. This technique is also known as constrained codes with unconstrained positions.

- Finally, the *lossless compression* technique is a variation of reverse concatenation in which an additional lossless compression code is used in order to reduce the codeword expansion due to the constrained code. This is not applicable in situations where soft decoding of the ECC is desired.

Schemes like reverse concatenation and the bit insertion method allow for the decoder to make direct use of the soft information from the channel, and also to reduce error-propagation. Methods for decoding the constraint for additional coding gain are also discussed.

7.2   Configurations

7.2.1   Definitions

Since constraints are typically imposed on words of finite length in data storage applications, the approach taken here is to consider the constraint to be defined by a set of valid words called the *constraint set* S_C. Note that this approach is more general than the notion of constrained systems, in which the valid sequences correspond to traversals through a constraint graph. More background about constrained coding and its applications can be found in [12–14,17].

Some preliminary definitions are useful for the description of constrained and error-control coding: For an alphabet A and block length n, the elements $\{w_1, w_2, \ldots, w_n\}$ of A^n are known as words. We define a *code* as a subset of words that are specified by an encoder function f which is a one-to-one function from A^k to A^n whose image is Im f. The number of information symbols is given by $k = log_{|A|}(|\text{Im } f|)$, where $|S|$ denotes the number of elements of a set S. The words in Im f are known as *codewords*, and the *rate* of the code is k/n. We focus our attention on the binary alphabet A consisting of the elements 0 and 1. (Note that these binary values are then mapped into magnetic polarizations for magnetic recording.)

The constraint set S_C of length n_C is a subset of A^{n_C} that defines the valid words that satisfy the constraint. In this context, the capacity of the constraint is defined by $cap(S_C) = \frac{1}{n_C} log_{|A|}(|S_C|)$. A *constrained code* is a code whose codewords all belong to the constraint set, and can be defined by an encoder function f_C, whose image satisfies Im $(f_C) \subset S_C$. This implies $k_C \leq log_{|A|}(|S_C|)$.

An *error-control code* S_{ECC} of length n_{ECC} is a subset of $A^{n_{ECC}}$. The error-control code can be defined by an encoder function f_{ECC} that is a one-to-one map from $A^{k_{ECC}}$ to $A^{n_{ECC}}$. Note that $k_{ECC} = log_{|A|}(|S_{ECC}|)$. The set of ECC codewords $S_{ECC} = \text{Im } f_{ECC}$ is chosen so that it is possible for a decoder to recover the original codeword from a corrupted version of it.

Typically, the set S_{ECC} possesses some geometric or mathematical structure that can be made use of in the decoding algorithm. We focus on linear codes, where the codewords form an additive group. In this case, it is possible to put the ECC into systematic form such that the encoder function f_{ECC} produces a word that is the concatenation of the input word with a parity word. In other words, the output is $f_{ECC}(u) = \{u, v^{(p)}\} \in A^{n_{ECC}}$, which is a concatenation of the input word $u \in A^{k_{ECC}}$ of length k_{ECC} and a parity word $v^{(p)}$ of length $n_{ECC} - k_{ECC}$. (The notation $\{x, y\}$ represents the concatenated sequence $\{x_1, x_2, \ldots, x_k, y_1, y_2, \ldots, y_l\}$, where $x = \{x_1, x_2, \ldots, x_k\}$ and $y = \{y_1, y_2, \ldots, y_l\}$.) In this context, it is useful to define a parity encoder function f_{ECC}^p whose output is just the parity portion: $f_{ECC}^p(u) = v^{(p)}$.

An important class of error-control codes for magnetic recording are the Reed-Solomon codes, which are linear codes that can be put into systematic form. Reed-Solomon codes are symbol-oriented ECCs, with symbols of size B bits (corresponding to the Galois field GF(2^B)), where a typical value is $B = 8$. The encoder takes k_B symbols and produces n_B symbols, which corresponds to $k_{ECC} = B \cdot k_B$ bits and

$n_{ECC} = B \cdot n_B$ bits. With Reed-Solomon codes, if any bit in a symbol is incorrect, the entire symbol is considered incorrect. The decoder for the Reed-Solomon code has the ability to correct a limited number of symbol errors (e.g., it can correct $\lfloor \frac{n_B - k_B}{2} \rfloor$ symbol errors).

In many data storage systems, it is necessary to satisfy simultaneously the needs of constrained coding and error-control coding. A conceptually simple solution to this problem is to choose the intersection of the constraint set S_C with an error-control code S_{ECC}. Assuming that the two sets are of the same length ($n_C = n_{ECC}$), then the words in the intersection set $S_C \cap S_{ECC}$ would be suitable for decoding by both the ECC decoder and constraint decoder since they are codewords in both codes. An encoding function f for this situation could be defined by an indexing function for an exhaustive list of the words in $S_C \cap S_{ECC}$. Suppose A is a binary alphabet. Let $k = \lfloor log_2(S_C \cap S_{ECC}) \rfloor$, and index 2^k words of the set $S_C \cap S_{ECC}$ using the numbers $\{0, 1, \ldots, 2^k - 1\}$ in binary representation. An encoder function f can then be defined by mapping this binary index $u \in A^k$ to the corresponding indexed word $w \in S_C \cap S_{ECC} \subset A^n$.

For practical implementations, it is important to choose a scheme that permits the encoding and decoding to be performed within the limits on complexity imposed by existing hardware. In particular, the direct approach of using an encoder function f for the intersection $S_C \cap S_{ECC}$ requires memory that is exponential in the codeword length to store an exhaustive encoder table. In contrast, most constrained codes and error-control codes in use today have associated encoders and decoders which offer efficient implementations; in particular, their complexity and storage requirements do not increase exponentially with the codeword length. This chapter considers a number of approaches to this problem of combined constrained and error-control coding based on configurations that make use of the existing encoders and decoders for the constrained code and the error-control code.

7.2.2   Standard Concatenation

The *standard concatenation* scheme, shown in Figure 7.1, consists of the serial concatenation of an ECC as the outer code and a constrained code as the inner code. In terms of the encoding functions, the user sequence u is mapped $v = f_{ECC}(u)$ and then to $w = f_C(f_{ECC}(u))$. Note that in this case, n_{ECC} is equal to k_C, and in particular, the output word w belongs to S_C, but does not belong to S_{ECC}. The overall rate of standard concatenation is $\frac{k_C}{n_C} \cdot \frac{k_{ECC}}{n_{ECC}}$.

Suppose that the transmitted word is w and the received word is $\hat{w}$. The decoding of the constrained code (also known as *demodulation*) is typically implemented as the inverse of the encoder function: $\hat{v} = f_C^{-1}(\hat{w})$. A problem known as *error-propagation* arises when a small number of errors in the received word becomes magnified by demodulation. Let $d(\cdot, \cdot)$ be a function that measures the Hamming distance between two words, that is, the number of symbols in which the two words differ. Generally speaking, error-propagation is said to occur when $d(w, \hat{w})$ is small but $d(f_C^{-1}(w), f_C^{-1}(\hat{w}))$ is large.

The constrained code is typically implemented in such a way as to reduce the error-propagation caused by the demodulation function f_C^{-1}. The impact of error-propagation depends on the details of the decoder for the error-control code. Common constructions of constrained codes involve block codes or sliding-block codes with short block lengths. As an example, consider a binary block code of length N_C with an encoder function F_C that maps from A^{K_C} to A^{N_C}. (Note that upper-case letters are used to represent parameters of the short block code, while lower-case letters are used to represent parameters of the full code.)

This encoder function F_C for the short block code can be used to implement the encoder function f_C for the entire constrained word. The image of F_C should be chosen such that the concatenation of any

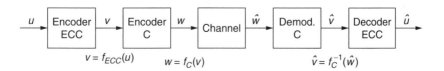

FIGURE 7.1   Standard concatenation scheme.

combination of words yields an overall word of length n_C that belongs to S_C (and the constraint is said to be maintained across blocks). If $u^{(1)}, u^{(2)}, \ldots, u^{(k_C/K_C)}$ are all words of length K_C, then the full constraint encoder f_C can be defined by the block encoder as follows

$$f_C\left(\left\{u^{(1)}, u^{(2)}, \ldots, u^{(k_C/K_C)}\right\}\right) = \left\{F_C\left(u^{(1)}\right), F_C\left(u^{(2)}\right), \ldots, F_C\left(u^{(k_C/K_C)}\right)\right\}$$

For the short block code, the encoding function F_C can be an arbitrary map from A^{K_C} to A^{N_C}. As a result, for the corresponding demodulation map F_C^{-1}, one or more bit errors in a word of N_C bits are assumed to result in an incorrect word upon demodulation, resulting in many bits in error in the demodulated output. The error-propagation effect, however, can be limited through the use of short block codes.

Suppose that a constrained code based on short block codes of rate K_C/N_C is used with an ECC with symbol size B. In the worst case, a single bit error can propagate into $\lceil K_C/B \rceil$ symbol errors, and a short burst of bit errors on the boundary of two blocks can result in $\lceil 2K_C/B \rceil$ symbol errors. The error-propagation effect becomes worse for large K_C, so that it is typical to choose a value of K_C that is matched to the symbol size B, such as $K_C = B$. This need to use short block lengths usually places a significant restriction on the design of the constrained code, making it difficult to use constrained codes whose rate approaches the capacity of the modulation constraint.

7.2.3   Reverse Concatenation

To allow for better combining of the error-control code and the constrained code so as to mitigate the effects of error-propagation, one promising approach is the *reverse concatenation* scheme, which reverses the order of the ECC and constrained code so that the encoding of the constraint takes place before the encoding of the error-control code, as shown in Figure 7.2. This method allows the decoder to perform decoding on most of the received word with the need for demodulation.

This method has the following benefits:

- It allows the use of arbitrary constrained codes
- It reduces error propagation
- It facilitates the use of soft decoding

Reverse concatenation is also known as "modified concatenation," or the "commuted configuration." Its history in the literature goes back to Bliss [4] and Mansuripur [16], and was analyzed in [7] and [11] as a method of preventing error-propagation in magnetic recording systems. It may also be viewed as related to specific schemes presented in [3,10,15,18,19] that combine (d, k) run-length constraints with error-control codes for detecting and correcting bit errors or bit shifts.

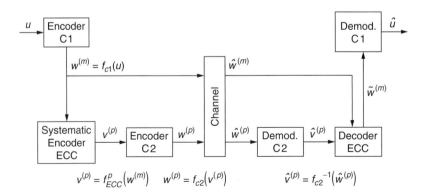

FIGURE 7.2   Reverse concatenation scheme.

Reverse concatenation involves a *main constrained code* $C1$ and an *auxiliary constrained code* $C2$, which have rates k_{C1}/n_{C1} and k_{C2}/n_{C2}, respectively. It also assumes the use of a systematic ECC of rate k_{ECC}/n_{ECC}. There are no restrictions on the design of the main constrained code $C1$ since the demodulation of this code takes place after its decoding by the ECC, so that it does not suffer from error-propagation. This allows the usage of constrained codes that have rates arbitrarily close to the capacity of the constraint. The auxiliary constrained code, however, must be designed in such a way as to prevent error propagation during demodulation, much as in the case of the constrained code in standard concatenation.

The encoding for reverse concatenation uses the encoding functions for $C1, C2$ and the ECC as follows: First, the main constrained code maps the user word u to $w^{(m)} = f_{C1}(u)$. This constrained codeword is then encoded by the systematic encoder for the ECC. The parity portion $v^{(p)} = f_{ECC}^{p}(f_{C1}(u)))$ of the ECC codeword is then encoded by the constrained code $C2$ to give $w^{(p)} = f_{C2}(v^{(p)})$. The end result is the word

$$w = \left\{ w^{(m)}, w^{(p)} \right\} \tag{7.1}$$

$$= \left\{ f_{C1}(u), f_{C2}\left(f_{ECC}^{p}(f_{C1}(u)) \right) \right\} \tag{7.2}$$

The code parameters for reverse concatenation are related as follows: $n_C = n_{C1} + n_{C2}, k_{C2} = n_{ECC} - k_{ECC}$, $k_{ECC} = n_{C1}$, and $k_{C1} = len(u)$. The two constrained codes $C1$ and $C2$ should be chosen such that for any input u and v, placing the two words $f_{C1}(u)$ and $f_{C2}(v)$ together gives a word $w = \{ f_{C1}(u), f_{C2}(v) \}$ that belongs to the target constraint set S_C.

Note that the overall rate is

$$\frac{k_{C1}}{n_{C1} + n_{C2}} = \frac{k_{C1}}{n_{C1} + \frac{n_{C2}}{k_{C2}} (n_{ECC} - k_{ECC})} \tag{7.3}$$

$$= \frac{k_{C1}}{n_{C1}} \left(1 + \frac{n_{C2}}{k_{C2}} \left(\frac{n_{ECC}}{k_{ECC}} - 1 \right) \right)^{-1} \tag{7.4}$$

If the constrained code $C2$ were not used (i.e., $k_{C2}/n_{C2} = 1$), then the overall rate would be the same as in standard concatenation. The constrained code $C2$ is necessary, however, to make sure that the parity bits also satisfy the constraint.

To reduce error-propagation during demodulation of the auxiliary constrained code $C2$, it is typically implemented using block codes or sliding-window block codes with short block length, as in the case of the constrained code in standard concatenation. For example, for the case of an ECC with symbol size B, the code $C2$ could be implemented by putting together words from a short block code of rate K_{C2}/N_{C2}, where the length K_{C2} is chosen to match B (e.g., it is equal to B, or a multiple of B).

The advantages of reverse concatenation over standard concatenation lie in the decoding process. Error-propagation occurs when demodulation must be performed before decoding takes place. With reverse concatenation, the error-propagation during demodulation is restricted to the parity portion of the ECC codeword. Note that reverse concatenation is most effective when the ECC code rate k_{ECC}/n_{ECC} is high.

The decoding procedure for reverse concatenation is as follows: (For simplicity in exposition in this section, we consider the ECC to be a hard-decision ECC, although this discussion generalizes in a straightforward manner to soft-decoding.) Suppose the channel decoder (e.g., a Viterbi decoder) produces a received sequence $\hat{w}$ that is a possibly erroneous copy of the transmitted word w. This sequence can be divided into a message portion $\hat{w}^{(m)}$ and a parity portion $\hat{w}^{(p)}$, based on the correspondence to $C1$ and $C2$, respectively. The parity portion $\hat{w}^{(p)}$ is first demodulated by the code $C2$, which has limited error-propagation by design (e.g., $K_{C2} = B$), to obtain the word $\hat{v}^{(p)}$. On the other hand, the message portion $\hat{w}^{(m)}$ can go directly to the ECC decoder without any need for demodulation. The ECC decoder performs decoding on the word $\{\hat{w}^{(m)}, \hat{v}^{(p)}\}$. If the ECC decoding is successful, then the output $\tilde{w}^{(m)}$ is an error-free version of the constrained message bits. In this case there is no risk of error-propagation during demodulation, and applying demodulation using f_{C1}^{-1} yields a replica $\tilde{u}$ of the original user bits u, completing the successful decoding.

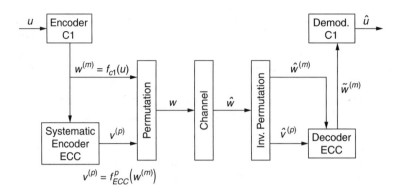

FIGURE 7.3 Bit insertion scheme.

To summarize, for reverse concatenation, the encoder's output to the channel satisfies the constraint, while the decoder sees only limited error propagation from demodulation. The main constrained code $C1$ can have arbitrary design, so there is no restriction on the function f_{C1}, and in particular, it is possible to use a constrained code with extremely long block length. This method is most effective when the ECC code rate is high, since the parity portion comprises a small part of the whole codeword. Reverse concatenation gives an effective method to meet a desired modulation constraint using a near-capacity constrained code.

7.2.4 Bit Insertion

For certain classes of constraints, it is possible to use a variation on reverse concatenation to insert the parity bits into a constrained message sequence in such a way that the resulting sequence does not violate the constraint. This *bit insertion scheme*, shown in Figure 7.3, entails choosing a modulation code $C1$ such that inserting bits into the sequence in some predetermined pattern results in a sequence that still meets the target constraint (e.g., it belongs to the constraint set S_C). In other words, instead of using a constrained code $C2$, the parity bits $v^{(p)}$ from the encoder are inserted directly into pre-specified locations in the sequence $w^{(m)} = f_{C1}(u)$.

The surprising advantage of this approach is that no error-propagation due to demodulation takes place at all. The entire sequence received from the channel decoder can be directly used by the ECC decoder (after possibly some permutation of the order), without the need for demodulation. (This bit insertion method is also beneficial for soft decoding as discussed in Section 7.3, since soft information for all bits is directly usable by the ECC decoder.) As in Section 7.2.3, if the ECC decoding is successful, then the decoder output $\tilde{w}^{(m)}$ is error-free and can be demodulated by f_{C1}^{-1} to yield the original user bits u.

While simple and effective, the bit insertion method is only useful for certain classes of constraints. It has been considered by Anim-Appiah and McLaughlin [1] for using $(0, k)$ modulation codes with Turbo Codes, and has been considered by van Wijngaarden and Immink [20,21] for the $(0, G/I)$-RLL constraint. An extensive analysis of this technique is given by Campello et al. in [5], which considers the "unconstrained positions" in a constrained code, referring to the locations that are suitable for placing parity bits with arbitrary values.

7.2.5 Lossless Compression

One issue with reverse concatenation is that the input to the ECC encoder is increased by a factor of n_{C1}/k_{C1} compared with standard concatenation since f_{ECC} is applied to $f_{C1}(u)$ instead of u directly. For constrained codes with low rate (e.g., $k_{C1}/n_{C1} \approx 0.5$), this poses a problem as the length of the ECC codeword expands proportionally, which leads to increased complexity in the ECC encoder and decoder.

To reduce codeword expansion in reverse concatenation, Immink proposed in [11] the use of a lossless compression code to compress the input to the ECC encoder. This technique is typically used in a scenario with a symbol-oriented hard-decision decoder such as a Reed-Solomon code. In this context, the compression should have the following characteristics:

- The sequence to be compressed satisfies a modulation constraint.
- The compression map should be lossless, so that it is possible to exactly recover the original sequence.
- The compression map should cause little error propagation. (This can be accomplished using short block lengths.)

The lossless compression code can be defined by an encoding function f_L whose image is a *superset* of the constraint set: $Im(f_L) \supset S_C$. The rate is lower-bounded by the constraint capacity: $cap(S_C) \leq k_L/n_L$. (In contrast, with constrained coding, the encoding function f_C has an image that is a *subset* of the constrained set, $Im(f_C) \subset S_C$, and constrained code rate is upper-bounded by the constraint capacity, $k_C/n_C \leq cap(S_C)$.) This map f_L can also be called an *expanding coder*, or "excoder" (in analogy to "encoder"). This is an invertible mapping from A^{k_L} to A^{n_L}. The inverse map f_L^{-1} then gives the corresponding *compression map*, which is a bijection from $Im(f_L)$ to A^{k_L} [9].

Applying this idea to reverse concatenation, the excoder map f_L should have an image that is a superset of S_{C1}, where $n_L = n_{C1}$. Then as shown in Figure 7.4, the compression map f_L^{-1} is applied to the constrained word $w^{(m)} = f_{C1}(u)$ to obtain t. Then the ECC encoder produces parity corresponding to t using a systematic ECC encoder to obtain $v^{(p)} = f_{ECC}^p(t)$. Finally, the parity portion is modulated by the constraint $C2$, and the transmitted word has the form:

$$w = \left\{ f_{C1}(u), f_{C2}\left(f_{ECC}^p\left(f_L^{-1}(f_{C1}(u))\right)\right) \right\}$$

As in reverse concatenation, $C1$ and $C2$ are chosen such that the resulting word w belongs to the constraint set S_C. In this case, $n_L = n_{C1}$, while $k_L = k_{ECC}$.

As for the decoder for this configuration, if the received word is $\hat{w}$, the message portion $\hat{w}^{(m)}$ must be first compressed by f_L^{-1} before it can be used by the ECC decoder. Hence, the compression map f_L^{-1} should be designed to have limited error-propagation. In particular, the goal is to create an excoder function f_L such that the compression map f_L^{-1} has limited error-propagation. In terms of practical implementation, this can be accomplished through short block codes or sliding-block codes.

The rest of the decoding procedure goes as follows, as shown in Figure 7.4. Let $\hat{t} = f_L^{-1}(\hat{w}^{(m)})$ represent the compressed version of the message portion of the received word. Meanwhile, the parity portion $\hat{w}^{(p)}$

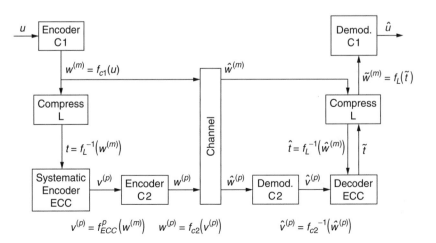

FIGURE 7.4 Lossless compression scheme for reverse concatenation.

must be demodulated by $C2$, so that the input to the ECC decoder is $\{\hat{t}, f_{C2}^{-1}(\hat{w}^{(p)})\}$. Upon successful decoding, a corrected version of the message portion is obtained, which is denoted by $\tilde{t}$. Next, the excoder f_L performs decompression on $\tilde{t}$ to recover the corrected $\tilde{w}^{(m)} = f_L(\tilde{t})$, and finally, demodulation is performed for the constrained code $C1$ to recover the user data $\tilde{u} = f_{C1}^{-1}(\tilde{w}^{(m)})$.

For the design of the lossless compression code, the goal is to decrease the codeword expansion, so that it is desirable to have as low a compression rate K_L/N_L as possible. On the other hand, error propagation must be avoided, which limits the selection of lossless compression codes (e.g., to ones with short block length). At one extreme, the compression could be trivial, so that the output t is exactly the same as the input $w^{(m)}$; this situation corresponds to reverse concatenation (Figure 7.2). At the other extreme, the lossless compression could compress $w^{(m)}$ back to u (demodulating the code $C1$), corresponding to standard concatenation (Figure 7.1). Choosing a compression code in between these two extremes allows for the benefits of reverse concatenation, while minimizing its codeword expansion relative to standard concatenation.

(It should be noted that this lossless compression technique is not generally applicable for soft decoding, since the compression step tends to further obscure the reliability information. In general, the compression code may be an arbitrary assignment of codewords, so that the computation of postcompression reliability information is a difficult task, similar to computing postdemodulation reliability information for an arbitrary constrained code.)

Block codes for lossless compression are given in [7] and [11]. The basic construction is a block code of rate K_L/N_L and an invertible function F_L that maps from K_L bits to N_L bits. Then the excoder f_L can be constructed using F_L (similarly to how f_C is built from F_C):

$$f_L\left(\left\{v^{(1)}, v^{(2)}, \dots, v^{(k_L/K_L)}\right\}\right) = \left\{F_L\left(u^{(1)}\right), F_L\left(u^{(2)}\right), \dots, F_L\left(u^{(k_L/K_L)}\right)\right\}$$

When any sequence from the constraint set S_C is divided into words of length N_L, these words should all lie in the image of F_L and so that they can be mapped via the compression map F_L^{-1} to a word of length K_L.

Sliding-block lossless compression codes are discussed in [9]. The compression map is a sliding-block decoder from sequences of N_L-codewords of the constraint set S_C to unconstrained sequences of length K_L; that is, a N_L-codeword $w^{(i)}$ is compressed into a K_L-frame $t^{(i)}$ as a deterministic function of the current N_L-codeword and perhaps some m preceding and a following N_L-codewords. The *sliding-block window length* is defined by the sum m + a + 1, and the code is said to be *sliding-block compressible*. The excoder, on the other hand, takes the form of a finite-state machine.

The existence of sliding-block compressible codes was shown in [9] for constraints defined as finite-memory constrained systems: if $\mathsf{cap}(S_C) \leq k_L/n_L$, then for this constrained system S_C, there exists a lossless compression code of rate K_L/N_L that is sliding-block compressible. Recall that the state-splitting algorithm for designing encoders for constrained codes (Ref. [17]) is guided by an "approximate eigenvector" that satisfies a certain inequality. For sliding-block lossless compression codes, this reversed inequality is used in a variant of the state-splitting algorithm for constructing finite-state excoders.

7.3   Reverse Concatenation and Soft Iterative Decoding

A very important benefit of reverse concatenation is the ability for the ECC decoder to make use of the full information available from the channel. Many channels provide *soft information*, which is typically a real-valued metric on each bit indicating its reliability, for example, a probability or a log-likelihood ratio (LLR). For many situations, the ability to obtain this soft information on the channel output allows the decoder for the error-control codes to improve its performance.

With standard concatenation, the usual implementations of constrained codes make it difficult to associate soft information to the bits of the demodulated output. In other words, given bit-wise probabilities for the word w, it is not straightforward to obtain bit-wise probabilities for the demodulated word v. With reverse concatenation, however, it is possible to directly use soft information for the message portion ($w^{(m)}$).

This is possible because no demodulation step is necessary for the message portion. (The bit insertion method is even better, allowing the use of soft information on the whole received word! The bit insertion method, however, can only be used on a limited class of constraints.) This ability of reverse concatenation to use soft information directly from the channel is critical for using constrained codes with ECCs such as Turbo codes and low-density parity-check (LDPC) codes whose iterative decoders rely on soft information for their superior performance over other codes. Of the extensive literature on applying these soft iterative ECCs in the context of magnetic and optical storage, a number of papers (e.g., [1,2,8]) also consider their application in the context of a constrained code.

In addition, with reverse concatenation, a soft decoder for the constraint can be used in conjunction with the ECC decoder to obtain additional coding gain from the constraint, as presented in [8]. The basic idea is that the constraint imposes restrictions on the valid codewords, as defined by the constrained set S_C, and the constraint decoder uses knowledge of these restrictions to improve the soft information (e.g., the bit-wise probabilities) for use in the ECC decoder.

While for arbitrary constraint sets S_C, it may be difficult to perform this soft decoding, for certain constraints, it is possible to take advantage of the structure of the constraint to perform a soft-in, soft-out (SISO) decoding to make useful updates to the soft information. For example, as shown in [6,8], for the case of the (d, k)-RLL constraints (and other constraints whose codewords can be represented by traversals on a trellis), it is possible to use the structure of the constraint to define a SISO decoder for the constraint using the BCJR algorithm.

With a SISO decoder for the constraint, it is possible to iterate between the ECC decoder and the constraint decoder as follows: Assume the parity portion has been demodulated according to the auxiliary constrained code C2. Then the ECC decoder receives the message portion and performs a SISO decoding (e.g., for LDPC or Turbo codes) to obtain updated soft output. This is used to create a new estimate of soft information that is passed to the constraint decoder. The SISO decoder for the constraint then uses the structure of the constraint to produce soft output that can then be combined with the channel soft information to create an updated soft input for the ECC decoder. By iterating back and forth between the ECC decoder and the constraint decoder, it is possible to gain additional coding gain by making use of the redundancy that is inherent in the constraint.

Note that the effectiveness of the soft constraint decoder requires the use of reverse concatenation (or the bit insertion method), as opposed to standard concatenation, in which the process of demodulation can distort the soft information. Also, note that the gains from decoding the constraint are more significant for lower rate constraints (e.g., k_C/n_C less than 2/3) since there is more redundancy imposed by the constraint that can be exploited for error-correction purposes. As a result, this technique of decoding the constraint may be more applicable to channels that use lower-rate constraints (e.g., optical storage). Finally, the channel decoder has been largely ignored in this discussion. As intersymbol interference is present in both magnetic and optical recording channels, it is common to use a trellis-based (e.g., Viterbi) detector for the channel. This is an additional issue that needs to be considered in combination with the constraint decoder and the ECC decoder.

References

[1] K. Anim-Appiah and S. McLaughlin, Turbo codes cascaded with high-rate block codes for $(0, k)$ constrained channels, *IEEE Journal on Selected Areas of Communication* vol. 19, no. 4. pp. 677–685, April 2001.

[2] W. Ryan, S. McLaughlin, K. Anim-Appiah, and M. Yang, Turbo, LDPC, and RLL codes in magnetic recording, *Proc. 2nd Int. Symp. on Turbo Codes and Related Topics,* September 2000.

[3] A. Bassalygo, Correcting codes with an additional property, *Prob. Inform. Trans.,* vol. 4, no. 1, pp. 1–5, Spring 1968.

[4] W.G. Bliss, Circuitry for performing error correction calculations on baseband encoded data to eliminate error propagation, *IBM Techn. Discl. Bul.,* vol. 23, pp. 4633–4634, 1981.

[5] J. C. de Souza, B.H. Marcus, R. New, and B.A. Wilson, Constrained systems with unconstrained positions, Presented at the DIMACS Workshop on Theoritical Advances in Recording of Information, New Brunswik, NJ, April 2004. Available at http://rutgers.edu

[6] J.L. Fan, *Constrained Coding and Soft Iterative Decoding,* Kluwer Academic Publishers, Boston, 2001.

[7] J.L. Fan and A.R. Calderbank, A modified concatenated coding scheme, with applications to magnetic storage, *IEEE Trans. Inform. Theory,* vol. 44, no. 4, pp. 1565–1574, July 1998.

[8] J.L. Fan and J.M. Cioffi, Constrained coding techniques for soft iterative decoders, *Proc. Globecom* (Rio de Janeiro), 1999.

[9] J.L. Fan, B.H. Marcus, and R.M. Roth, Lossless compression in constrained coding, *Proc. 37th Allerton Conf. on Commun., Control, and Computing,* 1999.

[10] H.M. Hilden, D.G. Howe, and E.J. Weldon, Jr., Shift error correcting modulation codes, *IEEE Trans. Magnetics,* vol. 27, no. 6, pp. 4600–4605, November 1991.

[11] K.A.S. Immink, A practical method for approaching the channel capacity of constrained channels, *IEEE Trans. Inform. Theory,* vol. 43, no. 5, pp. 1389–1399, September 1997.

[12] K.A.S. Immink, *Codes for Mass Data Storage,* Shannon Foundation Press, 1999.

[13] K.A.S. Immink, P.H. Siegel, and J.K. Wolf, Codes for digital recorders, *IEEE Trans. Inform. Theory,* vol. 44, no. 6, pp. 2260–2299, October 1998.

[14] R. Karabed, P.H. Siegel, and E. Soljanin, Constrained coding for binary channels with high inter-symbol interference, *IEEE Trans. Inform. Theory,* vol. IT-45, pp. 1777–1797, September 1999.

[15] W.H. Kautz, Fibonacci codes for synchronization control, *IEEE Trans. Inform. Theory,* pp. 284–292, April 1965.

[16] M. Mansuripur, Enumerative modulation coding with arbitrary constraints and post-modulation error correction coding and data storage systems, *Proc. SPIE,* vol. 1499, pp. 72–86, 1991.

[17] B.H. Marcus, R. Roth, and P.H. Siegel, Constrained systems and coding for recording channels, in *Handbook of Coding Theory,* Pless, V.S. and Huffman W.C., Eds., Elsevier, Amsterdam, pp. 1635–1764, 1998.

[18] A. Patapoutian and P.V. Kumar, The (d, k) subcode of a linear block code, *IEEE Trans. Inform. Theory,* vol. 38, no. 4, pp. 1375–1382, July 1992.

[19] P.N. Perry, Runlength-limited codes for single error detection in the magnetic recording channel, *IEEE Trans. Inform. Theory,* vol. 41, no. 3, pp. 809–814, May 1995.

[20] A.J. van Wijngaarden and K.A.S. Immink, Efficient error control schemes for modulation and synchronization codes, *Proc. IEEE ISIT* (Boston), p. 74, 1997.

[21] A.J. van Wijngaarden and K.A.S. Immink, Maximum run-length limited codes with error control capabilities, *IEEE J. Sel. Areas Commn.,* vol. 19, no. 4, April 2001.

8

Convolutional Codes for Partial-Response Channels

Bartolomeu F. Uchôa-Filho
Federal University of Santa Catarina
Florianopolis, Brazil

Mark A. Herro
University College
Dublin, Ireland

Miroslav Despotović
University of Novi Sad
Novi Sad, Yugoslavia

Vojin Šenk
University of Novi Sad
Novi Sad, Yugoslavia

8.1 Introduction

Partial-response equalization plays an important role in magnetic recording. In particular, the partial-response system described by the polynomial $P_n(D) = (1 - D)(1 + D)^n$, where n is a nonnegative integer and D denotes a symbol delay, represents an interesting and largely adopted model for the high-density magnetic recording channel [2, 3]. The excellent performance of systems based on partial-response models sparked interest in the search for compatible coding techniques. In this chapter, we describe the application of convolutional coding to partial-response channels, a technique that was introduced by Wolf and Ungerboeck [1] for the $(1 \pm D)$ channel in 1986. Many authors have subsequently elaborated on this coded system, such as Zehavi and Wolf [4], Hole [14], Hole and Ytrehus [15], and Siala and Kaleh [16]. The extension of this coding technique to the more general case of the $P_n(D)$ channel has been considered by Uchôa-Filho and Herro [12, 13], and refined by Despotović et al. [19, 20].

The key point in dealing with the convolutionally coded partial-response systems is to view the $P_n(D)$ channel as a finite state machine, as is usually done for convolutional encoders. Assuming binary channel inputs, the trellis associated with the channel has 2^{n+1} states. This channel trellis is then combined with the trellis describing the convolutional code, resulting in a trellis for the overall system that describes all possible coded sequences. Decoding is provided by the Viterbi algorithm operating on this trellis. The properties of this trellis are discussed in detail in this chapter. The combination of a precoder and a convolutional encoder drawn from a restricted set of generator matrices plays a key role in reducing the number of required decoder states. Other issues taken into consideration are the limitation of the length of equal symbol runs, which is addressed through the use of cosets of convolutional codes, and a code search for good codes based upon the (squared Euclidean) distance spectrum criterion.

This chapter is organized as follows. The description of the coded system and some preliminaries are given in Section 8.2. In Section 8.3, we derive a lower bound on the minimum free squared Euclidean distance of the trellis corresponding to the overall system. This bound, which is related to the minimum free Hamming distance of a subcode of the convolutional code, is then shown to be weak (and hence of little significance) when $n > 0$, which suggests the need for a computer-aided search for good codes. In Section 8.4, we point out several structural properties of the coded system and establish the rules which guide the computer search. In this section, we introduce the important theory of trellis matching. Section 8.5 addresses the problem of limiting the length of equal symbol runs for synchronization purposes. Section 8.6 deals with the problem of avoiding flawed codewords. Flawed codewords result whenever two infinitely long paths in the trellis of the overall system have the same labels (symbols from the output of the partial-response channel). In Section 8.7, the distance spectrum criterion for trellis codes is presented, which we adopt for finding good codes. These codes are tabulated for the $(1 - D)(1 + D)^2$ and $(1 - D)(1 + D)^3$ channels in Section 8.8.

8.2 Encoding System Description and Preliminaries

The block diagram of the encoded system that we consider in this chapter is shown in Figure 8.1. In this section we describe the components of this system and introduce some needed definitions. Unless otherwise stated, the signals and transfer functions of the blocks in Figure 8.1 are represented by their D-transforms. Beginning at the left, the input sequence, $\mathbf{U}(D) = [U^1(D), \ldots, U^k(D)]$, is first encoded by the (m, k) convolutional encoder of rate $R_c = k/m$, represented by the generator matrix $\mathbf{G}(D)$. The ith input sequence is represented by the polynomial $U^i(D) = u_0^i + u_1^i D + \cdots$, where $u_t^i \in GF(2)$ for $1 \leq i \leq k$, and for each time instant $t \geq 0$. The input sequence produces the encoded sequence, $\mathbf{V}(D) = \mathbf{U}(D)\mathbf{G}(D) = [V^1(D), \ldots, V^m(D)]$. The representation of the jth encoded sequence is $V^j(D) = v_0^j + v_1^j D + \cdots$, where $v_t^j \in GF(2)$ for $1 \leq j \leq m$, and for each time instant $t \geq 0$. The generator matrix $\mathbf{G}(D) \triangleq (G_i^j(D))$, where $G_i^j(D) = \sum_{l=0}^{v} g_{l,i}^j D^l$, and the elements $g_{l,i}^j \in GF(2)$, $1 \leq i \leq k$, and $1 \leq j \leq m$, represent the tap connections in the encoder. We define the ith input constraint length of the encoder by $v_i = \max_j\{\deg G_i^j(D)\}$ and the overall constraint length of the encoder by $v = \sum_{i=1}^{k} v_i$. The jth encoded

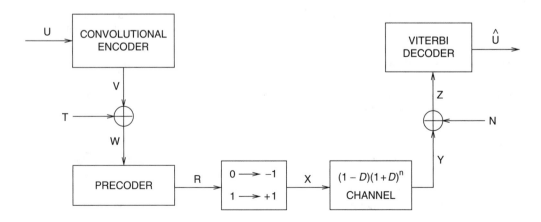

FIGURE 8.1 The convolutionally coded partial-response system. (Copyright © 2001 IEEE. Reproduced with permission.)

sequence, $V^j(D)$, can be written in terms of $\mathbf{U}(D)$ and $\mathbf{G}(D)$ as follows:

$$V^j(D) = \sum_{i=1}^{k} U^i(D)G_i^j(D), \quad 1 \le j \le m \tag{8.1}$$

The m encoded sequences are multiplexed into a single sequence, called the *codeword*, as follows:

$$V(D) = \sum_{j=1}^{m} D^{j-1}V^j(D^m) \tag{8.2}$$

A coset of the code is used to avoid the all zero sequence, since this sequence causes loss of clock synchronization. The coset sequence $\tilde{T}(D)$ is added to the codeword $V(D)$, giving

$$W(D) \equiv V(D) \oplus \tilde{T}(D) \tag{8.3}$$

where $\oplus$ denotes addition of polynomials over $GF(2)$. The sequence $W(D)$ is passed through a precoder. In the communication context, precoding is used as a method of avoiding error propagation in symbol-by-symbol detection of partial-response signals [11]. In this chapter, where maximum-likelihood sequence detection is adopted, the precoder has other purposes, as will become clear in the following sections. The transfer function of the precoder for the $P_n(D) = (1 - D)(1 + D)^n$ channel is chosen to be $(1 \oplus D)^{-n-1}$, which corresponds to $[1/P_n(D)]_{\mod 2}$. The precoded sequence, $R(D)$, is then given by

$$R(D) \equiv (1 \oplus D)^{-n-1} W(D) \tag{8.4}$$

where we consider the two sequences $R(D)$ and $W(D)$ to have the same length. The polar NRZ modulator (i.e., the map: $0 \to -1, 1 \to +1$) is introduced to make the system of Figure 8.1 compatible with saturation recording [6]. The output of the modulator is given by

$$X(D) = 2R(D) - \mathbf{1}(D) \tag{8.5}$$

where $\mathbf{1}(D)$ is the all one sequence with the same length as $R(D)$. The modulated sequence $X(D)$ is then sent to the $P_n(D) = p_0 + p_1 D + p_2 D^2 + \cdots + p_{n+1}D^{n+1}$ channel, which results in the multilevel channel output sequence

$$Y(D) = P_n(D)X(D) + \sum_{j=1}^{n+1}\left(\sum_{i=j}^{n+1}(-1)p_i\right)D^{j-1} - \left\{\begin{array}{l}\text{terms in } D^j, \text{ where}\\ j \ge \text{length of } W(D)\end{array}\right\} \tag{8.6}$$

where the second term is the response (independent of $X(D)$) due to the initial content of the unit-delay cells of the $P_n(D)$ channel (namely $-1, -1, \ldots, -1$). We illustrate this with an example.

Example 8.1

Consider the $P_2(D) = (1 - D)(1 + D)^2$ channel and the sequence $W(D) = (10110100) = 1 + D^2 + D^3 + D^5$. Then the precoder output is $R(D) = (11100000) = (1 \oplus D)^{-3}W(D) = 1 + D + D^2$. The modulated sequence is $X(D) = (+ + + - - - - -) = 2R(D) - \mathbf{1}(D) = 1 + D + D^2 - D^3 - D^4 - D^5 - D^6 - D^7$. Finally, the multilevel channel output sequence is $Y(D) = (+2, +4, +2, -2, -4, -2, 0, 0) = X(D)P_2(D) + (1 + 2D + D^2) - (D^8 + 2D^9 + D^{10}) = 2 + 4D + 2D^2 - 2D^3 - 4D^4 - 2D^5$.

A closed form for the number of output levels as a function of n is in general difficult to obtain. But for the most commonly used channels, namely for $n = 0, 1, 2$, or 3, these numbers are well known. For $n = 0$ and $n = 1$ there are three levels, namely 0 and ± 2. For $n = 2$ there are five levels, namely 0, ± 2, and ± 4. And for $n = 3$ there are seven levels, namely 0, ± 2, ± 4, and ± 6.

After passing through the partial response "channel," zero mean, i.i.d., Gaussian noise $N(D)$ is added to $Y(D)$, producing the noisy sequence $Z(D)$. A Viterbi detector is then used to find the maximum likelihood estimate $\hat{U}(D)$ of $U(D)$ from the noisy channel output $Z(D)$.

8.3 Trellis Codes for Partial-Response Channels Based Upon the Hamming Metric

For the trellis code formed at the output of the $1 - D$ channel, in the encoded system of Figure 8.1 (with $n = 0$), Wolf and Ungerboeck [1] derived a lower bound on the minimum free squared Euclidean distance, which is monotonically related to the minimum free Hamming distance of the convolutional code. From this result, it follows that convolutional codes with large free Hamming distance generate trellis codes with large free squared Euclidean distance. In this section, we present the extension of the result for the $1 - D$ channel to the $P_n(D)$ channel case, and conclude that similar guidelines exist for choosing a convolutional code that leads to a trellis code with large free squared Euclidean distance. We also introduce some concepts which will be useful in the forthcoming sections.

The set of all output sequences $V(D)$, as the input sequence $\mathbf{U}(D)$ ranges over all possible values, is called the *convolutional code*, denoted by C, and generated by the encoder $\mathbf{G}(D)$. Let $\omega_H(V(D))$ be the Hamming weight of the coded sequence $V(D)$, that is, the number of nonzero coefficients in $V(D)$. The minimum free Hamming distance of C, denoted by $d_H(C)$, is defined as the minimum of the set $\{\omega_H(V(D)) \mid V(D) \in C, V(D) \neq 0\}$.

We denote the *channel code* by the set of noiseless output sequences $Y(D)$ from the $P_n(D)$ channel in Figure 8.1. This code is multilevel and nonlinear trellis. The trellis for the channel code is called the *decoder trellis*, which is the trellis for the overall system modelled as the combination of the convolutional encoder, the precoder, and the $P_n(D)$ channel. A *channel codeword* of the channel code consists of the multilevel labels on any path in the decoder trellis that starts in any state and ends in any state (not necessarily the same as the starting state). We say that an *error event* of length l has occurred when two paths in the decoder trellis, say $Y(D)$ and $Y'(D)$, diverge from each other at time t and remerge at time $t + l - 1$. We may assume with no loss of generality that $t = 0$ and that the correct channel codeword $Y(D) = y_0 + y_1 D + \cdots + y_{l-1} D^{l-1}$ and the incorrect channel codeword $Y'(D) = y'_0 + y'_1 D + \cdots + y'_{l-1} D^{l-1}$. A typical error event is shown in Figure 8.2.

The squared Euclidean distance between $Y(D)$ and $Y'(D)$ is defined as

$$d_E^2(Y, Y') \stackrel{\Delta}{=} \sum_{i=0}^{l-1} (y_i - y'_i)^2 \tag{8.7}$$

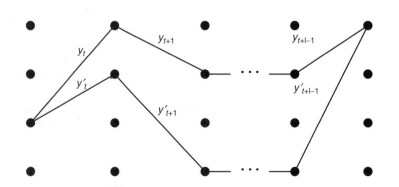

FIGURE 8.2 A typical error event in a decoder trellis. (Modified from Uchôa Filho, B. F. and Herro, M. A., *IEEE Trans. Inform. Theory*, vol. 43, No. 2, pp. 441–453, Mar. 1997. Copyright © 1997 IEEE. With permission.)

The minimum *free* squared Euclidean distance of the channel code is defined as

$$d_{\text{free}}^2 \overset{\Delta}{=} \min_{Y \neq Y'} \left(d_E^2 (Y, Y') \right) \tag{8.8}$$

where the minimization is over all possible error events.

We now present a lower bound on the minimum free squared Euclidean distance, d_{free}^2, of the channel code for the $P_n(D)$ channel, generated by the convolutional code C. This lower bound relates $d_H(C)$ to d_{free}^2. In the derivation we consider $\tilde{T}(D) \equiv 0$. However, it can easily be shown that adding $\tilde{T}(D) \not\equiv 0$ modulo 2 to every codeword of C, as given in Equation 8.3, will not change $d_H(C)$. Thus the bound is also valid for $\tilde{T}(D) \not\equiv 0$.

We define the function $|Y(D)|$ for polynomials $Y(D)$ with even coefficients as

$$|Y(D)| \overset{\Delta}{\equiv} \frac{y_0}{2} + \frac{y_1}{2} D + \frac{y_2}{2} D^2 + \cdots \pmod{2} \tag{8.9}$$

where $|Y(D)|$ reduces to a binary sequence because the (mod 2) reduction is carried out on each coefficient of $Y(D)$, after the division by 2. As an example, consider the sequence $Y(D) = 2 + 4D + 2D^2 - 2D^3 - 4D^4 - 2D^5$ given in Example 8.1. We have that $|Y(D)| \equiv 1 + 2D + D^2 - D^3 - 2D^4 - D^5 \pmod{2} \equiv 1 + D^2 + D^3 + D^5$.

It is shown in Appendix A that the correspondence between $Y(D)$ and $W(D)$ is given by

$$W(D) \equiv |Y(D)| \tag{8.10}$$

Note that the binary sequence obtained in the previous paragraph is the sequence $W(D)$ in Example 8.1. It can be seen from Equation 8.10 that the sequence $W(D)$ can be recovered from the noiseless channel output $Y(D)$ in a symbol-by-symbol fashion. Moreover, because of Equation 8.10, whenever a bit w is 0, the corresponding channel output y is divisible by 4, and whenever a bit w is 1, the corresponding channel output y is divisible by 4 with remainder 2. This can be expressed analytically as:

$$w \equiv 0 \pmod{2} \iff y \in 4\mathbb{Z} \overset{\Delta}{=} \{0, \pm 4, \pm 8, \dots\}$$
$$w \equiv 1 \pmod{2} \iff y \in 4\mathbb{Z} + 2 \overset{\Delta}{=} \{2, \pm 6, \pm 10, \dots\} \tag{8.11}$$

where $\mathbb{Z}$ is the set of all integers.

We now define the binary representation of an error event $E(D)$, as done in references [5, 15]. The binary representation of an error event $E(D)$ is the modulo 2 sum of the two binary sequences $W(D)$ and $W'(D)$ that produce, at the output of the channel, the sequences $Y(D)$ and $Y'(D)$, respectively. Using the function defined in Equation 8.9 we may write

$$E(D) \overset{\Delta}{=} (w_0 \oplus w_0') \oplus (w_1 \oplus w_1')D \oplus \cdots \oplus (w_{l-1} \oplus w_{l-1}')D^{l-1}$$
$$\overset{\Delta}{\equiv} (|y_0| \oplus |y_0'|) \oplus (|y_1| \oplus |y_1'|)D \oplus \cdots \oplus (|y_{l-1}| \oplus |y_{l-1}'|)D^{l-1}$$
$$\overset{\Delta}{\equiv} e_0 \oplus e_1 D \oplus \cdots \oplus e_{l-1}D^{l-1} \tag{8.12}$$

Note that $E(D)$ is independent of $\tilde{T}(D)$ since $E(D) \equiv W(D) \oplus W'(D) \equiv V(D) \oplus \tilde{T}(D) \oplus V'(D) \oplus \tilde{T}(D) \equiv V(D) \oplus V'(D)$. It also follows from these equivalences that $E(D) \equiv V(D) \oplus V'(D)$ is a codeword of C, since $V(D)$ and $V'(D)$ are two codewords of C, and C is a linear code. Zehavi and Wolf [5] observed that since $E(D)$ must be a codeword of the convolutional code, only sequences that are codewords can be potential error events, that is, not every sequence is the binary representation of an error event. In the next lemma we reduce even further the set of such possible binary representations.

Define the subset C_n of the convolutional code C as

$$C_n \overset{\Delta}{=} \{V(D) \in C \mid (1 \oplus D)^{n+1} \text{ divides } V(D)\}$$

Clearly C_n is a linear subcode of C and $C_{-1} = C \supseteq C_0 \supseteq C_1 \supseteq C_2 \supseteq \cdots$.

Lemma 8.1   *If $E(D)$ is the binary representation of an error event, then $E(D) \in C_n$.*

Proof 8.1   Consider the difference sequence $Y(D) - Y'(D)$. From Equation 8.6 we have that

$$Y(D) - Y'(D) = (X(D) - X'(D))P_n(D)$$

since the terms independent of $X(D)$ cancel out. If we apply the polynomial function defined in Equation 8.9 to the previous equation and use the results in Appendix B, we have the following equivalences:

$$\begin{aligned}
|Y(D) - Y'(D)| &\equiv |(X(D) - X'(D))P_n(D)| \\
&\equiv |X(D) - X'(D)|[P_n(D)]_{\mathrm{mod}\,2} \\
&\equiv |X(D) - X'(D)|(1 \oplus D)^{n+1} \\
&\equiv E(D)
\end{aligned}$$

where the last equivalence, namely $E(D) \equiv |Y(D) - Y'(D)|$, comes from Equation 8.12. Therefore $(1 \oplus D)^{n+1}$ divides $E(D)$ and consequently $E(D) \in C_n$ as claimed. □

We are now ready to state the lower bound on d_{free}^2.

Theorem 8.1   *The minimum free squared Euclidean distance of the channel code, for the precoded $P_n(D) = (1-D)(1+D)^n$ channel, generated by the convolutional code C, according to the encoded system of Figure 8.1, is lower bounded by:*

$$d_{\mathrm{free}}^2 \geq 4d_H(C_n)$$

Proof 8.2   First note that because of Equation 8.11 whenever two encoded bits, say w and w', differ from each other, the squared Euclidean distance between their corresponding channel outputs, namely y and y', is lower bounded by the intersubset squared Euclidean distance between the subsets $4\mathbb{Z}$ and $4\mathbb{Z} + 2$, which is 4. The rest of the proof follows from Lemma 8.1. The binary representation of any error event in the decoder trellis is a codeword in C_n, and this has at least $d_H(C_n)$ ones. Therefore d_{free}^2 is lower bounded by $4d_H(C_n)$. □

Remark 8.1   For $n = 0$ this bound is equivalent to the bound of Wolf and Ungerboeck (Lemma 8.2, [1]), namely $d_{\mathrm{free}}^2 \geq 4d_H^{(e)}$, where $d_H^{(e)}$ is the weight of the lowest even weight codeword in C. To see this equivalence note that, for $n = 0$, C_0 is the set of all codewords of C that have even weight, since a polynomial $V(D)$ is divisible by $(1 \oplus D)$ in $GF(2)$ if and only if $V(D)$ has even weight. Thus, for $n = 0$, $d_H(C_0) = d_H^{(e)}$.

The immediate consequence of Theorem 8.1 is that convolutional codes whose subcode C_n has large minimum free Hamming distance $d_H(C_n)$ are good candidates to generate channel codes with large d_{free}^2. The advantage of this construction is that many convolutional codes designed for the Hamming metric have already been found and are tabulated in references such as [10]. Many of these channel codes for the $1 - D$ channel were found in [1]. For this simpler channel, the bound given in Theorem 8.1 is always tight [1]. However, for the $P_n(D)$ channel ($n > 0$), the actual d_{free}^2 of the channel code is often much larger than the lower bound. We can see this by means of an example. Assume $n = 2$, so that the output levels are $0, \pm 2$, and ± 4. The bit $w = 0$ may be converted to the output level $y = -4$, and the bit $w' = 1$ may be converted to $y' = +2$. Then $d_E^2(y, y') = d_E^2(-4, +2) = 36$, as opposed to 4 given by the bound in Theorem 8.1. This discrepancy between the actual d_{free}^2 and the bound becomes even more prominent when $\tilde{T}(D) \neq 0$, as we shall verify in Section 8.4. We now give an example of channel codes designed using this method.

Example 8.2

Consider the $(4, 1)$ convolutional code C with constraint length $\nu = 1$ generated by $\mathbf{G}(D) = [1+D, 1, 0, 1]$. This code has $d_H(C) = d_H(C_0) = d_H(C_2) = 6$. Hence the bound given in Theorem 8.1 becomes $d_{free}^2 \geq 4 \times 6 = 24$. For $n = 0$, the channel code generated by C has $d_{free}^2 = 24$. But for $n = 2$, $d_{free}^2 = 56 > 24$.

Since the bound given in Theorem 8.1 is weak when $n > 0$, basing a search for channel codes with large d_{free}^2 solely on convolutional codes with large d_H appears to be too restrictive and does not exploit the channel memory in an efficient way. A computer-aided search is thus required to find good channel codes. The theoretical elements related to this search are presented next in Section 8.4.

8.4 Trellis-Matched Codes for Partial-Response Channels

In this section we point out some structural properties of channel codes and establish the rules (for choosing $\mathbf{G}(D)$ and $\tilde{T}(D)$) which guided the computer search that led to the channel codes given in Section 8.8.

The first concern is with the state complexity of the decoder trellises for the channel codes generated by the convolutional encoders. As mentioned in the introductory section, under binary inputs the $P_n(D)$ channel has 2^{n+1} states. The precoder does not increase the number of states since it shares the same (up to a one-to-one map — the polar NRZ map) states with the $P_n(D)$ channel. In other words, *once the state of the precoder is known, the state of the channel trellis can be determined.* On the other hand, since the decoder trellis is obtained from the product of the trellis of the convolutional code (which has 2^ν states) and the channel trellis, the number of states of the decoder trellis is in general $2^{n+\nu+1}$. However, depending on a structural property of the encoder, the decoder trellis can have fewer states. We use the following classification as we refer to the number of states of the decoder trellis.

We say that a channel code is unmatched (UM) to the $P_n(D)$ channel if the decoder trellis has $2^{n+\nu+1}$ states; that it is partially trellis matched (PTM) to the $P_n(D)$ channel if the decoder trellis has $2^{n'+\nu+1}$ states, where $0 \leq n' < n$; and that it is totally trellis matched (TTM) (or simply trellis matched (TM)) to the $P_n(D)$ channel if the decoder trellis has at most 2^ν states. Since reduced-complexity decoder trellises are preferred, this chapter focuses on channel codes that are TTM to the $P_n(D)$ channel.

In Theorem 8.2 we will present a sufficient condition for a convolutional encoder to generate a channel code that is TTM to the $P_n(D)$ channel. To do this we require the following definitions.

Definition 8.1   We define the *input generator vector*, $\mathbf{G}_{in}(D)$, as another representation for an (m, k) convolutional encoder which is equivalent to and derived from $\mathbf{G}(D)$ as given below:

$$\mathbf{G}_{in}(D) = \left[G_{in}^1(D), \ldots, G_{in}^k(D) \right]$$

where $G_{in}^i(D) = \sum_{j=1}^m D^{j-1} G_i^j(D^m)$, $1 \leq i \leq k$. Note that $G_{in}^i(D)$ is the result of multiplexing the polynomials in the ith row of $\mathbf{G}(D)$.

The input generator vector, for a particular encoder, is given in the following example.

Example 8.3

Consider the (5,2) convolutional encoder with constraint length $v = 2$ represented by the following generator matrix:

$$\mathbf{G}(D) = \begin{pmatrix} D & 1+D & 1 & 0 & 0 \\ 1 & 0 & 0 & D & 0 \end{pmatrix}$$

Then $m = 5$, $k = 2$, and $\mathbf{G_{in}}(D) = [G_{in}^1(D), G_{in}^2(D)] = [D + D^2 + D^5 + D^6, 1 + D^8]$.

Definition 8.2   An encoder is said to be *feedback free* if none of the $G_i^j(D)$ are rational polynomials.

Clearly the encoder of Example 8.3 is a feedback free encoder. All encoders considered in this chapter are feedback free encoders. In the following theorem, $\tilde{T}(D) \equiv 0$. The case $\tilde{T}(D) \neq 0$ will be treated later in this section.

Theorem 8.2   *Let n' be the largest nonnegative integer (if it exists) such that the input generator vector $\mathbf{G_{in}}(D)$ for C may be factored as:*

$$\mathbf{G_{in}}(D) = (1 \oplus D)^{n'+1} \, \mathbf{G'_{in}}(D)$$

Then the channel code generated by C is TTM to the $P_n(D)$ channel if $n' \geq n$.

Proof 8.3   We begin this proof by writing the overall output sequence $V(D)$ in terms of $\mathbf{G_{in}}(D)$. Substituting Equation 8.1 into Equation 8.2 yields

$$V(D) = \sum_{j=1}^{m} D^{j-1} \left(\sum_{i=1}^{k} U^i(D^m) G_i^j(D^m) \right)$$

$$= \sum_{i=1}^{k} U^i(D^m) \sum_{j=1}^{m} D^{j-1} G_i^j(D^m)$$

$$= \sum_{i=1}^{k} U^i(D^m) G_{in}^i(D)$$

From the assumption of the theorem we may write $V(D)$ as

$$V(D) = (1 \oplus D)^{n'+1} \sum_{i=1}^{k} U^i(D^m) G_{in}'^i(D) \tag{8.13}$$

Then the precoded output in equivalence (8.4), for $\tilde{T}(D) \equiv 0$, becomes

$$R(D) = (1 \oplus D)^{-n-1} V(D) = (1 \oplus D)^{n'-n} \sum_{i=1}^{k} U^i(D^m) G_{in}'^i(D)$$

Now we note that if $n' \geq n$ the feedback free modified encoder

$$(1 \oplus D)^{n'-n} \mathbf{G'_{in}}(D) \tag{8.14}$$

may be used to produce $R(D)$ directly from the input $U(D)$, and it has at most 2^v states. To see this note that since the degree of $G_{in}^i(D)$ satisfies $mv_i \leq \deg G_{in}^i(D) \leq mv_i + m - 1$, the input constraint length of the encoder, v_i, can be determined by:

$$v_i = \left\lfloor \frac{\deg G_{\text{in}}^i(D)}{m} \right\rfloor$$

where $[x]$ is the largest integer less than or equal to x. From Equation 8.14, the corresponding degree for the modified encoder is equal to $n' - n + \deg G_{\text{in}}^{\prime i}(D) = n' - n + (\deg G_{\text{in}}^i(D) - n' - 1) = \deg G_{\text{in}}^i(D) - n - 1$. Then the input constraint length of the modified encoder, $v_i(\text{modified})$, can be similarly determined by:

$$v_i(\text{modified}) = \left\lfloor \frac{\deg G_{\text{in}}^i(D) - n - 1}{m} \right\rfloor$$
$$\leq \left\lfloor \frac{mv_i + m - n - 2}{m} \right\rfloor$$
$$\leq v_i,$$

which shows that the modified encoder has at most 2^v states (since $v = \sum_{i=1}^{k} v_i$).

It follows from these facts that all states of the precoder can be determined by at most 2^v states, the states of the modified encoder. We know, from the second paragraph of this section, that the states of the precoded $P_n(D)$ channel trellis can be determined from the states of the precoder. Therefore the states of the precoded $P_n(D)$ channel trellis can be determined from the states of the modified encoder, and consequently the decoder trellis has at most 2^v states if $n' \geq n$. □

In the following we define a class of convolutional encoders for which an encoder, combined with the precoder for the $P_n(D)$ channel, generates a decoder trellis with exactly 2^v states.

Definition 8.3 There can exist many encoders $\mathbf{G}(D)$ that generate the same code C. An encoder $\mathbf{G}(D)$ of constraint length v (realized with 2^v states) that generates C is said to be a *minimal encoder* for C if no other encoder with constraint length smaller than v (realized with fewer than 2^v states) generates C.

Lemma 8.2 *If the convolutional encoder $\mathbf{G}(D)$ is minimal and satisfies the condition in Theorem 8.2, then it generates a decoder trellis with* exactly 2^v *states.*

Proof 8.4 Let C be the convolutional code generated by a minimal convolutional encoder $\mathbf{G}(D)$ (or $\mathbf{G_{in}}(D)$), with constraint length v, satisfying the condition in Theorem 8.2. Let $\mathcal{C}$ be the channel code generated by C. Suppose that a decoder trellis for $\mathcal{C}$ with fewer than 2^v states exists. If we apply the equivalence in Equation 8.10 to each channel codeword $Y(D)$ of $\mathcal{C}$, the resulting set of sequences is clearly the convolutional code C. Hence, by this procedure, we have found a trellis for C with fewer than 2^v states, which contradicts the assumption on the minimality of $\mathbf{G}(D)$. Therefore, from Theorem 8.2, the decoder trellis has exactly 2^v states. □

A criterion for determining whether or not a convolutional encoder is minimal was established by Forney [8]. He showed that an encoder $\mathbf{G}(D)$ is minimal if and only if each k by k subdeterminant of $\mathbf{G}(D)$ has degree not exceeding v, and the greatest common divisor of all such subdeterminants is equal to 1. We treat the case where the encoder is nonminimal later in this section.

In Theorem 8.2, we note that since $n' \geq n$, Equation 8.13 implies that *every codeword $V(D)$ of C is divisible by $(1 \oplus D)^{n+1}$*, which further implies that $C_n = C$. Hence, from Lemma 8.1, *every codeword $V(D)$ of C is the binary representation of some error event*. From this fact, and from Lemma 8.2, it follows that the decoder trellis and the trellis of the convolutional code are the same, except for the labels. We illustrate this by means of an example.

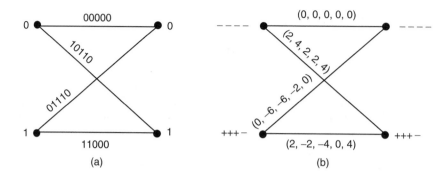

FIGURE 8.3 (a) The trellis of the convolutional code and; (b) the decoder trellis of Example 8.4. (Copyright © 1997 IEEE. Reproduced with permission.)

Example 8.4

Consider the minimal (5,1) convolutional encoder with constraint length $v = 1$, represented by the following generator matrix $\mathbf{G}(D) = [1, D, 1 + D, 1 + D, 0]$. The input generator vector can be factored as $\mathbf{G_{in}}(D) \equiv [1 \oplus D^2 \oplus D^3 \oplus D^6 \oplus D^7 \oplus D^8] \equiv (1 \oplus D)^5[1 \oplus D \oplus D^3]$. The trellis of the convolutional code and the decoder trellis when the $P_3(D) = (1 - D)(1 + D)^3$ channel is used are shown in Figure 8.3.

Remark 8.2 In this chapter, we label the states of the decoder trellis with the contents of the unit-delay cells of the precoded $(1 - D)(1 + D)^n$ partial-response system when the inputs are the coset branch bits. In the decoder trellis of Figure 8.3(b), for example, the state labels are "$- - - -$" and "$+ + + -$".

In Example 8.4, $n' = 4 > 3 = n$. Note that both trellises are the same. Also note the correspondence between the bits in the trellis of the convolutional code and the multilevel labels in the decoder trellis. It satisfies the equivalence in Equation 8.10. Every codeword $V(D)$ of this convolutional code is divisible by $(1 \oplus D)^5$, which implies that they are also divisible by $(1 \oplus D)^4$, where $4 = n + 1$. Recall that we can also represent a codeword of a convolutional code by the labels on a path in the trellis of the convolutional code that starts and ends in the zero state. For example, consider the codeword "101101100001110," which corresponds to the path in the trellis of Figure 8.3(a) that leaves state zero, goes to state one and remains there for one time instant, and then returns to state zero again. The D-transform of this codeword is $V(D) \equiv 1 \oplus D^2 \oplus D^3 \oplus D^5 \oplus D^6 \oplus D^{11} \oplus D^{12} \oplus D^{13} \equiv (1 \oplus D)^5(1 \oplus D \oplus D^3 \oplus D^5 \oplus D^6 \oplus D^8)$, that is, $V(D)$ is divisible by $(1 \oplus D)^5$.

The convolutional code in Example 8.4 has $d_H(C) = d_H(C_3) = 6$, so that the bound of Theorem 8.1 becomes $d_{\text{free}}^2 \geq 4 \times 6 = 24$. However the corresponding channel code has $d_{\text{free}}^2 = 108$, more than four times larger than the bound. This example also supports the fact that the bound of Theorem 8.1 is rather weak.

8.5   Run-Length Limited Trellis-Matched Codes

8.5.1   Cosets of Convolutional Codes

Now we begin to consider nontrivial cosets of convolutional codes, that is, we start to analyze the case $\tilde{T}(D) \not\equiv 0$. Assume that $\tilde{T}(D)$ is not a codeword of C. Then the set

$$C \oplus \tilde{T} \triangleq \{V(D) \oplus \tilde{T}(D) | V(D) \in C\}$$

is a *nontrivial coset* of the convolutional code C. The trellis for the coset $C \oplus \tilde{T}$ is called the *coset trellis* [15]. The sequence $\tilde{T}(D)$ is called the *coset representative* and is not unique. The binary labels on any path in the coset trellis that starts and ends in the all zero state is called a *coset word*. $\tilde{T}(D)$ is chosen to be a periodic sequence of period m, the number of output lines of the convolutional encoder. Let $T(D)$ be the polynomial of degree at most $m - 1$ which corresponds to a period of $\tilde{T}(D)$. With this choice the binary addition in Equation 8.3, which generates the coset of C, can be implemented by simply inverting the bits on the output lines of the convolutional encoder corresponding to ones in $T(D)$. The periodic sequence $\tilde{T}(D)$ can be written in terms of $T(D)$ as below:

$$\tilde{T}(D) = \sum_{\ell=1}^{n_b} D^{(\ell-1)m} T(D) \tag{8.15}$$

where n_b is the number of branches in the path whose labels form the codeword $V(D)$ in Equation 8.3.

We denote the *maximum zero-run length* of the channel code, that is, the maximum run of zeros between two consecutive nonzero symbols in any path of the decoder trellis, by $L_{\max}$. We can see from Equation 8.11 that the channel output $y = 0$ only if the coset bit $w = 0$. Hence $L_{\max}$ can be upper bounded by the maximum zero-run length of the coset. Focusing on the $(1 - D)$ channel, Hole and Ytrehus [15] have observed that a good choice of $T(D)$ may not only limit $L_{\max}$, but may also yield a trellis code with d_{free}^2 larger than the lower bound given in Theorem 8.1. It was shown in [13] that this assertion remains true for the $P_n(D)$ channel. We verify this by means of an example.

Example 8.5

Consider the convolutional encoder of Example 8.4. The coset trellis and the decoder trellis for the choice $T(D) = [10001] = 1 + D^4$ are given in Figure 8.4.

The channel code given in Example 8.5 has $d_{\text{free}}^2 = 264$. This is 11 times larger than the bound of Theorem 8.1. Note also that the maximum zero-run length of the coset is 3. However, we can see by inspection that the decoder trellis of Figure 8.4(b) has $L_{\max} = 2 < 3$. This is because some of the bits equal to "0" in the coset are converted into either -4 or 4 in the decoder trellis.

It is important to note that the output lines of the convolutional encoder can not be permuted. This would not necessarily imply the corresponding permutation of the multilevel labels in the branches of the decoder trellis. This is due to the fact that $P_n(D)$ channels have memory. The coset representative cannot be changed either. For example, if we choose $T(D) = [01111] = D + D^2 + D^3 + D^4$ instead of the $T(D)$ given in Example 8.5, the channel code generated by this coset can not be represented by a trellis with only two states. Four states are needed instead. To see this consider the four input sequences: $U_1^{(1)} = [0]$,

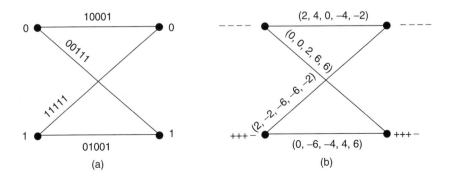

FIGURE 8.4 (a) The coset trellis and; (b) the decoder trellis of Example 8.5. (Copyright © 1997 IEEE. Reproduced with permission.)

$U_1^{(2)} = [1]$, $U_1^{(3)} = [1,1]$, and $U_1^{(4)} = [1,1,0]$. The branch labels in the coset trellis for these inputs are: $R^{(1)} = [01111]$, $R^{(2)} = [11001]$, $R^{(3)} = [11001, 10111]$ and $R^{(4)} = [11001, 10111, 00001]$, respectively. These branch labels in turn drive the channel state from "$----$" to "$----$", "$---+$", "$+++-$" and "$+++$", respectively. Any other sequence will drive the channel state from "$----$" to one of the channel states above. Therefore, the decoder trellis has four states. After these observations we can see that a good choice of the coset $C \oplus \tilde{T}$ is necessary to achieve good results. We devote the remainder of this section to establishing the rules that lead to good choices of $T(D)$.

In the last paragraph, we noted that the choice of $T(D)$ may also have an effect on the structure of the channel code. In particular, while C generates a channel code that is TTM to the $P_n(D)$ channel, the coset $C \oplus \tilde{T}$ may not do so. It is therefore important to know for which polynomials $T(D)$ the channel code generated by the coset $C \oplus \tilde{T}$ is TTM to the $P_n(D)$ channel. This is addressed in the theorem below.

Theorem 8.3 *Consider an (m,k) convolutional encoder that generates C and satisfies the condition in Theorem 8.2, that is, C generates a channel code that is TTM to the $P_n(D)$ channel. Let $C \oplus \tilde{T}$ be a coset of C, where $T(D)$ is a polynomial of degree at most $m-1$, and $\tilde{T}(D) \notin C$. Then the channel code generated by $C \oplus \tilde{T}$ is TTM to the $P_n(D)$ channel if $(1 \oplus D)^{n+1}$ divides $T(D)$, which in turn requires that $m \geq n+2$.*

Proof 8.5 From Equation 8.15, since $(1 \oplus D)^{n+1}$ divides $T(D)$, $(1 \oplus D)^{n+1}$ certainly divides $\tilde{T}(D)$. By assumption, $(1 \oplus D)^{n+1}$ divides $V(D)$. Consequently $(1 \oplus D)^{n+1}$ also divides the coset sequence $W(D) \equiv V(D) \oplus \tilde{T}(D)$ in Equation 8.3. If we replace $V(D)$ by $W(D) \equiv V(D) \oplus \tilde{T}(D)$ in Theorem 8.2, and repeat the proof of that theorem, we have proved the first part of Theorem 8.3. The requirement on the minimum number of encoder output lines m can be seen from the fact that $T(D)$ must have degree at least $n+1$ for $(1 \oplus D)^{n+1}$ to divide $T(D)$. Since the degree of $T(D)$ is at most $m-1$, the result follows immediately. □

Note that in Example 8.5, $T(D) \equiv 1 \oplus D^4 \equiv (1 \oplus D)^4$, which satisfies the condition in Theorem 8.3.

The remainder of this section discusses the search for good channel codes performed in [13, 20], whether adding constraints to $G(D)$ and $T(D)$ can lead to good channel codes while reducing search time, or whether these constraints overly restrict the choices of $G(D)$ and $T(D)$, making it difficult to find good channel codes.

Note that the requirement on the minimum number of encoder output lines $m \geq n+2$, from Theorem 8.3, imposes a constraint on the choice of code rates. For example, run-length limited time-invariant channel codes of rate $R_c = k/3$, TTM to the $(1-D)(1+D)^2$ channel, do not exist. Neither do run-length limited time-invariant channel codes of rate $R_c = k/4$, TTM to the $(1-D)(1+D)^3$ channel. The choice of $T(D)$ as in Theorem 8.3 always leads to a time-invariant decoder trellis. If $T(D)$ is not so restricted, however, there may still exist TTM channel codes with a time-varying decoder trellis. In this chapter, only the time-invariant case, that is, only the case of cosets that satisfy the condition in Theorem 8.3, is considered.

8.5.1.1 Bit Stuffing

A very simple technique to limit L_{max} is the so-called *bit stuffing* technique. In this technique we form the (m,k) convolutional code as follows. First a convolutional code with parameters $(m-b,k)$, with $1 \leq b < m-k$, is used. Then b "ones" are "stuffed" into the b bit positions (left uncoded), and this leads to the desired (m,k) code. A simple bound on L_{max} can be derived when this technique is used. For example, let "1" represent a "stuffed" bit and "x" an encoded bit. Suppose that $m=8$ and $b=2$, and we choose "$xx1x1xxx$". Since "x" can be either "0" or "1," the worst scenario would be the following: $\ldots xx1x1000001x1xxx\ldots$. Thus L_{max} can be bounded by $L_{max} \leq 5$. Another equivalent representation for the encoder in Figure 8.1 which is particularly helpful to visualize the bit stuffing technique is as follows.

Definition 8.4 *We define the output generator vector as*

$$\mathbf{G}_{\text{out}}(D) = \left[G_{\text{out}}^1(D), \ldots, G_{\text{out}}^m(D) \right]$$

where $G_{\text{out}}^j(D) = \sum_{i=1}^k D^{i-1} G_i^j(D^k)$, $1 \le j \le m$.

The representation given in Definition 8.4, for the encoder in Example 8.3, is $\mathbf{G}_{\text{out}}(D) = [G_{\text{out}}^1(D), G_{\text{out}}^2(D), G_{\text{out}}^3(D), G_{\text{out}}^4(D), G_{\text{out}}^5(D)] = [D + D^2, 1 + D^2, 1, D^3, 0]$. Note that $\mathbf{G}_{\text{out}}(D)$ is the modified 1 by m generator matrix $\mathbf{G_m}(D)$ introduced for the same purpose in [15]. We can see that an encoder $G(D)$ with b all zero columns, or equivalently, an encoder $\mathbf{G}_{\text{out}}(D)$ with b coordinates equal to zero, along with a choice of $T(D)$ having ones in at least one of the b positions that are equal to zero in $\mathbf{G}_{\text{out}}(D)$, may be used to accomplish the bit stuffing technique. In other words, the encoded system of Figure 8.1 is general enough to realize bit stuffing by an appropriate choice of $\mathbf{G}_{\text{out}}(D)$ and $T(D)$. Despite its simplicity, this technique can generate channel codes with large d_{free}^2 and very low values of L_{max}. However, the bit stuffing technique considerably limits the number of possible choices of encoders. As a result, no good codes, and in some cases no code at all, exist for certain code rates. In a code search, the bit stuffing technique is used whenever it leads to good channel codes. When this is not the case, L_{max} is determined by inspection.

8.6 Avoiding Flawed Codewords

A channel codeword is said to be a *flawed* channel codeword if the initial state is not uniquely determined from the multilevel labels alone [7, 15]. Such a codeword exists in a channel code if and only if there are two infinitely long paths in the decoder trellis with the same multilevel labels. This is an undesirable situation since it may lead to ambiguity in the decoding process. For the $1 - D$ channel, Hole and Ytrehus [15] showed that minimal encoders generate channel codes that do not contain flawed codewords. Their proof was based on a property of the convolutional codes generated by minimal encoders derived by Forney [9]. It can be similarly shown that the same is true for the $P_n(D)$ channels. However, in this case, some nonminimal encoders also generate channel codes not containing flawed codewords. This is due to the fact that each of the coset bits "0" and "1" can be converted, at the output of the $P_n(D)$ channel, into more than one multilevel symbol, if $n > 0$. Consequently, two infinitely long paths in the trellis of the convolutional code with the same binary labels may be converted by the $P_n(D)$ channel into two paths in the decoder trellis with different labels. In other words, a convolutional code containing flawed codewords may generate a channel code free of flawed channel codewords. In addition, there exist many nonminimal encoders that satisfy the condition in Theorem 8.2. Note that minimality of convolutional encoders by itself does not imply the total match of a channel code, i.e., the minimum number of states for the decoder trellis. In fact minimality is not a condition in Theorem 8.2. Therefore some nonminimal encoders can lead to trellis matched channel codes, whereas some minimal encoders may not. In the next example, we give a nonminimal (4,2) convolutional encoder with constraint length $\nu = 2$ for the $(1 - D)(1 + D)^2$ channel. For the same parameters, no minimal encoder exists that leads to a good channel code TTM to the $(1 - D)(1 + D)^2$ channel.

Example 8.6

Consider the nonminimal (4,2) convolutional encoder with constraint length $\nu = 2$ represented by the following generator matrix:

$$\mathbf{G}(D) = \begin{pmatrix} 0 & 1 + D & 0 & 1 + D \\ 0 & 1 + D & 1 + D & 1 + D \end{pmatrix}$$

The decoder trellis for the channel code TTM to the $(1 - D)(1 + D)^2$ channel is shown in Figure 8.5.

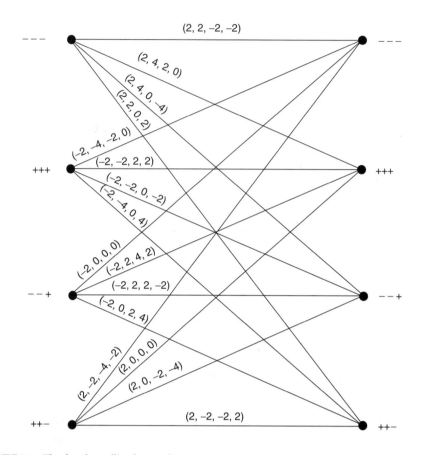

FIGURE 8.5   The decoder trellis of Example 8.6. (Copyright © 1997 IEEE. Reproduced with permission.)

This channel code has $d_{\text{free}}^2 = 40$ and $L_{\max} = 3$. It can be seen from Figure 8.5 that this channel code does not contain any flawed codewords. In a code search, when minimal encoders do not lead to good channel codes, the minimality constraint is relaxed and the best channel code generated by a nonminimal encoder (when this exists) is presented.

We end this section with a summary of the conditions that limit the choices of $\mathbf{G}(D)$ and $T(D)$ according to the results of this section.

Requirements on $\mathbf{G}(D)$ and $\mathbf{T}(D)$:

(i) $m \geq n + 2$
(ii) $1 \leq k < m$
(iii) $\mathbf{G}(D)$ is a feedback free encoder
(iv) $\mathbf{G}(D)$ is a minimal encoder
(v) $\mathbf{G}_{\text{in}}(D) = (1 \oplus D)^{n'+1} \mathbf{G}'_{\text{in}}(D)$, $n' \geq n$
(vi) $\mathbf{G}_{\text{out}}(D)$ has at least one coordinate equal to zero, for bit stuffing
(vii) $(1 \oplus D)^{n+1}$ divides $T(D)$
(viii) $T(D)$ has "1" in at least one of the positions where $\mathbf{G}_{\text{out}}(D)$ is zero, for bit stuffing
(ix) $1 \leq \nu \leq 5$
(x) $\nu_i \geq 1$, for $1 \leq i \leq k$, to avoid parallel transitions in the decoder trellis

If the bit stuffing technique is too restrictive, then conditions (vi) and (viii) are eliminated. If minimality is too restrictive, then condition (iv) is eliminated.

8.7 The Distance Spectrum Criterion for Trellis Codes

The *distance spectrum* of a trellis code is defined as the collection of ordered pairs (d_i, M_i), $i = 1, 2, \ldots,$ of all distances d_i, where $d_i < d_{i+1}$, together with the average number (multiplicity), M_i, of paths in the trellis at distance d_i from a given (reference) path in the trellis, under the condition that the two paths diverge from the same node at $t = 0$ and remerge only once at some later time. The average is taken over all paths as reference paths. The pair (d_i, M_i) is called the ith *spectral line*.

For the trellis codes considered in this chapter, the distance d_i represents the squared Euclidean distance between sequences and, by convention, $d_{\text{free}}^2 \triangleq d_1$. It is widely known that, under the assumption of AWGN and high SNR, higher d_{free}^2 implies lower probability of error. At moderate to low SNR, however, this need not be true since the whole distance spectrum influences the upper bound on the probability of error of the code [17], and a code optimized for some specific SNR may not be optimal if the SNR is changed. Nevertheless, it is easily observable that a higher d_{free}^2 code should have better distance spectrum altogether, since codewords at distance d_i from the reference codeword should be mutually separated by at least d_{free}^2. Thus, if d_{free}^2 is smaller, then there is more space available for a larger number of codewords at d_i, potentially increasing M_i. Naturally, counterexamples can be constructed with low d_{free}^2 and good spectrum at d_i, but high d_{free}^2 codes surely cannot be bad in this regard. The same argument may be used for the number M_j of codewords at distance d_j, with regard to distance d_i, $j > i$. Consequently, it is reasonable to assume, and it has been confirmed by all the examples we encountered, that the cumulative distance spectra [18] of two codes, defined as $\sum_{d_i < d_{\text{lim}}} M_i$, almost never intersect as d_{lim} increases. Consequently, the error performance curves of these two codes almost never intersect if all the other parameters of the two codes remain the same [20, 21].

In the next section, we use a simple yet reasonably robust criterion for achieving good codes, called the (squared Euclidean) *distance spectrum* (DS) criterion, defined as follows. In comparing (say) code $\mathcal{C}$ with code $\mathcal{C}'$ with distance spectra (d_i, M_i) and (d_i, M_i'), $i = 1, 2, \ldots$, respectively, the distance spectrum criterion declares that code $\mathcal{C}$ is better than code $\mathcal{C}'$ if there exists a positive integer j such that $M_i = M_i'$ for all $i < j$ and $0 \leq M_j < M_j'$. Note that the code with larger d_{free}^2 is considered better according to this criterion, regardless of the other spectral lines.

8.8 Good Trellis-Matched Codes for the Partial-Response Channels Based on the Distance Spectrum Criterion

Regarding the coded system of Figure 8.1, the best channel codes [13, 20], along with their respective distance spectra (first five spectral lines), are shown in Table 8.1 for the EPR4 channel and in Table 8.2 for the E^{2}PR4 channel. Similar tables for the $1 - D$ channel can be found in [15]. In Figure 8.1, the column $T(D)$ refers to the coset representative (see [6, Equation 8.15 and tables description] for details). The input generator vector, $\mathbf{G}_{\text{in}}(D)$, and $T(D)$ are given in the standard octal notation. For example, $\mathbf{G}_{\text{in}}(D) = 6^4$ [24, 34] and $T(D) = 74$ denote: $\mathbf{G}_{\text{in}}(D) = (110)^4$ [010100, 011100] $= (1+D)^4[D+D^3, D+D^2+D^3]$ and $T(D) = 111100 = 1 + D + D^2 + D^3$. In the $L_{\max}$ column, the maximum identical symbol-run length is listed. Most frequently, this is the zero-run length, but in two cases (shown in Table 8.2), the maximum run-length channel output symbol is nonzero, and is listed in parenthesis following the length of the run. For instance, 4(2) indicates that the run "2222" of length 4 is the longest identical symbol-run. All codes shown in Table 8.1 and Table 8.2 are free of flawed codewords.

Appendix A

In this appendix we show the equivalence in Equation 8.10. First we observe that the coefficients of $P_n(D)$ satisfy $p_i = -p_{n+1-i}$. If we substitute Equation 8.5 into Equation 8.6, and disregard the terms in D^j

TABLE 8.1　Distance Spectra for Channel Codes TTM to the EPR4 Channel

R_c	ν	$\mathbf{G_{in}}$	$T(D)$	L_{max}	Distance spectrum
1/5	1	$6^3[67]$	36	1	(88,0.25)(104,0.25)(120,0.25)(136,0.25)(144,0.125)
1/5	1	$6^3[76]$	42	2	(128,1)(224,0.5)(288,0.5)(320,0.25)(384,0.5)
1/5	2	$6^3[7754]$	42	3	(200,0.25)(216,0.25)(240,0.25)(256,0.75)(264,0.25)
1/5	3	$6^3[55777]$	42	2	(224,0.25)(240,0.5625)(256,0.25)(272,0.25)(288,0.25)
1/5	4	$6^3[5574076]$	42	2	(288,0.375)(304,0.5)(312,0.03125)(320,0.38281)(328,0.09375)
1/5	5	$6^5[63155246]$	42	2	(328,0.15625)(344,0.28125)(360,0.25)(376,0.29688)(392,0.21094)
1/4	1	$6^6[1]$	74	1	(80,1)(144,0.5)(176,0.5)(208,0.25)(240,0.5)
1/4	2	$6^6[26]$	74	1	(104,1)(184,0.5)(192,0.5)(224,0.5)(232,0.5)
1/4	3	$6^6[2134]$	74	2	(168,0.25)(184,0.25)(200,0.5)(216,0.75)(232,0.5)
2/5	2	$6^3[174,514]$	74	4	(48,1.5)(56,0.25)(64,0.75)(72,0.375)(80,1.03125)
2/5	3	$6^3[74,5124]$	42	2	(80,0.75)(88,0.5)(96,1.5)(104,0.5)(112,2)
2/4	2	$6^4[24,34]$	74	3	(40,2)(56,1)(72,1.5)(80,1)(88,3.25)
2/4	3	$6^4[5,436]$	74	8	(48,1.28125)(56,0.63281)(64,1.30469)(72,1.90039)(80,3.67847)
2/4[a]	4	$6^4[202,552]$	74	3	(56,0.875)(64,1.35156)(72,2.18945)(80,3.16211)(88,3.92285)
3/5	3	$6^3[26,15,404]$	74	11	(32,2.0625)(40,2.00391)(48,4.2544)(56,7.86763)(64,12.7282)
3/5	4	$6^3[14,61,4044]$	74	14	(40,2.875)(48,1.75879)(56,5.93607)(64,9.51839)(72,18.3108)

[a] Code found in [20]. All other codes were obtained from [13].

Source: Despotović, M., Šenk, V., and Uchôa-Filho, B. F., *IEEE Trans. Commun.*, vol. 49, no. 7, pp. 1121–1124, July 2001. Copyright © 2001 IEEE. Reproduced with permission.

TABLE 8.2　Distance Spectra for Channel Codes TTM to the E^2PR4 Channel

R_c	ν	$\mathbf{G_{in}}$	$T(D)$	L_{max}	Distance spectrum
1/5	1	$6^5[64]$	42	2	(264,1)(512,0.5)(544,0.5)(760,0.25)(792,0.5)
1/5	2	$6^4[7734]$	42	2	(464,0.125)(480,0.125)(496,0.25)(504,0.125)(512,0.25)
1/5	3	$6^8[7416]$	42	2	(536,0.25)(552,0.25)(616,0.25)(632,0.25)(656,0.0625)
1/5	4	$6^5[5253524]$	42	3	(664,0.0625)(680,0.0625)(688,0.01563)(720,0.01563)(728,0.0625)
1/5	5	$6^5[42116304]$	42	3	(792,0.01563)(800,0.03125)(808,0.125)(824,0.125)(848,0.03516)
2/5[a]	2	$6^5[12,16]$	42	3	(112,2)(184,1)(192,0.5)(216,0.5)(224,1)
2/5[a]	2	$6^4[7,47]$	42	4	(120,0.25)(152,1.5)(160,0.5)(200,0.28125)(224,0.1875)
2/5[a]	3	$6^4[3004,74]$	42	3	(152,0.375)(200,0.375)(216,0.0625)(224,1.375)(232,0.51563)
2/5[a]	3	$6^4[37,7074]$	42	11	(160,0.46875)(192,1.5)(200,0.30469)(232,0.03125)(240,0.66406)
2/5[a]	4	$6^4[74,30601]$	42	3	(216,0.5)(224,2)(232,0.5)(240,1)(256,1)
3/5	3	$6^4[3,24,62]$	42	4	(48,1.5)(64,1.25)(72,0.875)(80,0.34375)(88,1.25781)
3/5	3	$6^4[54,42,21]$	42	8	(72,1)(80,1.25)(88,1.5625)(96,0.35156)(104,0.76563)
3/5	4	$6^5[2,3,43]$	42	4	(48,1.5)(64,0.125)(72,0.85156)(80,0.49414)(88,0.70313)
3/5	4	$6^4[7,26,201]$	42	4(2)	(64,0.0625)(72,0.28125)(80,1.28125)(88,1.41016)(96,1.04297)
3/5	4	$6^4[3,43,4604]$	42	11	(80,1.6875)(88,0.6875)(96,0.625)(104,0.57813)(112,1.55469)
4/5	4	$6^4[2,1,02,05]$	42	4(-2)	(40,3.0625)(48,2.75)(56,1.75)(64,1.47656)(72,2.51758)

[a] Code found in [20]. All other codes were obtained from [13].

Source: Despotović, M., Šenk, V., and Uchôa-Filho, B. F., *IEEE Trans. Commun.*, vol. 49, no. 7, pp. 1121–1124, July 2001. Copyright © 2001 IEEE. Reproduced with permission.

where $j \geq$ length of $W(D)$, the channel output sequence becomes:

$$Y(D) = 2P_n(D)R(D) - \underbrace{\mathbf{1}(D)P_n(D)}_{a} + \underbrace{\sum_{j=1}^{n+1}\left(\sum_{i=j}^{n+1}(-1)p_i\right)D^{j-1}}_{b}$$

Expanding a and b we have:

$$a = p_0 + (p_1 + p_0)D + (p_2 + p_1 + p_0)D^2 + \cdots + (p_{n+1} + \cdots + p_1 + p_0)D^{n+1}$$
$$+ (p_{n+1} + \cdots + p_1 + p_0)D^{n+2} + (p_{n+1} + \cdots + p_1 + p_0)D^{n+3} + \cdots$$

and

$$b = (-p_1 - p_2 - \cdots - p_{n+1}) + (-p_2 - p_3 - \cdots - p_{n+1})D + \cdots + (-p_{n+1})D^n$$

Thus,

$$-a + b = (-p_{n+1} - \cdots - p_1 - p_0) \sum_i D^i = 0$$

since $p_i = -p_{n+1-i}$. Then the channel output sequence becomes:

$$Y(D) = 2P_n(D)R(D) = 2 \underbrace{P_n(D)(1 \oplus D)^{-n-1}}_{\equiv 1 (\mathrm{mod}\, 2)} W(D)$$

and we find that

$$W(D) \equiv \frac{Y(D)}{2} \ (\mathrm{mod}\ 2)$$

Appendix B

In this appendix we show the equivalence $|(X(D) - X'(D))P_n(D)| \equiv |X(D) - X'(D)|(1 \oplus D)^{n+1}$, used in the proof of Lemma 8.1. Note that $x_i - x'_i \in \{0, \pm 2\}$ and $p_i \in \mathbb{Z}$. Then the division by 2 in the function defined in Equation 8.9 can be carried out in the polynomial $(X(D) - X'(D))$. Thus, we can have

$$|(X(D) - X'(D))P_n(D)| \equiv \frac{(X(D) - X'(D))P_n(D)}{2} \ (\mathrm{mod}\ 2)$$

$$\equiv \frac{(X(D) - X'(D))}{2} P_n(D) \ (\mathrm{mod}\ 2)$$

$$\equiv |(X(D) - X'(D))|[P_n(D)]_{\mathrm{mod}\, 2}$$

$$\equiv |(X(D) - X'(D))|(1 \oplus D)^{n+1}$$

References

[1] J. K. Wolf and G. Ungerboeck, Trellis coding for partial-response channels, *IEEE Trans. Commun.*, vol. COM-34, pp. 765–773, August 1986.

[2] H. Kobayashi and D. T. Tang, Application of partial-response channel coding to magnetic recording systems, *IBM J. Des. Dev.*, vol. 14, pp. 368–375, July 1970.

[3] H. K. Thapar and A. M. Patel, A class of partial response systems for increasing storage density in magnetic recording, *IEEE Trans. Magn.*, vol. MAG-25, pp. 3666–3668, September 1987.

[4] E. Zehavi and J. K. Wolf, On saving decoder states for some trellis codes and partial response channels, *IEEE Trans. Commun.*, vol. COM-36, pp. 222–224, February 1988.

[5] E. Zehavi and J. K. Wolf, On the performance evaluation of trellis codes, *IEEE Trans. Inform. Theory*, vol. IT-33, pp. 196–202, March 1988.

[6] P. H. Siegel and J. K. Wolf, Modulation and coding for information storage, *IEEE Commun Mag.*, vol. 29, pp. 68–86, December 1991.

[7] T. A. Lee and C. Heegard, An inversion technique for the design of binary convolutional codes for the $1 - D^N$ Channel, *Proceedings IEEE regional meeting*, Johns Hopkins University, Baltimore, MD, February 1985.

[8] G. D. Forney, Jr., Convolutional codes I: algebraic structure, *IEEE Trans. Inform. Theory*, vol. IT-16, pp. 720–738, November 1970.

[9] G. D. Forney, Jr., Structural analysis of convolutional codes via dual codes, *IEEE Trans. Inform. Theory*, vol. IT-19, pp. 512–518, July 1973.

[10] S. Lin and D. J. Costello, Jr., *Error Control Coding: Fundamentals and Applications*, Prentice Hall, Englewood Cliffs, NJ, 1983.

[11] J. G. Proakis and M. Salehi, *Communication Systems Engineering*, Prentice Hall, Englewood Cliffs, NJ, 1994.

[12] B. F. Uchôa Filho and M. A. Herro, Convolutional codes for the high density $(1-D)(1+D)^n$ magnetic recording channel, *Proceedings IEEE Int. Symp. Inform. Theory*, pp. 211, Trondheim, Norway, June-July 1994.

[13] B. F. Uchôa Filho and M. A. Herro, Good convolutional codes for the precoded $(1 - D)(1 + D)^n$ partial response channels, *IEEE Trans. Inform. Theory*, vol. 43, no. 2, pp. 441–453, March 1997.

[14] K. J. Hole, Punctured convolutional codes for the $1 - D$ partial-response channel, *IEEE Trans. Inform. Theory*, vol. I-37, pp. 808–817, May 1991.

[15] K. J. Hole and Ø. Ytrehus, Improved coding techniques for precoded partial-response channels, *IEEE Trans. Inform. Theory*, vol. I-40, pp. 482–493, March 1994.

[16] M. Siala and G. K. Kaleh, Block and trellis codes for the binary $(1 - D)$ partial response channel with simple maximum likelihood decoders, *IEEE Trans. on Commun.*, vol. 44, no. 12, pp. 1613–1615, December 1996.

[17] M. Rouanne and D. J. Costello, Jr., An algorithm for computing the distance spectrum of trellis codes," *IEEE J. Select. Areas Commun.*, vol. SAC-7, pp. 929–940, August 1989.

[18] M. Despotović and V. Šenk, Distance spectrum of channels trellis codes on precoded partial-response $1 - D$ channel, FACTA UNIVERSITATIS (NIS), series: Electronics and Energetics vol. 1 (1995), pp. 57–72, http://factaee.elfak.ni.ac.yu/.

[19] M. Despotović, V. Šenk, and B. F. Uchôa-Filho, Convolutional codes with optimized distance spectrum for the EPR4 and EEPR4 channels, *Proceedings the 1999 IEEE Int. Conference Commun.*, vol. 3, pp. 1658–1662, June 1999.

[20] M. Despotović, V. Šenk, and B. F. Uchôa-Filho, Distance spectra of convolutional codes over partial-response channels, *IEEE Trans. Commun.*, vol. 49, no. 7, pp. 1121–1124, July 2001.

9

Capacity-Approaching Codes for Partial Response Channels

Nedeljko Varnica
Harvard University
Boston, MA

Xiao Ma
City University of Hong Kong
Kowloon, Hong Kong

Aleksandar Kavčić
Harvard University
Boston, MA

9.1 Introduction

A partial response (PR) channel is an intersymbol interference (ISI) channel with a binary input alphabet and additive white Gaussian noise (AWGN). The capacity of a PR channel is strictly greater than its i.u.d. capacity, which is defined as the information rate when the PR channel inputs are independent and uniformly distributed (i.u.d.) random variables.

The computations of the capacity C and the i.u.d. capacity $C_{\text{i.u.d.}}$ of a PR channel have been subjects of research for some time [1–3]. The information rates and the capacities of finite-state machines and Markov chains are closely related to the capacities of PR channels and have been studied in [4–7]. For a summary of capacity computation methods, see [8].

Recently, a Monte Carlo method for computing the information rate of a finite-state machine channel whose inputs are Markov processes was proposed independently by Arnold and Loeliger [9] and Pfister et al. [10]. This method can be used to compute $C_{\text{i.u.d.}}$, which is a lower bound on C. In [11], Kavčić proposed a Markov process optimization algorithm to tighten the lower bounds, and in [12] Vontobel and Arnold used this algorithm to compute tight upper bounds. The feedback capacity of PR channels computed in [13] is also an upper bound on C, and is in some cases tighter than the Vontobel-Arnold bound [12]. These methods (summarized in [8]) give bounds that are so tight that, for all practical purposes, we consider the capacities of PR channels to be known.

Error-correction and modulation codes for PR channels are numerous. Here, we only consider error-correcting codes that are aimed at achieving the capacities of PR channels. Before the invention of turbo codes [14] the codes for PR channels were dominated by trellis codes [15–17]. Matched spectral null (MSN) trellis codes whose binary codeword sequences have nulls in specific frequency locations matching the spectral nulls of the channels were proposed in [18].

The advent of turbo [14] and low-density parity-check (LDPC) [19, 20] codes has sparked research in iteratively decodable codes for PR channels. Typically the BCJR algorithm [21] is used to perform the maximum a posteriori symbol detection on the channel, while iterative decoding algorithms are used to provide constituent decoders in the iterative decoding schemes. Early iteratively decodable codes for PR channels were parallel and serially concatenated turbo codes [22–25]. Subsequent schemes concentrated on simplifying turbo equalization [26, 27]. The LDPC codes and turbo product codes over PR channels have been investigated in [28–30]. Luby et al. [31] introduced irregular LDPC codes and proposed a method to analyze the asymptotic performance of these codes over erasure channels. The analysis was soon thereafter expanded to various other memoryless channels by Richardson and Urbanke [32]. They used the *density evolution* analysis tool for the computation of noise thresholds [32], and the optimization of code degree sequences [33] that produced the capacity-achieving codes for memoryless channels [34]. Kavčić et al. [35] modified the density evolution to fit the PR channels, computed noise thresholds of *random* LDPC codes over PR channels and proved that these codes can achieve rates that approach the i.u.d. capacity, but not higher. This was supported by evidence of code constructions that approach $C_{\text{i.u.d.}}$ very closely [36].

Here, we construct codes that achieve rates higher than $C_{\text{i.u.d.}}$. Our approach is to (1) compute a (maximized) channel information rate achievable by an input Markov process of reasonably low complexity, (2) construct an inner trellis code that mimics this Markov process and therefore achieves a comparable information rate over the channel of interest, and (3) construct an outer LDPC code that ensures that the information rate of the inner code is approached. The strategy is therefore to compute an achievable information rate, and then construct a concatenation of two codes whose overall code rate matches the computed information rate. Hence the name *matched information rate* (MIR) code.

This chapter is organized into six sections. Section 9.2 and Section 9.3, are tutorial-like expositions of the channel capacity and trellis code terminologies. In Section 9.4 and Section 9.5, we design inner MIR trellis codes and outer LDPC codes, respectively. In Section 9.6 the success of the methodology is demonstrated with two examples of constructed codes (one for a channel with a spectral null, and the other for a channel without a spectral null).

9.2 The Channel Model and Capacity Definitions

9.2.1 The Channel Model

Let $t \in \mathbb{Z}$ denote a discrete-time variable. Denote the channel input as a sequence of random variables X_t, and the channel output as a sequence of random variables Y_t. The channel law is

$$Y_t = \sum_{k=0}^{J} h_k X_{t-k} + W_t \tag{9.1}$$

where $J > 0$ is the ISI length and $h_0, h_1, \ldots, h_J$ capture the memory in the channel. Often the channel memory is represented by a partial response polynomial $h(D) = \sum_{k=0}^{J} h_k D^k$. For example, the partial response polynomial of the *dicode* channel ($Y_t = X_t - X_{t-1} + W_t$) is $h(D) = 1 - D$. We assume that the noise process W_t is white and Gaussian, with mean 0 and variance σ^2. We also assume that the realizations x_t of the random variables X_t take values in a binary alphabet $\mathcal{X} = \{-1, 1\}$ which is of practical relevance in many binary recording and communications channels. We will interchangeably use the input alphabets $\mathcal{X} = \{-1, 1\}$ and $\mathcal{A} = \{0, 1\}$. Thereby, we will use the standard conversion $X_t = 1 - 2A_t$, where A_t is a discrete-time random process with binary realizations $a_t \in \mathcal{A} = \{0, 1\}$.

The signal-to-noise ratio (in dB) of a PR channel is defined as

$$\text{SNR} = 10 \log_{10} \frac{\sum_{k=0}^{J} h_k^2}{\sigma^2} \tag{9.2}$$

The most comprehensive known model to which our methods apply is the finite-state model with autoregressive observations described in [37, 38]. However, since all methods for the model in Equation 9.1 canonically extend to the channel model in [37, 38], without loss of generality, it suffices to consider the model in Equation 9.1 with binary inputs and AWGN.

9.2.2 The Channel Capacity

The channel capacity of a PR channel is defined as [39]

$$C = \lim_{N \to \infty} \frac{1}{N} \left[\max_{P_{X_1^N}(x_1^N)} I\left(X_1^N; Y_1^N \mid X_{-J+1}^0 = x_{-J+1}^0\right) \right] = \lim_{N \to \infty} \frac{1}{N} \left[\max_{P_{X_1^N}(x_1^N)} I\left(X_1^N; Y_1^N\right) \right] \tag{9.3}$$

where $I(X_1^N; Y_1^N \mid X_{-J+1}^0 = x_{-J+1}^0)$ is the mutual information rate between the sequence of channel inputs $X_1^N = (X_1, X_2, \ldots, X_N)$ and the sequence of channel outputs $Y_1^N = (Y_1, Y_2, \ldots, Y_N)$, given that the symbols transmitted over the channel at time $t = -J + 1$ through $t = 0$ are x_{-J+1}^0. Following [1], we can drop the conditioning on the symbols $X_{-J+1}^0 = x_{-J+1}^0$ because the channel is indecomposable [39].

Another capacity of interest is the i.u.d. capacity, defined as the information rate when the input sequence X_1^N consists of i.u.d. random variables X_t

$$C_{\text{i.u.d.}} = \lim_{N \to \infty} \frac{1}{N} \cdot I\left(X_1^N; Y_1^N\right) \Big|_{P_{X_1^N}(x_1^N) = 2^{-N}} \tag{9.4}$$

9.2.3 Trellis Representations

We introduce a general notion of a channel trellis. At time t, the channel trellis has 2^L states (where $L \geq J$), indexed by $\mathcal{S} = \{0, 1, \ldots, 2^L - 1\}$. Exactly 2^n branches emanate from each trellis state, where $n \geq 1$. A branch at time t is determined by a 4-tuple $b_t = (s_{t-1}, u_t, v_t, s_t)$, where $s_{t-1} \in \mathcal{S}$ and $s_t \in \mathcal{S}$ are the starting state and the ending state, respectively. The symbol u_t is an n-tuple of input symbols, that is, $u_t \in \mathcal{A}^n = \{0, 1\}^n$ or $u_t \in \mathcal{X}^n = \{-1, 1\}^n$. The symbol $v_t \in \mathbb{R}^n$ is an n-tuple of real-valued *noiseless* channel outputs. Every branch is uniquely determined by its starting state s_{t-1} and its input n-tuple u_t. That is, v_t and s_t are (deterministic) functions of s_{t-1} and u_t. The random 4-tuple $B_t = (S_{t-1}, U_t, V_t, S_t)$ stands for a random branch, whose realizations are the 4-tuples b_t.

To characterize a time-invariant channel trellis, we need only specify one trellis section. We distinguish 3 trellis types determined by the properties of their trellis sections:

- A minimal channel trellis is a channel trellis whose trellis section has the smallest possible number of states (i.e., 2^J states) and corresponds to a single input and a single output (i.e., $n = 1$).
- An original channel trellis is a channel trellis whose trellis section has $2^L \geq 2^J$ states and corresponds to a single input and a single output (i.e., $n = 1$).
- An extended channel trellis is a channel trellis whose trellis section corresponds to n channel inputs and n channel outputs, where $n > 1$. The number of states is $2^L \geq 2^J$.

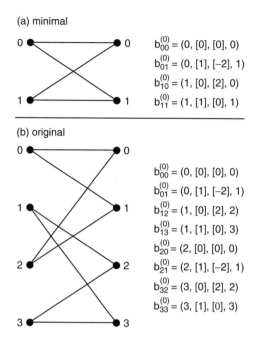

(a) minimal

0 •——————————• 0

1 •——————————• 1

$b_{00}^{(0)} = (0, [0], [0], 0)$

$b_{01}^{(0)} = (0, [1], [-2], 1)$

$b_{10}^{(0)} = (1, [0], [2], 0)$

$b_{11}^{(0)} = (1, [1], [0], 1)$

(b) original

0 •——————————• 0

1 •——————————• 1

2 •——————————• 2

3 •——————————• 3

$b_{00}^{(0)} = (0, [0], [0], 0)$

$b_{01}^{(0)} = (0, [1], [-2], 1)$

$b_{12}^{(0)} = (1, [0], [2], 2)$

$b_{13}^{(0)} = (1, [1], [0], 3)$

$b_{20}^{(0)} = (2, [0], [0], 0)$

$b_{21}^{(0)} = (2, [1], [-2], 1)$

$b_{32}^{(0)} = (3, [0], [2], 2)$

$b_{33}^{(0)} = (3, [1], [0], 3)$

FIGURE 9.1   (a) The minimal trellis for $1 - D$ channel. (b) An original trellis for $1 - D$ channel with $L = 2$ and $n = 1$.

The minimal trellis for the dicode channel with $2^J = 2$ states and $n = 1$ input (and output) symbol per section is shown in Figure 9.1(a). An original trellis for the dicode channel with $2^L = 4$ states and $n = 1$ input (and output) symbol per section is shown in Figure 9.1(b). The 3rd order extension of the minimal trellis is shown in Figure 9.2(a). This trellis is obtained by concatenating $n = 3$ sections of the minimal trellis in Figure 9.1(a). The 2nd order extension of the original trellis from Figure 9.1(b) is shown in Figure 9.2(b).

9.2.4   The Markov Channel Capacity

Let $i \in S$ and $j \in S$ denote the starting and the ending state of a branch in the trellis section, respectively. If the trellis is an original channel trellis, then there exists at most one branch connecting state i to state j. If, however, the trellis is an extended channel trellis, then there may be several branches connecting a pair of states i and j. Let $\mathcal{L}$ be the number of distinct branches that connect states i and j. These branches are denoted by $b_{ij}^{(\ell)} = (i, u_{ij}^{(\ell)}, v_{ij}^{(\ell)}, j)$, where $0 \leq \ell \leq \mathcal{L} - 1$, $u_{ij}^{(\ell)} \in \mathcal{A}^n$ is the binary channel input n-tuple and $v_{ij}^{(\ell)} \in \mathbb{R}^n$ is the real-valued noiseless channel output n-tuple.

Denote by τ the set of all triples (i, ℓ, j) for which a branch $b_{ij}^{(\ell)}$ exists in the trellis section. Since the trellis section uniquely determines the trellis, we say that τ represents a channel trellis. We call a branch $b_{ij}^{(\ell)}$ valid if $(i, \ell, j) \in \tau$. By assigning a probability $P_{ij}^{(\ell)} = \Pr(B_t = b_{ij}^{(\ell)} \mid S_{t-1} = i) = \Pr(S_t = j, B_t = b_{ij}^{(\ell)} \mid S_{t-1} = i)$ to every branch $b_{ij}^{(\ell)}$ of the channel trellis section we define an input Markov process on τ. Denote the probability of state i by $\mu_i = \Pr(S_t = i)$. The Markov process is stationary if for every $j \in S$,

$$\mu_j = \sum_{i,\ell:(i,\ell,j)\in\tau} \mu_i P_{ij}^{(\ell)} \qquad (9.5)$$

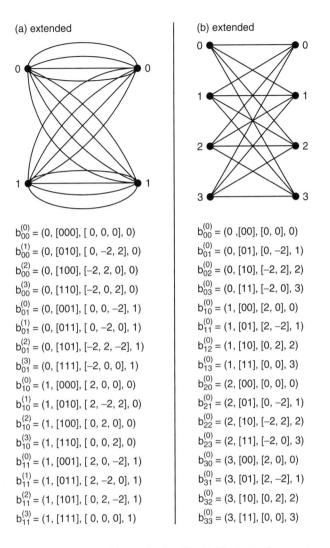

(a) extended

$b_{00}^{(0)} = (0, [000], [0, 0, 0], 0)$
$b_{00}^{(1)} = (0, [010], [0, -2, 2], 0)$
$b_{00}^{(2)} = (0, [100], [-2, 2, 0], 0)$
$b_{00}^{(3)} = (0, [110], [-2, 0, 2], 0)$
$b_{01}^{(0)} = (0, [001], [0, 0, -2], 1)$
$b_{01}^{(1)} = (0, [011], [0, -2, 0], 1)$
$b_{01}^{(2)} = (0, [101], [-2, 2, -2], 1)$
$b_{01}^{(3)} = (0, [111], [-2, 0, 0], 1)$
$b_{10}^{(0)} = (1, [000], [2, 0, 0], 0)$
$b_{10}^{(1)} = (1, [010], [2, -2, 2], 0)$
$b_{10}^{(2)} = (1, [100], [0, 2, 0], 0)$
$b_{10}^{(3)} = (1, [110], [0, 0, 2], 0)$
$b_{11}^{(0)} = (1, [001], [2, 0, -2], 1)$
$b_{11}^{(1)} = (1, [011], [2, -2, 0], 1)$
$b_{11}^{(2)} = (1, [101], [0, 2, -2], 1)$
$b_{11}^{(3)} = (1, [111], [0, 0, 0], 1)$

(b) extended

$b_{00}^{(0)} = (0 ,[00], [0, 0], 0)$
$b_{01}^{(0)} = (0, [01], [0, -2], 1)$
$b_{02}^{(0)} = (0, [10], [-2, 2], 2)$
$b_{03}^{(0)} = (0, [11], [-2, 0], 3)$
$b_{10}^{(0)} = (1, [00], [2, 0], 0)$
$b_{11}^{(0)} = (1, [01], [2, -2], 1)$
$b_{12}^{(0)} = (1, [10], [0, 2], 2)$
$b_{13}^{(0)} = (1, [11], [0, 0], 3)$
$b_{20}^{(0)} = (2, [00], [0, 0], 0)$
$b_{21}^{(0)} = (2, [01], [0, -2], 1)$
$b_{22}^{(0)} = (2, [10], [-2, 2], 2)$
$b_{23}^{(0)} = (2, [11], [-2, 0], 3)$
$b_{30}^{(0)} = (3, [00], [2, 0], 0)$
$b_{31}^{(0)} = (3, [01], [2, -2], 1)$
$b_{32}^{(0)} = (3, [10], [0, 2], 2)$
$b_{33}^{(0)} = (3, [11], [0, 0], 3)$

FIGURE 9.2   (a) The 3rd order extension of the minimal trellis. (b) The 2nd order extension of the original trellis in 9.1(b).

Denote by $\mathcal{P}(\tau)$ the collection of Markov transition probabilities $P_{ij}^{(\ell)}$ for all branches $(i, \ell, j) \in \tau$. For this Markov source, the Markov channel information rate is defined as

$$\mathcal{I}_{\mathcal{P}(\tau)} = \lim_{N \to \infty} \frac{1}{nN} I\left(X_1^{nN}; Y_1^{nN}\right) = \lim_{N \to \infty} \frac{1}{nN} I\left(B_1^{N}; Y_1^{nN}\right) \tag{9.6}$$

where the branch sequence B_1^N is a random sequence defined by the probabilities $\mathcal{P}(\tau)$.
   We define the Markov channel capacity for a trellis τ as

$$C_\tau = \max_{\mathcal{P}(\tau)} \mathcal{I}_{\mathcal{P}(\tau)} \tag{9.7}$$

where the maximization in Equation 9.7 is conducted over all Markov processes $\mathcal{P}(\tau)$ defined over τ. Since the set of stationary Markov processes is a subset of the set of all stationary discrete-time processes, it is clear that $C_\tau \leq C$. Let $\tau_o(L)$ be an *original* channel trellis with $2^L \geq 2^J$. Denote the Markov channel

TABLE 9.1   **Algorithm 1** Iterative Optimization of Markov Chain Transition Probabilities

---

Initialization Pick a channel trellis τ and an arbitrary probability mass function $P_{ij}^{(\ell)}$ defined over τ that satisfies: (1) $0 \le P_{ij}^{(\ell)} \le 1$ if $(i, \ell, j) \in \tau$; otherwise $P_{ij}^{(\ell)} = 0$ and (2) $\sum_{j, \ell:(i,\ell,j)\in\tau} P_{ij}^{(\ell)} = 1$ for any i.

Repeat until convergence

1. For N large (say, $N > 10^6$), generate a realization of a sequence of N trellis branches b_1^N according to the Markov probabilities $P_{ij}^{(\ell)}$. Determine the channel input sequence x_1^{nN} that corresponds to the branch sequence b_1^N. Pass x_1^{nN} through the PR channel to get a realization of the channel output y_1^{nN}.

2. Run the forward-backward sum-product (BCJR) algorithm and for all $1 \le t \le N$ compute the a posteriori probabilities $R_{ij}^{(\ell)}(t, y_1^{nN}) = \Pr(B_t = b_{ij}^{(\ell)} \mid Y_1^{nN} = y_1^{nN})$ and $R_i(t, y_1^{nN}) = \Pr(S_t = i \mid Y_1^{nN} = y_1^{nN})$.

3. Compute the estimate of the expectation term $\hat{T}_{ij}^{(\ell)} = \dfrac{1}{N} \sum_{t=1}^{N} \log_2 \dfrac{\dfrac{R_{ij}^{(\ell)}(t, y_1^{nN})}{\mu_i P_{ij}^{(\ell)}}}{\dfrac{R_i(t, y_1^{nN})}{\mu_i}}$ $\dfrac{R_{ij}^{(\ell)}(t, y_1^{nN})}{R_i(t, y_1^{nN})}$.

4. Compute the estimate of the noisy adjacency matrix, with the entries

$$\hat{A}_{ij} = \begin{cases} \sum_{\ell:(i,\ell,j)\in\tau} 2^{\hat{T}_{ij}^{(\ell)}} & \text{if states } i \text{ and } j \text{ are connected by at least one branch} \\ 0 & \text{otherwise} \end{cases},$$

and find its maximal eigenvalue $\hat{W}_{\max}$ and the corresponding right eigenvector $[\hat{e}_1, \hat{e}_2, \ldots, \hat{e}_M]^{\mathsf{T}}$.

5. For $(i, \ell, j) \in \tau$, compute the new transition probabilities as $P_{ij}^{(\ell)} = \dfrac{\hat{e}_j}{\hat{e}_i} \cdot \dfrac{2^{\hat{T}_{ij}^{(\ell)}}}{\hat{W}_{\max}}$ and go back to 1.

Terminate the algorithm and set $\hat{C}_\tau = \frac{1}{n} \log_2 \hat{W}_{\max}$. The input distribution $\mathcal{P}(\tau)$ that achieves $\hat{C}_\tau$ is given by the collection of probabilities $P_{ij}^{(\ell)}$.

---

capacity defined on this trellis by $C_{\tau_o(L)}$, or simply C_L. Then an alternative expression for the channel capacity is

$$C = \lim_{L\to\infty} C_{\tau_o(L)} = \lim_{L\to\infty} C_L \tag{9.8}$$

9.2.5   Computing the Markov Channel Capacity

Table 9.1 gives a method to estimate the capacity C_τ. Since the method in Table 9.1 is a Monte Carlo algorithm, we denote its capacity estimate by $\hat{C}_\tau$.

In Figure 9.3, we show the Markov rates $\hat{C}_\tau$ computed by the algorithm in Table 9.1 for the dicode channel trellises given in Figure 9.1 and Figure 9.2. Also shown in Figure 9.3 are the i.u.d. capacity $C_{\text{i.u.d.}}$, the *numerical capacity* $\hat{C}_6$ (which is the tightest known lower bound on the channel capacity C) and the minimum of all known upper bounds on C (the water-filling bound [39, 40], the feedback capacity [13] and the Vontobel-Arnold bound [12]).

9.3   Trellis Codes, Superchannels and Their Information Rates

We consider *binary* time-invariant trellis codes that map k input bits to n output bits. Each trellis section has N_S states, indexed by $\{0, 1, \ldots, N_S - 1\}$. A branch of a trellis code is described by a 4-tuple $b_t = (s_{t-1}, u_t, v_t, s_t)$, where $s_{t-1} \in \{0, 1, \ldots, N_S - 1\}$ is the starting state, $s_t \in \{0, 1, \ldots, N_S - 1\}$ is the ending state, $u_t \in \mathcal{A}^k$ is the input k-tuple, and $v_t \in \mathcal{A}^n$ is the output n-tuple. The code rate is $r = k/n$. Figure 9.4 shows an example of a Wolf-Ungerboeck (W-U) trellis code [15] with rate 2/3.

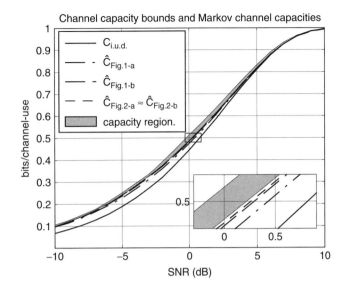

FIGURE 9.3 Markov Capacities for the trellises in Figure 9.1(a,b) and 9.2(a,b). The region between the tightest known lower and upper bound is shaded.

We refer to a superchannel as a concatenation of a trellis code and a PR channel and describe it by a joint code/channel trellis. The number of states in a superchannel trellis may be greater than the number of states in the constituent trellis code. Figure 9.5 gives an example of the superchannel trellis obtained by concatenating the trellis code in Figure 9.4 with the dicode channel in Figure 9.1(a). Note that exactly 2^k branches emanate from each state of the superchannel trellis, that is, one branch for every binary input k-tuple.

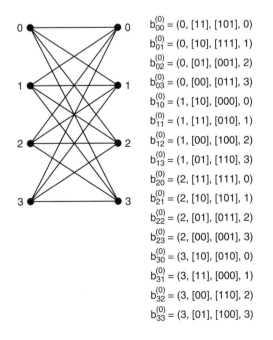

$$b_{00}^{(0)} = (0, [11], [101], 0)$$
$$b_{01}^{(0)} = (0, [10], [111], 1)$$
$$b_{02}^{(0)} = (0, [01], [001], 2)$$
$$b_{03}^{(0)} = (0, [00], [011], 3)$$
$$b_{10}^{(0)} = (1, [10], [000], 0)$$
$$b_{11}^{(0)} = (1, [11], [010], 1)$$
$$b_{12}^{(0)} = (1, [00], [100], 2)$$
$$b_{13}^{(0)} = (1, [01], [110], 3)$$
$$b_{20}^{(0)} = (2, [11], [111], 0)$$
$$b_{21}^{(0)} = (2, [10], [101], 1)$$
$$b_{22}^{(0)} = (2, [01], [011], 2)$$
$$b_{23}^{(0)} = (2, [00], [001], 3)$$
$$b_{30}^{(0)} = (3, [10], [010], 0)$$
$$b_{31}^{(0)} = (3, [11], [000], 1)$$
$$b_{32}^{(0)} = (3, [00], [110], 2)$$
$$b_{33}^{(0)} = (3, [01], [100], 3)$$

FIGURE 9.4 Wolf-Ungerboeck trellis code of rate $r = 2/3$.

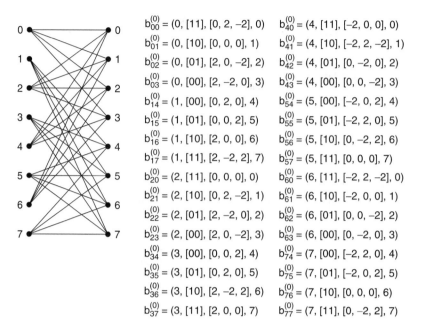

$b_{00}^{(0)} = (0, [11], [0, 2, -2], 0)$ $b_{40}^{(0)} = (4, [11], [-2, 0, 0], 0)$

$b_{01}^{(0)} = (0, [10], [0, 0, 0], 1)$ $b_{41}^{(0)} = (4, [10], [-2, 2, -2], 1)$

$b_{02}^{(0)} = (0, [01], [2, 0, -2], 2)$ $b_{42}^{(0)} = (4, [01], [0, -2, 0], 2)$

$b_{03}^{(0)} = (0, [00], [2, -2, 0], 3)$ $b_{43}^{(0)} = (4, [00], [0, 0, -2], 3)$

$b_{14}^{(0)} = (1, [00], [0, 2, 0], 4)$ $b_{54}^{(0)} = (5, [00], [-2, 0, 2], 4)$

$b_{15}^{(0)} = (1, [01], [0, 0, 2], 5)$ $b_{55}^{(0)} = (5, [01], [-2, 2, 0], 5)$

$b_{16}^{(0)} = (1, [10], [2, 0, 0], 6)$ $b_{56}^{(0)} = (5, [10], [0, -2, 2], 6)$

$b_{17}^{(0)} = (1, [11], [2, -2, 2], 7)$ $b_{57}^{(0)} = (5, [11], [0, 0, 0], 7)$

$b_{20}^{(0)} = (2, [11], [0, 0, 0], 0)$ $b_{60}^{(0)} = (6, [11], [-2, 2, -2], 0)$

$b_{21}^{(0)} = (2, [10], [0, 2, -2], 1)$ $b_{61}^{(0)} = (6, [10], [-2, 0, 0], 1)$

$b_{22}^{(0)} = (2, [01], [2, -2, 0], 2)$ $b_{62}^{(0)} = (6, [01], [0, 0, -2], 2)$

$b_{23}^{(0)} = (2, [00], [2, 0, -2], 3)$ $b_{63}^{(0)} = (6, [00], [0, -2, 0], 3)$

$b_{34}^{(0)} = (3, [00], [0, 0, 2], 4)$ $b_{74}^{(0)} = (7, [00], [-2, 2, 0], 4)$

$b_{35}^{(0)} = (3, [01], [0, 2, 0], 5)$ $b_{75}^{(0)} = (7, [01], [-2, 0, 2], 5)$

$b_{36}^{(0)} = (3, [10], [2, -2, 2], 6)$ $b_{76}^{(0)} = (7, [10], [0, 0, 0], 6)$

$b_{37}^{(0)} = (3, [11], [2, 0, 0], 7)$ $b_{77}^{(0)} = (7, [11], [0, -2, 2], 7)$

FIGURE 9.5   Superchannel obtained by concatenating the trellis code in Figure 9.4 and the dicode channel in Figure 9.1(a).

We assume that the trellis code inputs are i.u.d. symbols. This means that the conditional probability $P_{ij}^{(\ell)} = \Pr(B_t = b_{ij}^{(\ell)} \mid S_t = i)$ of each superchannel trellis branch equals 2^{-k}. Under this i.u.d. assumption, we define the *superchannel information rate* as the information rate between the superchannel input sequence U_1^N and the superchannel output sequence Y_1^{nN}

$$\mathcal{I}_S = \lim_{N \to \infty} \frac{1}{nN} I\left(U_1^N; Y_1^{nN}\right) \Bigg|_{P_{U_1^N}\left(u_1^N\right) = 2^{-kN}} \tag{9.9}$$

9.3.1 Coding Theorems for Superchannels

Theorem 9.1   *[achievability of $\mathcal{I}_S$ with a linear code] The i.u.d. superchannel information rate $\mathcal{I}_S$ in Equation (9.9) can be achieved by an outer linear (coset) code.*

Proof 9.1   The proof is given in [41].   □

Theorem 9.1 suggests that we could use a concatenation scheme to surpass the i.u.d. capacity $C_{i.u.d.}$ of the PR channel if we construct a superchannel whose i.u.d. rate is $\mathcal{I}_S > C_{i.u.d.}$. Consider the block diagram shown in Figure 9.6. The binary vector D_1^M first enters a linear encoder represented by an

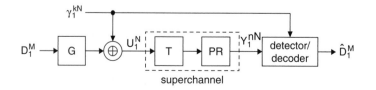

FIGURE 9.6   A concatenation scheme for partial response channels.

$M \times N$ generator matrix G. Before the codeword enters the trellis encoder 'T', a coset-defining binary vector γ_1^N is modulo-2-added on. Different choices of the vector γ_1^N may cause different error rates since the channel has memory. Although we cannot claim that the achievable rate of the best linear (coset) code is upper-bounded by $\mathcal{I}_S$, we have the following "random converse" theorem.

Theorem 9.2 (Random Code Converse) *If $R > \mathcal{I}_S$, there exists $\delta > 0$ such that for any given $M \times N$ generator matrix G (of a linear code) with $M/N > R$ and any given detection/decoding method, the average error rate over all the coset codes is greater than δ.*

Proof 9.2 The proof is given in [41]. □

Theorem 9.2 exposes the difficulties in finding a linear (coset) code that surpasses the superchannel i.u.d. rate $\mathcal{I}_S$. Namely, a randomly chosen linear (coset) code will not achieve a rate higher than $\mathcal{I}_S$. Theorem 9.2 also reveals that without the inner code, the average rate achievable by a linear (coset) code is upper-bounded by the i.u.d. channel capacity $C_{\text{i.u.d.}}$. To surpass $C_{\text{i.u.d.}}$, we separate the complex code construction problem into two smaller problems: designing an inner trellis code, and an outer linear (coset) code. We first construct an inner trellis code whose superchannel i.u.d. rate $\mathcal{I}_S$ satisfies $C \approx \mathcal{I}_S > C_{\text{i.u.d.}}$.

9.4 Matched Information Rate (MIR) Trellis Codes

We design a trellis code for a PR channel, such that the i.u.d. rate $\mathcal{I}_S$ of the resulting superchannel is as close as possible to the (numerical) channel capacity. An interesting feature of our design is that the trellis code is constructed for a specific target code rate r. Our general strategy is to first choose an extended channel trellis τ, and use the algorithm in Table 9.1 to find the optimal Markov probabilities for this trellis at the code rate r. We then construct a superchannel to mimic the optimal Markov process on the trellis τ.

We next specify the design rules to construct MIR superchannels. These rules are derived to formalize the design methodology. We adopt these rules because they satisfy our intuition and deliver good codes. We make no claim regarding their optimality.

9.4.1 Choosing the Extended Channel Trellis and the Superchannel Code Rate

We construct a superchannel trellis with n output symbols per every k binary input symbols. Our first task is to pick k and n. Let r be the target code rate. Pick an integer $n > 0$ and an nth order extended channel trellis τ. For this trellis, run the algorithm in Table 9.1. Denote by $P_{ij}^{(\ell)}$ the optimized probabilities of the trellis τ for which $\hat{C}_\tau = r$.

Rule 1: The rate r_{in} of the inner trellis code should satisfy the constraint

$$r < r_{\text{in}} = \frac{k}{n} \leq \frac{1}{n} \min_{(i,\ell,j)\in\tau} \left[\log_2 \frac{1}{P_{ij}^{(\ell)}} \right] \tag{9.10}$$

The reason for obeying the lower bound $r_{\text{in}} = k/n > r$ is that $r_{\text{in}} = r$ would mean that we would not have the option of using a powerful outer code.

The upper bound $k \leq \min_{(i,\ell,j)\in\tau} [-\log_2 P_{ij}^{(\ell)}]$ is motivated by the unique decodability of the trellis code. To avoid non-unique decodability, we require that all 2^k branches emanating from each superchannel state have distinct noiseless output n-tuples[1]. The assumption that the input to the superchannel is i.u.d. implies that the conditional probability of each of the branches is 2^{-k}. Since the goal is to

---

[1]This is sufficient (but not always necessary) requirement for the construction of a uniquely decodable trellis.

TABLE 9.2   Optimized Transition Probabilities for the Dicode Channel at Rate $r = 1/2$ Using the Extended Channel Trellis in Figure 9.2(a): The Integer Values k, K_i and $n_{ij}^{(\ell)}$ are Determined using Rules 1-3

Branch $b_{ij}^{(\ell)}$	Branch Label $\left(i, u_{ij}^{(\ell)}, v_{ij}^{(\ell)}, j\right)$	Transition Probability	Optimized Integers $k=2, K_0=5, K_1=5$	Integer Probability
$b_{00}^{(0)}$	$(0, [000], [\ 0,\ 0,\ 0], 0)$	$P_{00}^{(0)} = 0.005$	$n_{00}^{(0)} = 0$	$n_{00}^{(0)}/(K_0 \cdot 2^k) = 0.00$
$b_{00}^{(1)}$	$(0, [010], [\ 0, -2,\ 2], 0)$	$P_{00}^{(1)} = 0.146$	$n_{00}^{(1)} = 3$	$n_{00}^{(1)}/(K_0 \cdot 2^k) = 0.15$
$b_{00}^{(2)}$	$(0, [100], [-2,\ 2,\ 0], 0)$	$P_{00}^{(2)} = 0.146$	$n_{00}^{(2)} = 3$	$n_{00}^{(2)}/(K_0 \cdot 2^k) = 0.15$
$b_{00}^{(3)}$	$(0, [110], [-2,\ 0,\ 2], 0)$	$P_{00}^{(3)} = 0.195$	$n_{00}^{(3)} = 4$	$n_{00}^{(3)}/(K_0 \cdot 2^k) = 0.20$
$b_{01}^{(0)}$	$(0, [001], [\ 0,\ 0, -2], 1)$	$P_{01}^{(0)} = 0.066$	$n_{01}^{(0)} = 1$	$n_{01}^{(0)}/(K_0 \cdot 2^k) = 0.05$
$b_{01}^{(1)}$	$(0, [011], [\ 0, -2,\ 0], 1)$	$P_{01}^{(1)} = 0.145$	$n_{01}^{(1)} = 3$	$n_{01}^{(1)}/(K_0 \cdot 2^k) = 0.15$
$b_{01}^{(2)}$	$(0, [101], [-2,\ 2, -2], 1)$	$P_{01}^{(2)} = 0.231$	$n_{01}^{(2)} = 5$	$n_{01}^{(2)}/(K_0 \cdot 2^k) = 0.25$
$b_{01}^{(3)}$	$(0, [111], [-2,\ 0,\ 0], 1)$	$P_{01}^{(3)} = 0.066$	$n_{01}^{(3)} = 1$	$n_{01}^{(3)}/(K_0 \cdot 2^k) = 0.05$
$b_{10}^{(0)}$	$(1, [000], [\ 2,\ 0,\ 0], 0)$	$P_{10}^{(0)} = 0.066$	$n_{10}^{(0)} = 1$	$n_{10}^{(0)}/(K_1 \cdot 2^k) = 0.05$
$b_{10}^{(1)}$	$(1, [010], [\ 2, -2,\ 2], 0)$	$P_{10}^{(1)} = 0.231$	$n_{10}^{(1)} = 5$	$n_{10}^{(1)}/(K_1 \cdot 2^k) = 0.25$
$b_{10}^{(2)}$	$(1, [100], [\ 0,\ 2,\ 0], 0)$	$P_{10}^{(2)} = 0.145$	$n_{10}^{(2)} = 3$	$n_{10}^{(2)}/(K_1 \cdot 2^k) = 0.15$
$b_{10}^{(3)}$	$(1, [110], [\ 0,\ 0,\ 2], 0)$	$P_{10}^{(3)} = 0.066$	$n_{10}^{(3)} = 1$	$n_{10}^{(3)}/(K_1 \cdot 2^k) = 0.05$
$b_{11}^{(0)}$	$(1, [001], [\ 2,\ 0, -2], 1)$	$P_{11}^{(0)} = 0.195$	$n_{11}^{(0)} = 4$	$n_{11}^{(0)}/(K_1 \cdot 2^k) = 0.20$
$b_{11}^{(1)}$	$(1, [011], [\ 2, -2,\ 0], 1)$	$P_{11}^{(1)} = 0.146$	$n_{11}^{(1)} = 3$	$n_{11}^{(1)}/(K_1 \cdot 2^k) = 0.15$
$b_{11}^{(2)}$	$(1, [101], [\ 0,\ 2, -2], 1)$	$P_{11}^{(2)} = 0.146$	$n_{11}^{(2)} = 3$	$n_{11}^{(2)}/(K_1 \cdot 2^k) = 0.15$
$b_{11}^{(3)}$	$(1, [111], [\ 0,\ 0,\ 0], 1)$	$P_{11}^{(3)} = 0.005$	$n_{11}^{(3)} = 0$	$n_{11}^{(3)}/(K_1 \cdot 2^k) = 0.00$

create a superchannel trellis that mimics the optimal Markov process on the extended trellis τ, the occurrence probabilities of the superchannel output n-tuples should match the occurrence probabilities of the noiseless output n-tuples of the extended trellis τ. However, if $P_{ij}^{(\ell)} > 2^{-k}$, this would not be possible.

Our task now becomes finding the smallest possible positive integers k and n such that an n-th order extended trellis τ satisfies Equation 9.10. The search for k and n can be conducted systematically starting with $n = 1$ and increasing n until such a trellis is found. This procedure delivers k, n, the extended trellis τ, the optimized branching probabilities $P_{ij}^{(\ell)}$, and the signal-to-noise ratio $\text{SNR}_{(\tau,r)}$ for which $\hat{C}_\tau = r$.

Consider the dicode channel in Figure 9.1(a). Suppose our target code rate is $r = 1/2$. The simplest extended trellis of this channel for which Rule 1 holds is the extended trellis in Figure 9.2(a) with $n = 3$. Hence, we pick the inner trellis code rate $r_{in} = k/n = 2/3$ which satisfies Equation 9.10. The corresponding branching probabilities are given in Table 9.2. We shall use this example throughout the chapter to illustrate the design method.

9.4.2   Choosing the Number of States in the Superchannel

We now want to design a superchannel with rate $r_{in} = k/n$. Let $P_{ij}^{(\ell)}$ be the branching probabilities of the extended trellis τ evaluated at $\text{SNR}_{(\tau,r)}$. Let $\mu_i = \Pr(S_t = i)$ denote the stationary probability of each state $0 \le i \le N_S - 1$ in the extended channel trellis τ.

Let K denote the number of states in the superchannel trellis ($K \le K_{\max}$, where $K_{\max}$ is a predefined maximal number of states in the superchannel trellis). Our strategy is to split each state i of the channel trellis τ into K_i states of the superchannel trellis. We say that these K_i states are of type t_i. Obviously, $K = \sum_{i=0}^{N_S-1} K_i \le K_{\max}$. Our goal is to find integers K_i such that the state types t_i in the superchannel trellis occur with the same probabilities as the state i in the extended channel trellis τ, that is, we desire $\mu_i \approx K_i/K$. Define a probability mass function (pmf) $\kappa = (\kappa_0, \kappa_1, \ldots, \kappa_{N_S-1})$, where $\kappa_i = K_i/K$. Denote by μ the pmf $\mu = (\mu_0, \mu_1, \ldots, \mu_{N_S-1})$.

Rule 2: Pick the pmf κ such that $D(\kappa \| \mu)$ is minimized under the constraint $N_S \leq K = \sum_{i=0}^{N_S-1} K_i \leq K_{max}$, where $D(\cdot \| \cdot)$ denotes the Kullback-Leibler distance [40].

Consider again the dicode channel example with $r = 1/2$. Rule 1 gave $k = 2$ and $n = 3$, and the probabilities $P_{ij}^{(\ell)}$ in Table 9.2. Solving Equation 9.5, we get $\mu_0 = \mu_1 = 0.5$. With $K_{max} = 12$, we get six solutions that satisfy Rule 2. They are $1 \leq K_0 = K_1 \leq 6$. This illustrates that there may not be a unique solution to the optimization problem in Rule 2. If this happens, then further refinement using Rule 3 (presented next) is needed.

9.4.3 Choosing the Branch Type Numbers in the Superchannel

We say that a branch of the superchannel trellis is of type $t_{ij}^{(\ell)}$ if

1. Its starting state is of type t_i and its ending state is of type t_j
2. Its noiseless output n-tuple matches the noiseless output n-tuple of the branch $b_{ij}^{(\ell)}$ in τ

Denote by $n_{ij}^{(\ell)}$ the number of branches in the superchannel trellis of type $t_{ij}^{(\ell)}$. For a fixed $i \in \{0, 1, \ldots, N_S - 1\}$, denote by v_i the pmf whose individual probabilities are $n_{ij}^{(\ell)} / (K_i \cdot 2^k)$, where i is fixed and ℓ and j are varied under the constraint $(i, \ell, j) \in \tau$. Obviously

$$\sum_{i:(i,\ell,j)\in\tau} \frac{n_{ij}^{(\ell)}}{K_i \cdot 2^k} = 1 \tag{9.11}$$

Similarly, for a fixed $i \in \{0, 1, \ldots, N_S - 1\}$, denote by π_i the pmf whose individual probabilities are $P_{ij}^{(\ell)}$, where i is fixed and ℓ and j are varied under the constraint $(i, \ell, j) \in \tau$.

Rule 3: Determine $0 \leq n_{ij}^{(\ell)} \in \mathbb{Z}$ such that $\sum_{i=0}^{N_S-1} \kappa_i D(v_i \| \pi_i)$ is minimized under the constraints $\sum_{j,\ell:(i,\ell,j)\in\tau} n_{ij}^{(\ell)} = K_i \cdot 2^k$ and $\sum_{i,\ell:(i,\ell,j)\in\tau} n_{ij}^{(\ell)} = K_j \cdot 2^k$.

We have established that Rule 2 may not deliver a unique solution. In this case, among all solution candidates for Rule 2, we pick the solution that minimizes the objective function in Rule 3. In the dicode channel example with the target rate $r = 1/2$, Rule 2 delivered a set of solutions $K_0 = K_1 \leq K_{max}/2 = 6$. Applying Rule 3, we get the integers $n_{ij}^{(\ell)}$ in Table 9.2 and $K_0 = K_1 = 5$, i.e., a superchannel trellis with $K = K_0 + K_1 = 10$ states.

9.4.4 Choosing the Branch Connections

Rules 1-3 guarantee that the marginal probability that the superchannel branch is of type $t_{ij}^{(\ell)}$ is very close to the value $\mu_i P_{ij}^{(\ell)}$. However, this does not guarantee that the resulting output process of the superchannel will mimic the output hidden Markov process of the channel trellis τ. Therefore, we need to choose a branch connection assignment with the following three requirements in mind:

1. Exactly $n_{ij}^{(\ell)}$ branches should be of type $t_{ij}^{(\ell)}$.
2. A branch of type $t_{ij}^{(\ell)}$ must start at a state of type t_i and end at a state of type t_j.
3. Branches emanating from a given state must have distinct types[2], that is, there cannot be two (or more) branches of the same type $t_{ij}^{(\ell)}$ emanating from a given state of type t_i.

Rule 4: Pick the superchannel branch connections that satisfy requirements (1)-(3) and deliver a superchannel with the maximal information rate $\mathcal{I}_S$ evaluated at $\text{SNR}_{(\tau,r)}$.

If the integers K_i and $n_{ij}^{(\ell)}$ are very small, Rule 4 can be satisfied by an exhaustive search. Very often the exhaustive search procedure is too complex and we soften our goal by finding a "good enough"

---

[2]This requirement can be removed if we choose not to obey Rule 1.

TABLE 9.3   Rate $r_{in} = 2/3$ Superchannel Trellis for the $1 - D$ Channel and Design Rate $r = 1/2$

Start State	Superchannel Input k-tuple (Rule 5)	Trellis-code Output n-tuple	Noiseless Superchannel Output n-tuple (Rule 4)			End State
0	0,0	1,0,1	-2,	2,	-2	9
0	0,1	1,0,0	-2,	2,	0	0
0	1,0	0,1,1	0,	-2,	0	5
0	1,1	1,1,0	-2,	0,	2	1
1	0,0	1,1,1	-2,	0,	0	9
1	0,1	1,0,1	-2,	2,	-2	6
1	1,0	0,1,1	0,	-2,	0	8
1	1,1	1,1,0	-2,	0,	2	2
2	0,0	0,1,1	0,	-2,	0	6
2	0,1	1,0,1	-2,	2,	-2	8
2	1,0	0,1,0	0,	-2,	2	3
2	1,1	1,1,0	-2,	0,	2	0
3	0,0	1,0,1	-2,	2,	-2	7
3	0,1	1,0,0	-2,	2,	0	2
3	1,0	0,1,0	0,	-2,	2	4
3	1,1	1,1,0	-2,	0,	2	4
4	0,0	0,0,1	0,	0,	-2	7
4	0,1	1,0,1	-2,	2,	-2	5
4	1,0	0,1,0	0,	-2,	2	1
4	1,1	1,0,0	-2,	2,	0	3
5	0,0	0,1,1	2,	-2,	0	6
5	0,1	1,0,1	0,	2,	-2	8
5	1,0	0,1,0	2,	-2,	2	4
5	1,1	1,1,0	0,	0,	2	2
6	0,0	0,0,1	2,	0,	-2	5
6	0,1	1,0,1	0,	2,	-2	5
6	1,0	0,1,1	2,	-2,	0	7
6	1,1	0,1,0	2,	-2,	2	2
7	0,0	0,0,1	2,	0,	-2	9
7	0,1	1,0,1	0,	2,	-2	6
7	1,0	0,1,0	2,	-2,	2	1
7	1,1	1,0,0	0,	2,	0	3
8	0,0	0,0,1	2,	0,	-2	7
8	0,1	1,0,0	0,	2,	0	1
8	1,0	0,1,0	2,	-2,	2	3
8	1,1	0,0,0	2,	0,	0	0
9	0,0	0,0,1	2,	0,	-2	8
9	0,1	1,0,0	0,	2,	0	4
9	1,0	0,1,1	2,	-2,	0	9
9	1,1	0,1,0	2,	-2,	2	0

superchannel using the following *ordinal optimization* [42] randomized search procedure. We randomly pick (say) 2000 superchannels that satisfy Rules 1-3. For each of these superchannels, we *coarsely* estimate their i.u.d. rates $\mathcal{I}_S$ (say by using a trellis length of 10^4 trellis sections in the Monte Carlo method of [9, 10]). We keep (say) only 10 superchannels that have the 10 highest (coarsely estimated) i.u.d. rates. For these 10 superchannels, we now make *fine* estimates (with long trellises, say 10^8 trellis stages) of the i.u.d. rates $\mathcal{I}_S$, and pick the superchannel with the highest i.u.d. rate $\mathcal{I}_S$.

Applying Rule 4 to the dicode channel example with the target rate $r = 1/2$ delivered the branch connections assignment presented in Table 9.3. The information rate of the constructed MIR super-channel (shown in Figure 9.7) is only 0.1dB away from the Markov capacity $\hat{C}_\tau$ of the extended channel trellis in Figure 9.2(a) at the target rate $r = 1/2$. At $r = 1/2$, the designed MIR super-trellis has a higher

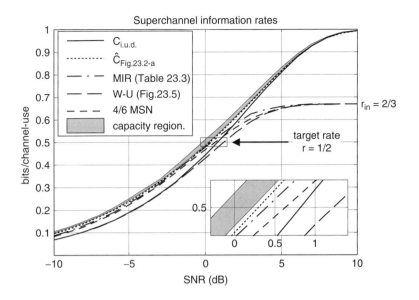

FIGURE 9.7 Information rates of the MIR (Table 9.3), W-U (Figure 9.5) and MSN superchannels with $r_{in} = 2/3$ (target rate $r = 1/2$). $\hat{C}_{Fig.\ 9.2-a}$ is the numerical Markov capacity of the trellis in Figure 9.2(a), whose optimal Markov probabilities were used to construct the MIR superchannel. The region between the tightest known lower and upper bound is shaded.

super-trellis information rate than other known trellis codes (the rate-2/3 W-U code [15] in Figure 9.5 and the rate-4/6 MSN code [18]). Much more importantly, we now have a *method* to construct superchannel trellises for *any* channel, even if the channel does not have spectral nulls.

9.5 Outer LDPC Codes

For the dicode channel, Figure 9.7 reveals that the i.u.d. capacity $C_{i.u.d.}$ equals the design rate $r = 1/2$ at SNR $= 0.82$ dB. The i.u.d. rate of the superchannel trellis shown in Table 9.3 equals $\mathcal{I}_S = 1/2$ at SNR $= 0.40$ dB (see Figure 9.7). Theorem 9.1 asserts that there exists at least one linear (coset) code such that, if we apply this code to the superchannel and utilize maximum likelihood (ML) decoding, we may gain 0.42 dB over $C_{i.u.d.}$. Since it is impractical to perform the ML decoding, we construct iteratively decodable outer codes.

9.5.1 Encoding/Decoding System

The normal graph [43] of the entire concatenated coding system is shown in Figure 9.8. We utilize k separate subcodes as the outer code (we explain the reason below), where k is the number of input bits at each stage of the MIR trellis code. Let the rate of the ith subcode be $r^{(i)} = K^{(i)}/N$, where $0 \le i \le k - 1$. Let $\underline{D}$ be an i.u.d. binary sequence of length $K^{(total)} = \sum_{i=0}^{k-1} K^{(i)}$ to be transmitted over the channel. The encoding algorithm can be described by the following three steps (see also Figure 9.8):

1. The sequence $\underline{D}$ is separated into k subsequences $\underline{D} = [\underline{D}^{(0)}, \underline{D}^{(1)}, \ldots, \underline{D}^{(k-1)}]$, where the ith subsequence $\underline{D}^{(i)}$ is of length $K^{(i)}$.
2. The ith subsequence $\underline{D}^{(i)}$ enters the ith LDPC encoder whose code-rate is $r^{(i)} = K^{(i)}/N$. The output sequence from the ith encoder is denoted by $[U^{(i)}]_1^N = [U_1^{(i)}, \ldots, U_N^{(i)}]$.
3. The whole sequence $U_1^N = [U_1, \ldots, U_t, \ldots, U_N]$ enters the MIR trellis encoder, where $U_t = (U_t^{(0)}, \ldots, U_t^{(k-1)})$. The sequence Y_1^{nN} is observed at the channel output.

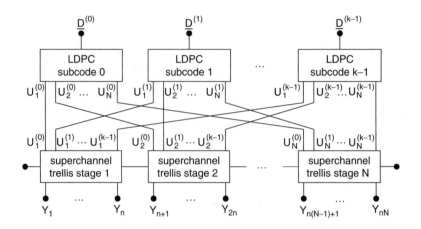

FIGURE 9.8   The normal graph of an outer LDPC code (consisting of k subcodes) and the inner superchannel trellis.

For the purpose of systematically designing good outer codes, we specify a serial multistage decoding schedule [44] depicted in Figure 9.9. To guarantee that the decoding algorithm of Figure 9.9 works well, we have to solve the following three problems:

- A trellis section can be viewed as a collection of branches $\{b_t = (s_{t-1}, u_t, v_t, s_t)\}$. In terms of optimizing the superchannel i.u.d. information rate $\mathcal{I}_S$, it is irrelevant how the input vector u_t of each branch is chosen. However, to construct a good and practically decodable LDPC code we need to choose the branch input-bit assignment judiciously. We develop the input-bit assignment design rule (Rule 5) in Section 9.5.2.

- Consider the soft-output random variables $L_t^{(i)} \overset{\text{def}}{=} \Pr(U_t^{(i)} = 0 \mid Y_1^{nN})$. For a general rate-$k/n$ inner trellis code, when $N \rightarrow \infty$, the statistical properties of $L_{t_1}^{(i)}$ and $L_{t_2}^{(j)}$ are the same if and only if $i = j$. In other words, different superchannel bit positions have different soft-output statistics, which is why we utilize k different subcodes. Now the question is: how to determine the rates $r^{(i)}$ of the constituent subcodes? Section 9.5.3 gives the answer.

- For large N, we need to optimize each of the k subcodes. In Section 9.5.4, we develop an optimization method along the lines of [33, 36].

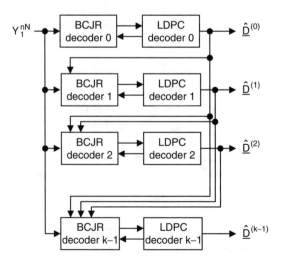

FIGURE 9.9   The iterative decoder. The k codes are successively decoded, each iteratively.

9.5.2 Choosing the Branch Input-Bit Assignment

We need to specify the input k-tuple $u_t = [u_t^{(0)}, u_t^{(1)}, \ldots, u_t^{(k-1)}]$ (where $u_t^{(i)} \in \mathcal{A}$) for every branch $b_t = (s_{t-1}, u_t, v_t, s_t)$ of the superchannel trellis. Generally, if the bit assignments for $u_t^{(0)}, \ldots, u_t^{(i-1)}$ are given, then only the bit assignment for $u_t^{(i)}$ determines the performance of Decoder i in Figure 9.9, irrespective of the bit assignment for $u_t^{(i+1)}, \ldots, u_t^{(k-1)}$. Therefore we use the following rule to determine the bit assignment.

Rule 5: (Greedy bit assignment algorithm)

For $0 \leq i \leq k - 1$

Assume that Decoders 0 through $i - 1$ decode without errors. Find the bit assignment for the ith location $u_t^{(i)}$ such that the ith BCJR decoder in Figure 9.9 delivers the lowest probability of bit error *in the first iteration*.

This algorithm guarantees a good decoding start which is typically sufficient to ensure good decoding properties for the concatenated LDPC decoder [35].

By applying Rule 5, we have selected one good input-bit assignment for the trellis code shown in Table 9.3. The input bit assignment is shown in the third column of Table 9.3.

9.5.3 Determining the Outer Subcode Rates

Ideally, the superchannel i.u.d. rate $\mathcal{I}_S$ should equal the target rate r

$$r = r_{\text{in}} \cdot r_{\text{out}} = \frac{k}{n} \cdot r_{\text{out}} = \lim_{N \to \infty} \frac{1}{nN} \cdot I\left(U_1^N; Y_1^{nN}\right) \Bigg|_{P_{U_1^N}(u_1^N) = 2^{-kN}} = \mathcal{I}_S$$

From the chain rule [40], we have

$$I\left(U_1^N; Y_1^{nN}\right) = I\left([U^{(0)}]_1^N; Y_1^{nN}\right) + \sum_{i=1}^{k-1} I\left([U^{(i)}]_1^N; Y_1^{nN} \mid [U^{(0)}]_1^N, \ldots, [U^{(i-1)}]_1^N\right)$$

where $[U^{(i)}]_1^N = [U_1^{(i)}, U_2^{(i)}, \ldots, U_N^{(i)}]$. Hence, the assumption that Decoders 0 through $i - 1$ perform error-less decoding before Decoder i starts the decoding process is valid only if

$$r^{(i)} \leq \lim_{N \to \infty} \frac{1}{N} \cdot I\left([U^{(i)}]_1^N; Y_1^{nN} \mid [U^{(0)}]_1^N, \ldots, [U^{(i-1)}]_1^N\right) \tag{9.12}$$

Therefore, a reasonable rate-assignment is

$$r^{(i)} = \lim_{N \to \infty} \frac{1}{N} \cdot I\left([U^{(i)}]_1^N; Y_1^N \mid [U^{(0)}]_1^N, \ldots, [U^{(i-1)}]_1^N\right) \tag{9.13}$$

where the sequences $[U^{(i)}]_1^N$ are i.u.d. for all $0 \leq i \leq k - 1$. The rates in Equation 9.13 can be computed by Monte Carlo simulations [9, 10]. Consequently, we get $r_{\text{out}} = \frac{1}{k} \sum_{i=0}^{k-1} r^{(i)}$.

To summarize, first we choose two integers $(K^{(total)}, N)$, where N is large enough, such that $K^{(total)}/(kN)$ is not greater than (and is as close as possible to) $r_{\text{out}} = r/r_{\text{in}}$. Then we choose $K^{(i)}$, for $0 \leq i \leq k - 1$ such that $\sum_{i=0}^{k-1} K^{(i)} = K^{(total)}$ and $K^{(i)}/N \approx r^{(i)}$, where $r^{(i)}$ is computed by Equation 9.13.

9.5.4 Subcode Optimization

9.5.4.1 Outer Code Optimization

To optimize the outer code for a given superchannel we generalize the *single* code optimization method for a PR channel (with no inner code) [33, 34, 36] to a *joint* optimization of k different subcodes. This generalization is summarized as follows:

1. Given a superchannel trellis of rate $r_{in} = k/n$, we optimize k different constituent subcodes shown in Figure 9.8 according to the decoding scenario shown in Figure 9.9. That is, while optimizing the degree sequences of the LDPC Subcode i, we assume that the symbols $[U^{(0)}]_1^N$, $[U^{(1)}]_1^N, \ldots,$ $[U^{(i-1)}]_1^N$ are decoded without errors and that no prior information on the symbols $[U^{(i+1)}]_1^N, \ldots,$ $[U^{(k-1)}]_1^N$ is available.

2. The ith constituent subcode of rate $r^{(i)}$ is optimized using the optimization from [36]. We make a change in the density evolution [32, 35] to reflect the fact that Decoders 0 through $i-1$ have decoded symbols $[U^{(0)}]_1^N$ through $[U^{(i-1)}]_1^N$ without errors. This is done by generating i.u.d. realizations of symbols $[u^{(0)}]_1^N$ through $[u^{(i-1)}]_1^N$, and running the BCJR algorithm (in the density evolution step) with prior knowledge of these symbols.

3. We obtain k pairs $(\underline{\lambda}^{(i)}, \underline{\rho}^{(i)}), 0 \leq i \leq k-1$, of the optimized edge degree sequences [31], each pair corresponding to one constituent subcode. The noise threshold [32] of the entire code is given by $\sigma^* = \min\{\sigma_0^*, \sigma_1^*, \ldots, \sigma_{k-1}^*\}$, where σ_i^* is the noise threshold for Subcode i.

9.5.4.2 LDPC Decoding

Figure 9.9 seems to suggest that we need k different BCJR decoders and k different LDPC decoders to decode our code. In fact we only need *one* BCJR detector and *one* LDPC decoder. The *single* LDPC code is constructed by interleaving the constituent subcodes.

The parity check matrix of Subcode i is constructed to be a low-density parity-check matrix with degree sequences $\underline{\lambda}^{(i)}$ and $\underline{\rho}^{(i)}$. Denote the parity-check matrix of Subcode i by

$$\mathbf{H}^{(i)} = \left[h_1^{(i)} \; h_2^{(i)}, \ldots, h_N^{(i)}\right]$$

where $\underline{h}_j^{(i)}$ represents the jth column of the matrix $\mathbf{H}^{(i)}$. The parity check matrix $\mathbf{H}$ of the entire code is obtained by interleaving the columns of the subcode parity check matrices

$$\mathbf{H} = \begin{bmatrix} \underline{h}_1^{(0)} & 0 & \cdots & 0 & \underline{h}_2^{(0)} & 0 & \cdots & 0 & & \underline{h}_N^{(0)} & 0 & \cdots & 0 \\ 0 & \underline{h}_1^{(1)} & \cdots & 0 & 0 & \underline{h}_2^{(1)} & \cdots & 0 & & 0 & \underline{h}_N^{(1)} & \cdots & 0 \\ \vdots & \vdots & \ddots & \vdots & \vdots & \vdots & \ddots & \vdots & \cdots & \vdots & \vdots & \ddots & \vdots \\ 0 & 0 & \cdots & \underline{h}_1^{(k-1)} & 0 & 0 & \cdots & \underline{h}_2^{(k-1)} & & 0 & 0 & \cdots & \underline{h}_N^{(k-1)} \end{bmatrix}. \quad (9.14)$$

The size of the matrix $\mathbf{H}$ is $(kN - K^{(total)}) \times (kN)$, where $K^{(total)} = \sum_{i=0}^{k-1} K^{(i)}$.

9.6 Optimization Results

We perform the optimization on the dicode $(1 - D)$ channel which has a spectral null at frequency $\omega = 0$, and on the $1 + 3D + D^2$ channel which does not have a spectral null.

9.6.1 Dicode Channel

The inner trellis code with code rate $r_{in} = k/n = 2/3$ for the dicode channel is given in Table 9.3. Since $k = 2$, we have 2 outer LDPC subcodes. Using Equation 9.13, we found their rates to be $r^{(0)} = 0.66$ and $r^{(1)} = 0.84$, respectively. The rate of the outer code is thus $r_{out} = \frac{1}{2} \cdot (r^{(0)} + r^{(1)}) = 3/4$. The resulting overall code rate is then $r = r_{in} \cdot r_{out} = \frac{2}{3} \cdot \frac{3}{4} = 1/2$, which is exactly our target code rate.

The optimized degree sequences together with their respective thresholds are given in Table 9.4. We constructed an outer LDPC code by interleaving the parity check matrices obtained from the optimized degree sequences, see Equation 9.14. The code block length was set to 10^6 binary symbols. The bit error rate (BER) simulation curve is shown in Figure 9.10. For comparison, Figure 9.10 also shows a tight lower bound $\hat{C}_6$ and an upper bound C_U on the capacity, the superchannel i.u.d. rate $\mathcal{I}_s$, the i.u.d. capacity $C_{i.u.d.}$

TABLE 9.4 Good Degree Sequences and the Noise Thresholds for Separately Coded Even and Odd Bits of the Outer LDPC Code on the Superchannel from Table 9.3

	Subcode 0 (even bits)			Subcode 1 (odd bits)	
	$r^{(0)} = 0.660$			$r^{(1)} = 0.840$	
x	$\lambda_x^{(0)}$	$\rho_x^{(0)}$	x	$\lambda_x^{(1)}$	$\rho_x^{(1)}$
2	0.2225		2	0.2191	
3	0.1611		3	0.1526	
5	0.1627		4	0.1057	
6	0.0164		15	0.0017	
10	0.0024		16	0.0003	
11		0.3325	17		0.0914
13		0.1406	23		0.3855
16		0.3929	49	0.1205	
24		0.1340	50	0.4001	0.1321
48	0.0035		51		0.0698
49	0.4314		59		0.3212
	threshold - Subcode 0 $\sigma_0^* = 1.322$			threshold - Subcode 1 $\sigma_1^* = 1.326$	
		threshold $\sigma^* = 1.322$			
		$\text{SNR}^* = 10\log_{10}\frac{2}{(\sigma^*)^2} = 0.59\,\text{dB}$			

and the noise tolerance threshold $\text{SNR}^* = 10\log_{10}\frac{\sum_j h_j^2}{(\sigma^*)^2}$ of the code in Table 9.4. We see that the threshold SNR^* surpasses the i.u.d. channel capacity $C_{\text{i.u.d.}}$ by 0.23 dB and is 0.19 dB away from the superchannel i.u.d. rate $\mathcal{I}_S$. The code simulation shows that a BER of 10^{-6} is achieved at an SNR that surpasses $C_{\text{i.u.d.}}$ by 0.14 dB.

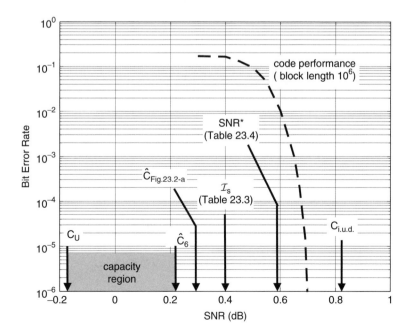

FIGURE 9.10 BER versus SNR for the Dicode channel. Code rate $r = 1/2$; code length 10^6 bits.

TABLE 9.5　Rate $r_{in} = 1/2$ Superchannel Trellis for the $1 + 3D + D^2$ Channel and Design Rate $r = 1/3$

Start State	Super-Channel Input k-tuple	Trellis-code Output n-tuple	Noiseless Superchannel Output n-tuple	End State
0	0	0,0	5,　5	1
0	1	0,1	5,　3	3
1	0	0,0	5,　5	2
1	1	1,1	3,−3	6
2	0	0,0	5,　5	0
2	1	1,1	3,−3	5
3	0	1,0	−3,−3	4
3	1	1,1	−3,−5	7
4	0	0,0	3,　5	0
4	1	0,1	3,　3	3
5	0	0,0	−3,　3	2
5	1	1,1	−5,−5	7
6	0	0,0	−3,　3	1
6	1	1,1	−5,−5	5
7	0	1,0	−5,−3	4
7	1	1,1	−5,−5	6

9.6.2　A Channel without Spectral Nulls

For the $1 + 3D + D^2$ channel we chose the design rate $r = 1/3$. Using Rules 1-5, we constructed an inner trellis code of rate $r_{in} = k/n = 1/2$ for the $1 + 3D + D^2$ channel; the superchannel trellis is given in Table 9.5. Since $k = 1$, we have only one outer LDPC subcode of rate $r_{out} = 2/3$, which gives the desired code rate $r = \frac{1}{2} \cdot \frac{2}{3} = 1/3$. The optimization of the outer LDPC code delivered the degree sequences and the threshold shown in Table 9.6. Based on these degree sequences we constructed an LDPC code of block length 10^6. The code's BER performance curve is shown in Figure 9.11. We observe from Figure 9.11 that

TABLE 9.6　Good Degree Sequences and the Noise Threshold for the Outer LDPC Code Designed for the Superchannel in Table 9.5

$r_{out} = r^{(0)} = 0.6667$		
i	λ_i	ρ_i
2	0.2031	
3	0.2195	
4	0.0022	
6	0.1553	
10		0.2974
13		0.3064
15		0.1906
27	0.1616	
28	0.1399	
38		0.2056
50	0.1184	

threshold
$$\sigma_0^* = \sigma^* = 4.475$$

$$\text{SNR}^* = 10 \log_{10} \frac{11}{(\sigma^*)^2}$$
$$= -2.60\text{dB}$$

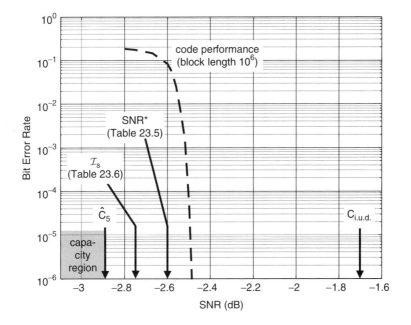

FIGURE 9.11   BER versus SNR for the $1 + 3D + D^2$ channel. Code rate $r = 1/3$; code length 10^6 bits.

the superchannel i.u.d. rate $\mathcal{I}_S$ is 0.14 dB away from the lower bound on the channel capacity $\hat{C}_5$. The noise threshold SNR* surpasses the i.u.d. channel capacity $C_{\text{i.u.d.}}$ by 0.90 dB and is 0.15 dB away from $\mathcal{I}_S$. The code simulation reveals that our design method yields a code that achieves a BER of 10^{-6} at an SNR that surpasses $C_{\text{i.u.d.}}$ by 0.77 dB.

9.7   Conclusion

We developed a methodology for designing capacity-approaching codes for partial response (PR) channels. Since the PR channel is a channel with memory, its capacity C is greater than its i.u.d. capacity $C_{\text{i.u.d.}}$. We showed that rates above $C_{\text{i.u.d.}}$ cannot be achieved by random linear codes. Given that our goal was to construct a code that surpasses $C_{\text{i.u.d.}}$, we chose a concatenated coding strategy where the inner code is a (generally nonlinear) trellis code, and the outer code is a low-density parity-check (LDPC) code.

The key step in the design of the inner code is to identify a Markov input process that achieves a high (capacity-approaching) information rate over the PR channel of interest. Then, we construct a trellis code that mimics the identified Markov process. Hence, we name it a *matched information rate* (MIR) trellis code. We choose an LDPC code as an outer code to show that the concatenation of the two codes approaches the computed information rate of the identified Markov process. The outer code is optimized by modified density evolution methods to fit our specific inner code and the PR channel.

MIR trellis code constructions are different from any previously known trellis code construction methods in that we do *not* base the code construction on an algebraic criterion. Instead, the code construction is purely *probabilistic*. We provided a set of rules to construct the inner MIR trellis codes. These rules apply to *any* PR channel. Using our design rules and the outer code optimization, we constructed examples of capacity-approaching codes for both channels with and without spectral nulls.

Acknowledgment

The authors would like to thank Michael Mitzenmacher for very helpful discussions over the course of this research.

References

[1] W. Hirt, *Capacity and Information Rates of Discrete-Time Channels with Memory.* Ph.D. thesis, Swiss Federal Institute of Technology (ETH), Zurich, Switzerland, 1988.

[2] S. Shamai (Shitz), L.H. Ozarow, and A.D. Wyner, Information rates for a discrete-time Gaussian channel with intersymbol interference and stationary inputs, *IEEE Trans. Inform. Theory,* vol. 37, pp. 1527–1539, 1991.

[3] S. Shamai (Shitz) and S. Verdú, Worst-case power-constrained noise for binary-input channels, *IEEE Trans. Inform. Theory,* vol. 38, pp. 1494–1511, 1992.

[4] C.E. Shannon, A mathematical theory of communications, *Bell Syst. Tech. J.,* vol. 27, pp. 379–423 (part I) and 623–656 (part II), 1948.

[5] A.S. Khayrallah and D.L. Neuhoff, Coding for channels with cost constraints, *IEEE Trans. Inform. Theory,* vol. 42, pp. 854–867, May 1996.

[6] E. Zehavi and J.K. Wolf, On runlength codes, *IEEE Trans. Inform. Theory,* vol. 34, pp. 45–54, January 1988.

[7] B. Marcus, K. Petersen, and S. Williams, Transmission rates and factors of Markov chains, *Contemporary Mathematics,* vol. 26, pp. 279–293, 1984.

[8] S. Yang and A. Kavčić, Capacity of partial response channels, *Chapter 13.*

[9] D. Arnold and H.-A. Loeliger, On the information rate of binary-input channels with memory, in *Proceedings' IEEE International Conference on Communications 2001,* Helsinki, Finland, June 2001.

[10] H.D. Pfister, J.B. Soriaga, and P.H. Siegel, On the achievable information rates of finite state ISI channels, in *Proceedings IEEE Globecom 2001,* San Antonio, Texas, pp. 2992–2996, November 2001.

[11] A. Kavčić, On the capacity of Markov sources over noisy channels, in *Proceedings IEEE Global Communications Conference 2001,* San Antonio, Texas, pp. 2997–3001, November 2001.

[12] P. Vontobel and D.M. Arnold, An upper bound on the capacity of channels with memory and constraint input, in *IEEE Information Theory Workshop,* Cairns, Australia, September 2001.

[13] S. Yang and A. Kavčić, Markov sources achieve the feedback capacity of finite-state machine channels, in *Proceedings of IEEE International Symposium on Inform. Theory,* Lausanne, Switzerland, July 2002.

[14] C. Berrou, A. Glavieux, and P. Thitimajshima, Near Shannon limit error-correcting coding and decoding: Turbo-codes, in *Proceedings IEEE International Conference on Communications,* pp. 1064–1070, May 1993.

[15] J.K. Wolf and G. Ungerboeck, Trellis coding for partial-response channels," *IEEE Trans. Commun.,* vol. 34, pp. 744–765, March 1986.

[16] A.R. Calderbank, C. Heegard, and T.A. Lee, Binary convolutional codes with application to magnetic recording, *IEEE J. Select. Areas Commun.,* vol. 32, pp. 797–815, November 1986.

[17] K.A.S. Immink, Coding techniques for the noisy magnetic recording channel: A state-of-the-art report, *IEEE Trans. Commun.,* vol. 35, pp. 413–419, May 1987.

[18] R. Karabed and P.H. Siegel, Matched spectral-null codes for partial response channels, *IEEE Trans. Inform. Theory,* vol. 37, pp. 818–855, May 1991.

[19] R.G. Gallager, *Low-Density Parity-Check Codes,* MIT Press, Cambridge, MA, 1962.

[20] D.J.C. MacKay and R.M. Neal, Near Shannon limit performance of low-density parity-check codes, *Electron. Lett.,* vol. 32, pp. 1645–1646, 1996.

[21] L.R. Bahl, J. Cocke, F. Jelinek, and J. Raviv, Optimal decoding of linear codes for minimizing symbol error rate, *IEEE Trans. Inform. Theory,* vol. 20, pp. 284–287, September 1974.

[22] W. Ryan, Performance of high-rate turbo codes on PR4-equalized magnetic recording channels, in *Proceedings of IEEE International Conference on Communications,* Atlanta, GA, pp. 947–951, June 1998.

[23] M.C. Reed and C.B. Schlegel, An iterative receiver for partial response channels, in *Proceedings IEEE International Symposiums Information Theory,* Cambridge, MA, p. 63, August 1998.

[24] T. Souvignier, A. Friedmann, M. Öberg, P. Siegel, R.E. Swanson, and J.K. Wolf, Turbo codes for PR4: Parallel versus serial concatenation, in *Proceedings of IEEE International Conference on Comm.,* pp. 1638–1642, June 1999.

[25] L.L. McPheters, S.W. McLaughlin, and K.R. Narayanan, Precoded PRML, serial concatenation, and iterative (turbo) decoding for digital magnetic recording, *IEEE Trans. Magn.*, vol. 35, September 1999.

[26] M. Tüchler, R. Kötter, and A. Singer, Iterative correction of ISI via equalization and decoding with priors, in *Proceedings of IEEE International Symposiums Information Theory*, Sorrento, Italy, p. 100, June 2000.

[27] J. Park and J. Moon, A new soft-output detection method for magnetic recording channels, in *Proceedings IEEE Global Communications Conference 2001*, San Antonio, Texas, pp. 3002–3006, November 2001.

[28] J. Fan, A. Friedmann, E. Kurtas, and S. McLaughlin, Low density parity check codes for magnetic recording, in *Proceedings Allerton Conference on Communications and Control*, 1999.

[29] M. Oberg and P.H. Siegel, Parity check codes for partial response channels, in *Proceedings IEEE Global Telecommunications Conference*, vol. 1, Rio de Janeiro, pp. 717–722, December 1999.

[30] J. Li, K.R. Narayanan, E. Kurtas, and C.N. Georghiades, On the performance of high rate turbo product codes and low density parity check codes for partial response channels, *IEEE Trans. Commun.*, vol. 50, pp. 723–734, May 2002.

[31] M. Luby, M. Mitzenmacher, M. A. Shokrollahi, and D. Spielman, Improved low-density parity-check codes using irregular graphs, *IEEE Trans. Inform. Theory*, vol. 47, pp. 585–598, February 2001.

[32] T. Richardson and R. Urbanke, The capacity of low-density parity check codes under message-passing decoding, *IEEE Trans. Inform. Theory*, vol. 47, pp. 599–618, February 2001.

[33] T. Richardson, A. Shokrollahi, and R. Urbanke, Design of capacity-approaching low-density parity-check codes, *IEEE Trans. Inform. Theory*, vol. 47, pp. 619–637, February 2001.

[34] S.-Y. Chung, G.D. Forney, T. Richardson, and R. Urbanke, On the design of low-density parity-check codes within 0.0045 dB of the Shannon limit, *IEEE Commun. Lett.*, February 2001.

[35] A. Kavčić, X. Ma, and M. Mitzenmacher, Binary intersymbol interference channels: Gallager codes, density evolution and code performance bounds, *IEEE Trans. Inform. Theory*, pp. 1636–1652, July 2003.

[36] N. Varnica and A. Kavčić, Optimized low-density parity-check codes for partial response channels, *IEEE Commun. Lett.*, vol. 7, pp. 168–170, April 2003.

[37] A. Kavčić and A. Patapoutian, A signal-dependent autoregressive channel model, *IEEE Trans. Magn.*, vol. 35, pp. 2316–2318, September 1999.

[38] E. Kurtas, J. Park, X. Yang, W. Radich, and A. Kavčić, Detection methods for data-dependent noise in storage channels, *Chapter 33*.

[39] R.G. Gallager, *Information Theory and Reliable Communication*, John Wiley and Sons, New York, 1968.

[40] T.M. Cover and J.A. Thomas, *Elements of Information Theory*, John Wiley and Sons, New York, 1991.

[41] A. Kavčić, X. Ma, and N. Varnica, Matched information rate codes for partial response channels, *submitted for publication in IEEE Trans. Inform. Theory*, July 2002.

[42] Y.-C. Ho, An explanation of ordinal optimization: Soft computing for hard problems, *Inform. Sci.*, vol. 113, pp. 169–192, February 1999.

[43] G.D. Forney Jr., Codes on graphs: Normal realizations, *IEEE Trans. Inform. Theory*, vol. 47, pp. 520–548, February 2001.

[44] H. Imai and S. Hirakawa, A new multilevel coding method using error correcting codes, *IEEE Trans. Inform. Theory*, vol. 23, pp. 371–377, May 1977.

10

Coding and Detection for Multitrack Systems

Bane Vasic
University of Arizona
Tucson, AZ

Olgica Milenkovic
University of Colorado
Boulder, CO

10.1   Introduction

In traditional magnetic disk drive systems user data is recorded on concentric tracks as a sequence of changes of the magnetization of small domains. The direction of the magnetization of a domain depends on the polarity of a data-driven write current. The areal density of data stored on a disk can be increased by either reducing the size of the magnetic domains along tracks (i.e., by increasing the linear density) and/or by reducing the track pitch (i.e., by increasing the track, or radial, density). The linear density is limited by several factors, including the finite sensitivity of the read head, properties of magnetic materials, head design [47, 51], as well as the ability to detect and decode recorded data in the presence of intersymbol interference (ISI) and noise. Most research in disk drive systems has focused on increasing linear density. Extremely high densities, for example, 10 Gbits/in^2, and data rates approaching 1 Gbits/s have been already demonstrated in commercially available systems. Recent progress in heads and media design has opened the possibility of reaching densities of 100 Gbits/in^2 and perhaps even 1 Terabit/in^2. However, the rate of increase of the linear density in future magnetic recording systems is not likely to be as high as in the past. This is due to the fact that as the linear density increases, the magnetic domains on the disk surface become smaller and increasingly thermally unstable. The so-called super-paramagnetic effect [6] represents a fundamental limiting factor for linear density.

Alternative approaches for increasing areal density are therefore required in order to meet the constant demand for increases in data rate and capacity of storage devices, largely driven by the Internet. Since the current linear densities are approaching the super-paramagnetic limit, the obvious alternative to an increase in linear density is an increase in radial density. In modern systems, the radial density is mostly limited by the mechanical design of the drive and the ability to accurately follow a track whose width is of the order of 1 μm. In order to further increase radial density, multiple-head arrays have been developed

[27, 36]. A head array is an arrangement of closely spaced heads that can read and write data simultaneously on multiple tracks. Such heads can potentially provide both high density and high speed, but they suffer from cross-talk or intertrack interference (ITI) [4]. This ITI is the consequence of a signal induced in the heads due to the superposition of magnetic transitions in neighboring tracks. Today's recording systems have a large track pitch, and therefore ITI has a negligible effect on their performance. However, significant advances in coding and signal processing for multitrack recording channels are required before potential large radial densities together with head arrays may be practically applied.

10.2   The Current State of Research in Multitrack Codes

A number of multitrack coding and detection schemes have been proposed in the last decade. They can be categorized as follows:

1. The first category is a class of multitrack codes which exploit the idea that the achievable areal density can be increased *indirectly*. This is done through relaxing the per-track maximum runlength constraint (the so-called k-constraint) in the recording codes and by imposing a constraint across multiple tracks. Such codes have been studied by Marcellin et al. [20–23], Swanson and Wolf [29], and Vasic et al. [39–43].

2. The second category is a class of techniques involving multitrack detection combined with partial-response (PR) equalization [12] and *maximum likelihood* (multiple) sequence detection (MLSD). Such techniques have computational complexity exponential in NM, where N is the number of tracks and M is the memory of the PR channel. Multitrack codes and reduced-complexity detectors for such systems have been studied by Soljanin et al. [31–33] and Kurtas et al. [12, 19]. More recently, a combination of equalization and *maximum a posteriori* (MAP) decoding, reducing both ISI and ITI, was considered by Wu et al. [52] (see also Chugg et al. [7] who first applied these ideas to page oriented holographic storage). The method presented in [12] combines a multitrack version of the BCJR algorithm [1] with iterative Wiener filtering, and performs very well in conjunction with low-density parity check (LDPC) codes. It's major drawback is very high complexity.

3. The third class of techniques uses the idea of imposing a constraint on a recorded bit sequence in such a way that ITI on each PR channel is either completely eliminated or reduced. Recently, Ahmed et al. [1] constructed a two-track runlength-limited code for a Class 4 partial response (PR4) channel. The code forbids any transitions of opposite polarity on adjacent tracks, and results in up to a 23% gain in areal density over an uncoded system. Similar two track schemes, but for a different multitrack constraint, have been proposed by Davey et al. [7, 11] and by Lee and Madisetti [20]. Due to their high complexity, these schemes can be used only for a small number of tracks and low order PR polynomials.

In order to achieve high linear densities, equalization with respect to higher-order PR polynomials is necessary. However, the complexity of read-channel chips increases exponentially with the order of the PR polynomial. Furthermore, the largest contribution to the complexity comes from MLSD, which is already the most complicated subsystem in the "read-channel" electronics and is a primary impediment to high data throughput. Thus, increasing the complexity of a detector by another factor (N) in the exponent, which is required for MLSD detection over N tracks, is not feasible.

10.3   Multitrack Channel Model

The magnetic recording channel is modeled by a discrete-time linear filter with a partial response polynomial typically of the form $h(D) = (1 - D)(1 + D)^M$, or $h(D) = (1 + D)^M, M \geq 1$, depending on whether longitudinal or perpendicular recording is employed [46]. User data is encoded by N separate error control encoders, and are organized in two-dimensional blocks of size $N \times n$ written on N adjacent tracks (see Figure 10.1). The sequence recorded in the kth track is denoted by $\{a_m^{(k)}\}_{m \in Z}$. Adjacent tracks

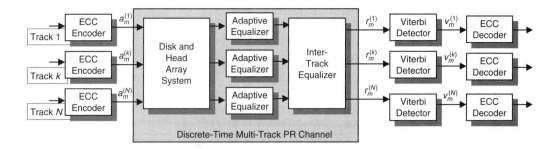

FIGURE 10.1   Discrete-time longitudinal multitrack recording channel model.

are read by an array of N heads. According to the channel model proposed by Vea and Moura [47], the signal read by the kth head is equalized to some partial response target and is given by

$$r^{(k)}(t) = \sum_{1 \le l \le N} \alpha_{|l-k|} \sum_{m \in Z} a_m^{(l)} \cdot h(t - mT) + n(t), \quad 1 \le k \le N$$

where $h(t)$ is the impulse response of the PR target, $n(t)$ is the colored noise process obtained by filtering additive white Gaussian noise (AWGN) by the equalizer; the real numbers $\alpha_d, 0 \le d \le N - 1$ specify the cross-talk between two tracks separated by d other tracks.

For longitudinal recording systems, we assume that the read head response to an isolated (positive-going) transition written on a disk is the Lorentzian pulse, $l(t) = (1 + (2t/PW_{50})^2)^{-1}$. For perpendicular recording, we assume that the transition response is the error function, $g(t) = (2/\sqrt{\pi}) \cdot \int_0^{S \cdot t} \exp(-x^2)\, dx = erf(St)$, where $S = 2\sqrt{\ln 2} \cdot T/P W_{50}$. The *dibit response* is defined as $(g(t + T/2) - g(t - T/2))$. The channel density is defined as $S_c = PW_{50}/T$, where PW_{50} represents the width of the channel impulse response at half of its peak value, and T is the channel bit interval. This models ignores the effects of timing jitter, that is, it assumes that the samples are taken at time instants $t = mT$, where the channel bit interval T is known and fixed. It also ignores the effects of track-following error, that is, it is assumed that the kth head is perfectly aligned with the kth track.

Another equalization scheme was recently proposed by Vasic and Venkateswaran [46] and shown in Figure 10.1. The intertrack interference block is described by the matrix $A = [\alpha_{|i-j|}]_{1 \le i, j \le N}$ [47]. The in-track equalizer is an adaptive least mean-square error equalizer, while the inter-track equalizer is a short zero-forcing equalizer. The equalized data are fed to a bank of N Viterbi detectors, and then to ECC decoders.

10.4   Multitrack Constrained Codes

10.4.1   ITI Reducing Codes for PR Channels

Consider an N-track system in which each track is equalized to a PR channel with polynomial $p(D) = h_0 + h_1 D + \cdots + h_L D^L$. The N-track constrained system is defined as an oriented, strongly connected graph $G = (V, E)$ with vertex set V and edge set E. The vertices are labeled by binary arrays Ψ of dimension $N \times L$, $\Psi = (\Psi_j)_{1 \le j \le L}$, where each Ψ_j is a column vector of length N. The edges are labeled by binary column vectors x of length N (more details on vector constrained systems can be found in [39] and [43].) The response of the multitrack channel to an array of input symbols (Ψ, x) is

$$y = h_0 \cdot x + \sum_{1 \le j \le L} h_j \cdot \Psi_{L+1-j}$$

TABLE 10.1   Shannon Capacities of ITI-Reducing Constraints

N	PR2 'sign'	PR2 'zero'	PR4 'sign'	PR4 'zero'	EPR4 'sign'	EPR4 'zero'
1	$C = 0.91625$	$C = 0.79248$	$C = 0.91096$	$C = 0.79248$	$C = 0.87472$	$C = 0.66540$
2	$C = 0.92466$	$C = 0.77398$	$C = 0.92190$	$C = 0.77398$	$C = 0.87703$	$C = 0.68491$
3	$C = 0.92668$	$C = 0.75000$	$C = 0.92444$	$C = 0.75000$	$C = 0.87680$	$C = 0.62652$
4	$C = 0.92910$	$C = 0.74009$	$C = 0.92731$	$C = 0.74009$	$C = 0.87976$	$C = 0.62696$
5	$C = 0.91625$	$C = 0.79248$	$C = 0.91096$	$C = 0.79248$	$C = 0.87472$	$C = 0.66540$

where the value $y_i, 1 \leq i \leq N$ in the vector y corresponds to the ith track. In this section, we will be interested in the following types of N-track ITI-reducing constraints imposed on elements of the vector y.

1. The number of zeros between two nonzero elements in y is at least $d, d > 0$ (the "zero" constraint)
2. Neighboring elements are either of the same sign or at least one of them is zero (the "sign" constraint)

It is not difficult to see that both of these constraints reduce ITI. The first constraint, also referred to as a perpendicular minimum runlength constraint or d-constraint, requires that the signals in d tracks neighboring the ith track all have zero crossings when the signal in track i is nonzero. This requirement completely cancels ITI. The second constraint does not completely eliminate ITI, but rather allows only "constructive" ITI, which in fact improves performance. This method is a generalization of the approaches taken by Ahmed et al. [1] and Davey et al. [11].

The Shannon noiseless capacities of such constraints for various PR targets of interest are given in Table 10.1. It can be seen that the rate penalty for ITI constraints (especially for the "sign" constraint) is not high. We were able to construct a 100% efficient rate 3/4 three-track code with 256 states for the PR2 and PR4 channels for the "zero" constraint.

10.4.2 Constrained Coding for Improved Synchronization

Multitrack codes that improve timing recovery and the immunity of synchronization schemes to media defects have been introduced and extensively studied by Vasic et al. [26, 40–42]. Synchronization immunity to media defects can be improved by allowing the clock recovery circuit to use any group of l tracks on which the k-constraint is satisfied. This new class of codes was named redundant multitrack (d, k) codes, or (d, k, N) codes, with N being the number of tracks. The redundancy $r = N - l$ is the number of bad tracks out of N tracks that can be tolerated while maintaining synchronization. Orcutt and Marcellin [23] considered the (d, k, N, l) constraint assuming $k \geq d$.

In [39], the starting point of the construction was a class of binary multitrack codes with very good clock recovery properties. These are (d, k, N, l) constrained codes with $k < d$. In the same paper, a reduced-state graph model of the constraint was defined, and based on it the Shannon capacities of the constraint were computed. The same approach was then extended for the case of multiamplitude, multitrack runlength limited (d, k) constrained channels with clock redundancy [41] (mainly for applications in optical recording channels). In [20] the Shannon capacities of these channels were computed and some simple, 100% efficient codes were constructed. The vertex labels of the graph of the constraint were modified to insure that they are independent on the number of tracks. This resulted in significant computational savings for the case when the number of tracks is large. In the same paper it was also shown that the increase of the number of tracks written in parallel provides a significant improvement of per-track capacity for a more restrictive clocking constraint case $k < d$.

10.4.3 Low-Complexity Encoder and Decoder Implementations

Substantial progress has been made in the theory of constrained codes using symbolic dynamics [23, 24], as well as in low complexity encoding and decoding algorithms [15]. Despite this progress, the design of constrained codes remains difficult, especially for large constraint graphs such as multitrack constraint graphs.

One class of codes that is not very difficult to implement is the class of multitrack enumerative codes. The idea of enumerative coding originates from the work of Labin and Aigrain [20], Kautz [16], and has been formulated as a general enumeration scheme for constrained sequences by Tang and Bahl [35] (see also Cover [9]). It was also used by Immink et al. [12] as a practical method for the enumeration of (single-track) (d, k) sequences and by Orcutt and Marcellin [29] for multitrack (d, k) block codes. For this type of codes, it is essential to design an encoder/decoder pair that does not require large memory for storing the constraint graph (because this would be prohibitively complex). This design criterion is met by creating only portions of the graph used in different stages of enumerative encoding/decoding.

10.5 Multitrack Soft Error-Event Correcting Scheme

Multitrack, soft error-event correcting schemes were recently introduced by Vasic and Venkateswaran [46]. This error-correcting scheme supports soft error-event decoding and has complexity slightly higher than the complexity of N MLSD detectors. The idea is to design a multitrack version of the "postprocessor" which has been discussed by Cideciyan et al. [5], Conway [8], and Sonntag and Vasic [34], and is employed in most of today's commercial disk drives. The generalization of the postprocessor concept to multiple tracks is nontrivial because the detector must be designed to mitigate the effect of errors caused both by ISI and ITI, as explained below. For more details, the reader is referred to [46].

As mentioned above, a magnetic recording channel is characterized by a PR polynomial. The appropriate PR polynomial depends on the recording density; this density unavoidably increases when going from the outer sectors of the disk towards the inner sectors. Implementation of hard drive subsystems such as tracking servo, timing recovery, and automatic gain control would be unacceptably costly if the PR polynomials were allowed to vary with recording density. Thus, practical systems typically use only two partial response polynomials, one for high-density regions and another for low-density regions.

As a consequence, the employed PR response polynomial is closely, but not completely, matched to the discrete-time channel response, and the noise samples are not independent. Moreover, the noise samples are neither Gaussian nor stationary because of media noise. The detector complexity, however, dictates the use of an MLSD detector with a squared Euclidean distance metric as opposed to a more complex detector with optimal metric. The metric inaccuracy is in practice compensated by a so-called *postprocessor* [34]. The idea of postprocessing can be generalized so as to apply for the multitrack scenario.

The Viterbi detector produces some error patterns more often than others and the most frequently occurring error patterns are a function of the PR polynomial and the noise coloration. The most frequent patterns, called the *dominant error sequences* or *error events* $E = \{e_i\}_{1 \leq i \leq I}$, and their probabilities can be obtained experimentally [34] or analytically [2]. The index i referrs to an *error type*. Note that the relative frequencies of error events strongly depend on the recording density.

The block diagram of a multitrack soft error-event correcting system is shown in Figure 10.2. User data is encoded by a high-rate error control code (ECC) as shown in Figure 10.1. The decoding algorithm combines syndrome decoding and soft-decision error correction. The error-event likelihoods needed for soft decoding are computed from the channel sequence by using an algorithm proposed by Conway [8] (see also [34].) By using the syndrome calculated for a received codeword, a list of all possible *positions* where error events might have occurred is created. Then the error-event likelihoods are used to select the most likely position and most likely type of the error event. Decoding is completed by finding the error event position and type [12].

Error detection is based on the fact that one can calculate the likelihoods of each of the dominant error sequences at each point in time. The parity bits introduced by the ECC serve to detect the errors, and to provide some localization of the error type and the position where the error ends. The likelihoods are then used to choose the most likely error events (type and position) for correction. The likelihoods in the k-th

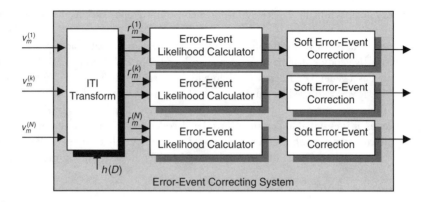

FIGURE 10.2   Block diagram of multitrack soft error-event correction scheme.

track are calculated using the following signals:

a. The read-back signal obtained from the kth head
b. The convolution of the signal obtained by combining the MLSD estimates of all tracks and the channel partial response
c. The convolution of an alternative data pattern (given by a particular error-event) and the channel partial response

The key idea of this approach is that not all candidate data patterns are considered, but only those whose differences with respect to MLSD estimates form the set E.

During each clock cycle, K error-events with largest likelihoods are chosen, and the syndromes for these error events are calculated. Throughout the processing of each block, a list of the K most likely error events, along with their associated error types, positions and syndromes, is maintained. At the end of the block, when the list of candidate error events is finalized, the likelihoods and syndromes are calculated for each of the $\binom{K}{s}$ combinations of s candidate error events that are possible. After disqualifying those s-sets of candidates that overlap in the time domain, and those candidates and s-sets of candidates which produced a syndrome which does not match the actual syndrome, the candidate or s-set which remains and which has the highest likelihood is chosen for correction.

Practical reasons (such as decoding delay and thermal asperities) dictate the use of short codes, and consequently, in order to keep the code rate high, only a relatively small number of parity bits is allowed, making the design of error-event detection codes nontrivial. If all errors from a list E were made up of contiguous bit errors and were shorter than m, then a cyclic code with $n - k = m$ parity bits could be used to detect a single error event [35]. In [44], Vasic introduced a systematic graph-based approach to construct codes for this purpose. This method was further developed in [45].

The performance of multitrack soft error-event decoding will be illustrated on the example of cyclic BCH codes; these codes have fairly large minimum distance, which allows for improvements of the performance of the detector. Single-track state of the art systems use different codes designed to perform well for a given set of dominant error events. However, these codes are not available in the public domain, and therefore will not be discussed here.

Figure 10.3 shows the performance of the multitrack $s = 2$ error-event correction scheme (denoted by 2-EE) based on the (255, 239) BCH code. These results have been obtained for a three-track Lorentzian channel, equalized to the E^2PR4 target, with user bit density of 2.5, and 10% ITI (i.e., $\alpha_1 = 0.1$). The channel bit density due to the [255, 239] BCH code is 2.67. This scheme provides a SNR gain of 3.5 dB at BER $= 10^{-6}$. An extension of the research regarding this multitrack scheme would include the investigation of various PR targets and ECC schemes, the characterization of effects of track misalignment, residual ISI and ITI on the performance of this scheme as well as low-complexity implementations.

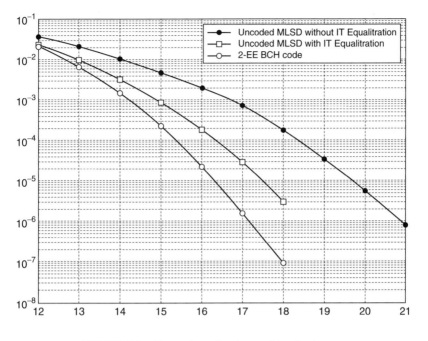

FIGURE 10.3 Comparison of various multitrack schemes.

Acknowledgement

This work is supported by the NSF under grant CCR-0208597.

References

[1] M.Z. Ahmed, T. Donnelly, T., P.J. Davey, and W.W. Clegg, Increased areal density using a two-dimensional approach, *IEEE Trans. Magn.*, vol. 37, no. 4, part: 1, pp. 1896–1898, July 2001.

[2] S.A. Altekar, M. Berggren, B.E. Moison, P.H. Siegel, and J.K. Wolf, Error-event characterization on partial-response channels, *IEEE Trans. Inform. Theory*, vol. 45, no. 1, pp. 241–247, January 1999.

[3] L.R. Bahl, J. Cocke, F. Jelinek, and J. Raviv, Optimal decoding of linear codes for for minimizing symbol error rate, *IEEE Trans. Inform. Theory*, vol. IT-20, no. 2, pp. 284–287, March 1974.

[4] L.C. Barbosa, Simultaneous detection of readback signals from interfering magnetic recording tracks using array heads, *IEEE Trans. Magn.*, vol. 26, pp. 2163–2165, September 1990.

[5] R. Cideciyan, J. Coker, E. Eleftheriou, and R.L. Galbraith, Noise predictive maximum likelihood detection combined with parity-based postprocessing, *IEEE Trans. Magn.*, vol. 37, no. 2 pp. 714–720, March 2001.

[6] S.H. Charrap, P.L. Lu, and Y. He, Thermal stability of recorded information at high densities, *IEEE Trans. Magn.*, part 2, vol. 33, no. 1, pp. 978–983, January 1997.

[7] K.M. Chugg, X. Chen, and M.A. Neifeld, Two-dimensional equalization in coherent and incoherent page-oriented optical memory, *J. Opt. Soc. Am.*, vol. 16, no. 3, pp. 549–562, March 1999.

[8] T. Conway, A new target response with parity coding for high density magnetic recording channels, *IEEE Trans. Magn.*, vol. 34, no. 4, pp. 2382–2386, July 1998.

[9] T.M. Cover, Enumerative source coding, *IEEE Trans. Inform. Theory*, vol. IT-19, pp. 73–77, January 1973.

[10] P.J. Davey, T. Donnelly, D.J. Mapps, and N. Darragh, Two-dimensional coding for multitrack recording system to combat inter-track interference, *IEEE Trans. Magn.* vol. 34, pp 1949–1951, July 1998.

[11] P.J. Davey, T. Donnelly, and D.J. Mapps, Two-dimensional coding for a multiple-track, maximum-likelihood digital magnetic storage system, *IEEE Trans. Magn.*, vol. 30, no. 6, Part: 1–2, pp. 4212–4214, November 1994.

[12] M. Despotovic and B. Vasic, Hard disk drive recording and data detection, *IEEE International Conference on Telecommunications Cable and Broadcasting Services*, Telsiks, Nis, vol. 2, pp. 555–561, October 8–13, 2001.

[13] I.J. Fair, W.D. Gover, W.A. Krzymien, and R.I. MacDonald, Guided scrambling: A new line coding technique for high bit rate fiber optic transmission systems, *IEEE Trans. Commun.*, vol. 39, pp. 289–297, February 1991.

[14] K.A.S. Immink, A practical method for approaching the channel capacity of constrained channels, *IEEE Trans. Inform. Theory*, vol. 43, no. 5, pp. 1389–1399, September 1997.

[15] K.A.S. Immink, *Coding Techniques for Digital Recorders*, Prentice-Hall Int., Englewood Cliffs, NJ (UK) Ltd., 1991.

[16] W.H. Kautz, Fibonacci codes for synchronization control, *IEEE Trans. Inform. Theory*, vol. IT-11, pp. 284–292, 1965.

[17] H. Kobayashi and D.T. Tang, Application of partial response channel coding to magnetic recording systems, *IBM J. Res. Develop*, vol. 14, pp. 368–375, July 1979.

[18] E. Kurtas, J. Proakis, and M. Salehi, Reduced complexity maximum likelihood sequence estimation for multitrack high density magnetic recording channels, *IEEE. Trans. Magn.* vol. 35, no. 4, pp. 2187–2193, July 1999.

[19] E. Kurtas, J.G. Proakis, and M. Salehi, Coding for multitrack magnetic recording systems, *IEEE Trans. Inform Theory*, vol. 43, no. 6, pp. 2020–2023, November 1997.

[20] E. Labin and P.R. Asgrain, Electric pulse communication system, U.K. Patent 713 614, 1951.

[21] J. Lee and V.K. Madisetti, Combined modulation and error correction codes for storage channels, *IEEE Trans. Magn.*, vol. 32, no. 2, pp. 509–514, March 1996.

[22] M.W. Marcellin and H.J. Weber, Two-dimensional modulation codes, *IEEE J. Select. Areas Commun.*, vol. 10, pp. 254–266, January 1992.

[23] B. Marcus, P. Siegel, and J.K. Wolf, Finite-state modulation codes for data storage, *IEEE J. Select. Areas Commun.*, vol. 10, no. 1, pp. 5–37, January 1992.

[24] B.H. Marcus, Sofic systems and encoding data, *IEEE Trans. Inform. Theory*, vol. IT-31, pp. 366–377, May 1985.

[25] O. Milenkovic and B. Vasic, Power spectral density of multitrack (0, G/I) codes, *Proceedings of the International Symposium on Information Theory*, p. 141, Ulm, Germany 1997.

[26] O. Milenkovic and B. Vasic, Permuation (d, k) codes: efficient enumeration coding and phrase length distribution shaping, *IEEE Trans. Inform. Theory*, vol. 46, no. 7, pp. 2671–2675, November 2000.

[27] H. Muraoka and Y. Nakamura, Multitrack submicron-width recording with a novel integrated single pole head in perpendicular magnetic recording, *IEEE Trans. Magn.*, vol. 30, no. 6, Part: 1–2, pp. 3900–3902, November 1994.

[28] E.K. Orcutt and M.W. Marcelin, Redundant multitrack (d, k) codes, *IEEE Trans. Inform. Theory*, vol. 39., no. 5., pp. 1744–1750, September 1993.

[29] E.K. Orcutt and M.W. Marcellin, Enumerable multitrack (d, k) block codes, *Trans. Inform. Theory*, vol. 39, no. 5, pp. 1738–1744, September 1993.

[30] R.E. Swanson and J.K. Wolf, A new class of two-dimensional RLL recording codes, *IEEE Trans. Magn.*, vol. 28, pp. 3407–3416, November 1992.

[31] E. Soljanin and C.N. Georghiades, Coding for two-head recording systems, *IEEE Trans. Inform. Theory*, vol. 41, no. 3, pp. 747–755, May 1995.

[32] E. Soljanin and C.N. Georghiades, A five-head, three-track, magnetic recording channel, *Proceedings, IEEE International Symposium on Information Theory*, p. 244, 1995.

[33] E. Soljanin and C. Georghiades, Multihead detection for multitrack recording channels, *IEEE Trans. Info. Theory*. vol. 44, no. 7, pp. 2988–2997, November 1998.

[34] J.L. Sonntag and B. Vasic, Implementation and bench characterization of a read channel with parity check post processor, *Digest of TMRC 2000*, Santa Clara, CA, August 2000.

[35] D.T. Tang and L.R. Bahl, Block codes for a class of constrained noiseless channels, *Inform. Contr.*, vol. 17, pp. 436–461, 1970.

[36] D.D. Tang, H. Santini, R.E. Lee, K. Ju, and M. Krounbi, A design concept of array heads, *IEEE Trans. Magn.*, vol. 33, no. 3, pp. 2397–2401, May 1997.

[37] H. Thapar and A. Patel, A class of partial response systems for increasing storage density in magnetic recording, *IEEE Trans. Magn.*, vol. 23, no. 5, September 1987.

[38] B. Vasic, G. Djordjevic, and M. Tosic, Loose composite constrained codes and their application in DVD, *IEEE J. Select. Areas in Communications*, vol. 19 no. 4, pp. 765 –773, April 2001.

[39] B. Vasic, Capacity of channels with redundant runlength costraints: The $k < d$ case, *IEEE Trans. Inform. Theory*, vol. 42, no. 5, pp. 1567–1569, September 1996.

[40] B. Vasic, O. Milenkovic, and S. McLaughlin, Power spectral density of multitrack (0, G/I) codes, *IEE Elect. Lett.*, vol. 33, no. 9, pp. 784–786, 1997.

[41] B. Vasic, S. McLaughlin, and O. Milenkovic, Channel capacity of M-ary redundant multitrack runlength limited codes, *IEEE Trans. Inform. Theory*, vol. 44, no. 2, March 1998.

[42] B. Vasic and O. Milenkovic, Cyclic two-dimensional IT reducing codes, *Proceedings of the International Symposium on Information Theory*, p. 414, Ulm, Germany 1997.

[43] B. Vasic, Spectral analysis of multitrack codes, *IEEE Trans. Inform. Theory*, vol. 44, no. 4, pp. 1574–1587, July 1998.

[44] B. Vasic, A graph based construction of high rate soft decodable codes for partial response channels, in *Proceedings of ICC-2001*, vol. 9, pp. 2716–2720, June 11–15, Helsinki, Finland.

[45] B. Vasic, Error event correcting codes for partial response channels, 5th-*IEEE International Conference on Telecommunications Cable and Broadcasting Services*, Telsiks, Nis, vol. 2, pp. 562–566, October 8–13, 2001.

[46] B. Vasic and V. Venkateswaran, Soft error-event decoding for multitrack magnetic recording channels, *IEEE Trans. on Magn.*, Vol. 40, No. 2, pp. 492–497, March 2004.

[47] M.P. Vea and J.M.F. Moura, Magnetic recording channel model with inter-track interference, *IEEE Trans. Magn.*, vol. 27, no. 6, pp. 4834–4836, November 1991.

[48] A. van Wijngaarden and K.A.S. Immink, Construction of constrained codes using sequence replacement techniques, *Proceedings of 1997 IEEE International Symposium on Information Theory*, p. 144, 1997.

[49] J.K. Wolf and D. Chun, The single burst error detection performance of binary cyclic codes, *IEEE Trans. Commun.*, vol. 42, no. 1, pp. 11–13, January 1994.

[50] R. Wood, Detection and capacity limits in magnetic media noise, *IEEE Trans. Magn.*, vol. MAG-34, No. 4, pp. 1848–1850, July 1998.

[51] R. Wood, The feasibility of magnetic recording at 1 terabit per square inch, *IEEE Trans. Magn.*, vol. 36, no. 1, pp. 36–42, January 2000.

[52] Y. Wu, J.A. O'Sullivan, R.S. Indeck, and N. Singla, Iterative detection and decoding for seperable two-dimensional intersymbol interference, *IEEE Trans. on Magn.*, Vol. 39, No. 4, pp. 2115–2120, July 2003.

11

Turbo Codes

Mustafa N. Kaynak
Arizona State University
Tempe, AZ

Tolga M. Duman
Arizona State University
Tempe, AZ

Erozan M. Kurtas
Seagate Technology
Pittsburgh, PA

Mustafa N. Kaynak: received the B.Sc. degree from Middle East Technical University, Ankara, Turkey, in 1999 and the M.Eng. degree from National University of Singapore, Singapore in 2001, both in electrical engineering, and is currently working toward the Ph.D. degree in electrical engineering at Arizona State University, Tempe, AZ.

His research interests are in digital and wireless communications, magnetic recording channels, iterative decoding algorithms, channel coding, for recording and wireless communication channels, turbo and low density parity check codes.

Tolga M. Duman: received the B.S. degree from Bilkent University in 1993, M.S. and Ph.D. degrees from Northeastern University, Boston, in 1995 and 1998, respectively, all in electrical engineering. Since August 1998, he has been with the Electrical Engineering Department of Arizona State University first as an assistant professor (1998–2004), and currently as an associate professor. Dr. Duman's current research interests are in digital communications, wireless and mobile communications, channel coding, turbo codes, coding for recording channels, and coding for wireless communications.

Dr. Duman is the recipient of the National Science Foundation CAREER Award, IEEE Third Millennium medal, and IEEE Benelux Joint Chapter best paper award (1999). He is a senior member of IEEE, and an editor for *IEEE Transactions on Wireless Communications*.

Erozan M. Kurtas: received the B.Sc. degree from Bilkent University, Ankara, Turkey, in 1991 and M.Sc. and Ph.D. degrees from Northeastern University, Boston, MA, in 1993 and 1997, respectively.

His research interests cover the general field of digital communication and information theory with special emphasis on coding and detection for inter-symbol interference channels. He has published over 75 book chapters, journal and conference papers on the general fields of information theory, digital communications and data storage. He has seven pending patent applications. Dr. Kurtas is currently the Research Director of the Channels Department at the research division of Seagate Technology.

Abstract

This chapter describes the parallel and serial concatenated convolutional codes, i.e., turbo codes. First, the use of these codes for additive white Gaussian noise (AWGN) channels is discussed. The iterative decoding algorithm that uses two soft output component decoders as building blocks is presented in detail. Then, the use of concatenated codes over recording channels is reviewed. Partial response channels (e.g., PR4, EPR4) with additive white Gaussian noise are used to illustrate the main ideas. The concepts of turbo decoding without turbo equalization and with turbo equalization as well as serial concatenation of a convolutional code with the channel are explained. As more accurate models, Lorentzian channels are also considered for longitudinal recording. Finally, the effects of precoding and media noise are studied in some depth, and the existence of burst errors is noted.

11.1   Principles of Turbo Coding

In his celebrated work, Shannon proved that, an arbitrarily low probability of error can be obtained in digital transmission over a noisy channel provided that the transmission rate is less than a certain quantity called "channel capacity" [1]. He also proved that randomly selected codes of very large block lengths can achieve this capacity. However, his proofs were not constructive, that is, he did not give any practical coding/decoding approaches. Since then, coding theorists have been attacking the problem of designing good channel codes with a lot of success (e.g., refer to the standard textbooks [2, 3] for a review of error correcting codes).

Since codes with very long block lengths are expected to perform very well, methods for constructing very long block length codes, that can be encoded and decoded relatively easily, have been investigated intensively. One way of obtaining a large block length code is concatenating two simple codes so that the encoding and the decoding of the overall code is less complex. For instance, Forney concatenated two simple codes in series to obtain a more powerful overall code [4].

Turbo codes proposed in 1993 by Berrou et al., represent a different form of concatenating two simple codes to obtain codes that achieve a near Shannon limit performance [5]. In turbo coding, two systematic recursive constituent convolutional encoders are concatenated in parallel via a long interleaver. For decoding, a suboptimal iterative decoding algorithm is employed. Let us now describe the turbo coding principle in detail.

11.1.1   Parallel Concatenated Convolutional Codes

Turbo codes generated an abundance of literature after their invention in 1993, mainly because of their exceptional performance for very low signal to noise ratios. For example, by using a rate $1/2$ turbo code with an interleaver size of 65536 and memory-4 component codes, Berrou et al. [5] demonstrated that a bit error rate of 10^{-5} can be obtained within 0.7 dB of the channel capacity over an AWGN channel. We note that the capacity for rate $1/2$ transmission is at 0 dB[1].

The idea in turbo coding is to concatenate two recursive systematic convolutional codes in parallel via an interleaver. For encoding, the information sequence is divided into blocks of a certain length. The input of the first encoder is the information block and the input of the second encoder is an interleaved version

---

[1]By the statement "the capacity is at 0 dB," we mean that the channel capacity is equal to the transmission rate, that is, $1/2$, when the signal to noise ratio is 0 dB.

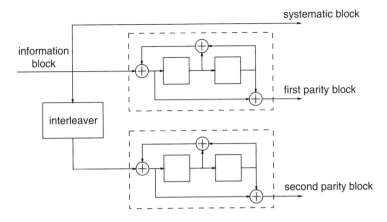

FIGURE 11.1 Rate 1/3 turbo code with 5/7 component codes.

of the information block. The systematic bits and the parity bits of the first and second encoders constitute the codeword corresponding to the information block. As an example, the block diagram of the rate 1/3 turbo code with 5/7 (in octal notation) component convolutional codes[2] is shown in Figure 11.1.

A code rate of 1/3 is not acceptable for many applications. Fortunately, there are means of obtaining higher rate turbo codes that still have a very good performance. To achieve higher code rates, different rate component convolutional codes can be used or certain puncturing schemes can be employed. For example, the rate 1/2 code example provided in the original turbo coding paper [5] is obtained by puncturing half of the parity bits.

For parallel concatenated convolutional codes, the component codes should be chosen as convolutional codes with feedback (hence, the term recursive), on the other hand, they can be nonsystematic. It is also common practice to choose identical component encoders. One can argue that any recursive convolutional encoder has an equivalent feedback-free representation, so the requirement about the recursiveness of the component codes may not be clear. To explain this, we note that, although the codewords generated by the two encoders are the same, they correspond to different input sequences. Therefore, for the purpose of constructing turbo codes, the recursive encoder and its feedback-free version are not equivalent.

The problem of selecting the feedback and the feedforward generator polynomials of the component codes to optimize the performance is studied in [6] and [7], and it is found that, the feedback polynomial should be primitive, whereas we have more flexibility in the selection of the feedforward polynomial. The reason is that, the number of problematic weight 2 information sequences (the sequences that terminate before the end of the block) are lowest if the feedback generator polynomial is primitive.

The other ingredient of the turbo coding scheme, the interleaver, can be chosen pseudo-randomly. When the interleaver is selected very large, in the order of several thousand bits or more, a very good bit error rate performance, usually within 1 dB or so of the Shannon limit is obtained, at the expense of the increased decoding delay. Although pseudo-random interleavers perform well, there are a number of interleaver design techniques that are useful. For instance, the S-random interleaver proposed in [8] provides a significant performance improvement over the pseudo-random interleavers, specifically, for relatively larger signal to noise ratios. For a review of different interleaver design techniques, the reader is referred to [9].

Maximum likelihood decoding of turbo codes is very difficult because of the pseudo-random interleaver used. In general, one has to consider all the possible codewords (there are 2^N possibilities, where N is the

---

[2]The term p/q convolutional code is used to indicate the feedforward and feedback connections in the convolutional code in octal notation. For example, 5/7, or in binary notation 101/111 means that the feedforward link is "connected, not connected and connected" and the feedback link is all connected.

interleaver size), compute the corresponding cost for each one (e.g., Euclidean distance with the received signal), and choose the one with the lowest cost as the transmitted codeword. Even for short interleavers, this is a tedious task. Fortunately, to perform this task practically, an iterative decoding algorithm is proposed in [5]. The iterative decoder offers a near optimal performance, and it is perhaps the most important contribution of the original turbo coding paper [5]. From Shannon's result, we know that codes with large block lengths chosen randomly usually have a very good performance. However, decoding such randomly chosen codes is nearly impossible. In essence, turbo coding is a way to obtain very large block length "random-like" (due to the existence of the interleaver) codes, yet we still have a near-optimal decoding algorithm. Since the iterative decoding algorithm is essential to understand turbo codes, we will describe it in some detail in a separate section.

Let us now give a brief explanation of why turbo codes perform so well. Turbo codes are linear block codes. Therefore, for the purposes of analysis, we can simply assume that the all-zero codeword is transmitted. The possible error sequences corresponding to this transmission are all the nonzero codewords. Consider a codeword with information weight 1, that is, a codeword obtained by encoding a weight one information sequence. Since the component encoders are recursive convolutional encoders, the parity sequences corresponding to this information sequence will not terminate until the end of the block is reached, because the component encoders are infinite impulse response filters. With a good selection of the interleaver, if the single "1" occurs towards the end of the information block for one of the component encoders, it will occur towards the beginning of the input block for the other component encoder. Therefore, the codewords with information weight 1, typically, have a large parity weight, hence a large total weight. Furthermore, the interleaver "breaks down" the sequences with information weight greater than one, that is, if the information block results in a lower weight parity sequence corresponding to one of the encoders, it will have a larger weight parity sequence for the other one. Therefore, most of the codewords will have large Hamming weights and they are less likely to be decoded as the correct codeword when the all-zero codeword is transmitted, at least, for an AWGN channel. On the other hand, the interleaver cannot break down all the sequences, and therefore, there will be some codewords with low weights as well, hence the overall free distance of a turbo code is usually small. For a discussion on the effective free distance of turbo codes, see [10]. Since the number of low weight codewords is typically small, although the asymptotic performance of the code (for large signal to noise ratios) is limited by its relatively low free distance, its performance is very good for low signal to noise ratios. We also note that, the most troublesome error sequences for turbo codes are the ones with information weight 2, since those are the most difficult ones for the interleaver to "break down". A more detailed distance spectrum interpretation of the turbo coding scheme can be found in [11].

To predict the performance of turbo codes, one can use Monte Carlo simulation results obtained by the suboptimal iterative decoding algorithm. However, it is also important to develop performance bounds and compare them with simulation results. In [12] and [13] the union bounding technique is applied to derive an average upper bound on the probability of error for turbo codes over an AWGN channel using maximum likelihood decoding. Other more sophisticated bounds on the performance are developed in [14, 15]. Simulation results show that the iterative turbo decoding algorithms perform very well. Particularly, the simulation results and the union bound are very close to each other (in the region where the union bound is tight) [13], which shows that the iterative decoding algorithm is near optimal. In [16], McEliece et al. demonstrated that, the iterative decoding algorithm is not always optimal and may not even converge. However, they also observed that it converges with a high probability.

Terminating the trellis of a recursive convolutional code (i.e., bringing the state of the encoder to the all-zero state) is not possible by appending a number of zeros at the end of the information sequence due to the existence of the interleaver and due the fact that, the states of the two component encoders are in general different from each other. Instead, depending on the current state of the encoder, a nonzero sequence should be appended. In [17], it has been demonstrated that, if the trellis is not terminated, the performance of the turbo code deteriorates. The trellis termination problem is studied in [18] and an algorithm which does not require the knowledge of the current state for either encoders is proposed. The problem of trellis termination for turbo codes is also studied in [19, 20].

FIGURE 11.2   Serial concatenation of convolutional and block codes.

11.1.2   Other Concatenation Schemes

Serial concatenation of convolutional and block codes are proposed in [21]. In this scheme, two component codes are concatenated serially with an interleaver between them as shown in Figure 11.2. The component codes can be chosen as convolutional or block codes. The input of the outer code is the original binary input sequence and the input of the inner code is the scrambled (interleaved) version of the codeword generated by the outer code. Generalization of this scheme to more than two component codes is straightforward. Convolutional codes are usually preferred over block codes because of the existence of simple soft input-soft output decoders and the possibility of greater interleaver gain. In [21], the code selection criteria are studied and it is shown that, for superior bit error rate performance, the inner encoder must be chosen as a recursive convolutional encoder. On the other hand, we have more flexibility to choose the outer code. However, for a better performance, it should have a large free distance.

For decoding, a suboptimal iterative decoding algorithm based on the information exchange between the two component decoders is used. The details of the decoding algorithm for serially concatenated convolutional codes will be given in a separate section.

There are some obvious generalizations of the standard turbo coding scheme. For instance, one can use linear block codes as the component codes instead of the recursive convolutional codes [22]. Or, one can concatenate more than two component encoders in parallel instead of only two [23].

Another concatenation scheme is the hybrid concatenation introduced in [24]. Basically, in this case, a third code is concatenated in parallel to the two serially concatenated component codes as shown in Figure 11.3.

11.1.3   Other Iteratively Decodable Codes

Another family of iteratively decodable capacity approaching codes is the low density parity check (LDPC) codes. These codes were first introduced in 1962 by Gallager [25] and after having been forgotten for almost 30 years, with the extensive research on "turbo-like" codes and iterative decoding algorithms; recently they were rediscovered by MacKay [26]. LDPC codes are linear block codes [2, 27], and they are represented by a large, randomly generated sparse parity check matrix **H**, that is, very few entries of the parity check matrix are ones and the rest are all zeros.

Similar to the turbo codes, LDPC codes can be decoded using a simple, practically implementable iterative decoding algorithm based on the message passing algorithm (or, belief propagation as it is named in the artificial intelligence community). They can be considered as the most serious competitors to turbo codes in terms of the offered performance and complexity. For example, recent results by Richardson et al.

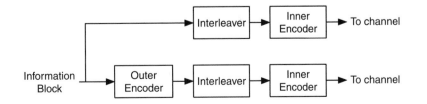

FIGURE 11.3   Hybrid concatenation of convolutional and block codes.

show that, irregular LDPC codes can achieve a performance within 0.13 dB of the channel capacity on AWGN channels [28] and thus outperform any code discovered to date including turbo codes.

11.2 Iterative Decoding of Turbo Codes

In this section, we describe the iterative turbo decoding algorithm in detail based on the presentation in [30]. Consider the standard rate 1/3 turbo coding scheme. Let us denote the input sequence by $d = (d_1, d_2, \ldots, d_N)$ where N is the interleaver length. The encoded sequence consists of three different parts: the systematic bits denoted by $x^s = (x_1^s, x_2^s, \ldots, x_N^s) = d$, the first parity part, that is, the parity bits produced by the first encoder, denoted by $x^{1p} = (x_1^{1p}, x_2^{1p}, \ldots, x_N^{1p})$, and the second parity part denoted by $x^{2p} = (x_1^{2p}, x_2^{2p}, \ldots, x_N^{2p})$, and generated by the second encoder from the interleaved information block.

We assume that the encoded sequence of bits is transmitted using BPSK modulation over an AWGN channel.[3] Therefore, the receiver observes the sequence $R = (R_1, R_2, \ldots, R_N)$ where

$$
\begin{aligned}
R_k &= \left(y_k^s, y_k^{1p}, y_k^{2p} \right) \\
&= \left(\sqrt{E_s}\left(2x_k^s - 1\right) + \eta_k^s, \sqrt{E_s}\left(2x_k^{1p} - 1\right) + \eta_k^{1p}, \sqrt{E_s}\left(2x_k^{2p} - 1\right) + \eta_k^{2p} \right)
\end{aligned}
$$

where E_s is the energy per symbol transmitted. If the overall code rate is r, we have $E_s = r E_b$ where E_b is the energy per information bit. The additive noise terms are distributed according to $\mathcal{N}(0, N_0/2)$, where $N_0/2$ is the noise variance. We refer to E_s/N_0 as the signal to noise ratio (SNR) per symbol, and E_b/N_0 as the signal to noise ratio per information bit.

If some of the parity bits are punctured to obtain a turbo code with a different rate, the decoder operates by inserting "0"s in the observation sequence for the bits that are punctured.

The rest of the section is divided into four parts. In the first part, we present the maximum a-posteriori (MAP) decoding algorithm for convolutional codes [31]. We describe the iterative decoding algorithm of turbo codes that uses the MAP decoders for the component codes as its building blocks in the second part. Then, in the third part, we describe the iterative decoding of serially concatenated convolutional codes. Finally, a brief review of other iterative decoding algorithms, including several simplified turbo decoders, is given.

11.2.1 Maximum A-Posteriori (MAP) Decoder

There are different algorithms for the decoding of convolutional codes. The Viterbi algorithm is the optimal decoder that minimizes the sequence error probability. However, minimizing the sequence error probability does not directly translate into minimizing the bit error probability. In other words, if the performance criteria is the minimum bit error probability, Viterbi algorithm is not optimal. In this case, the MAP decoding algorithm derived in [31] is optimal. Let us now describe the MAP decoder in some detail. Note that, we describe decoder structure with respect to the first component code, the decoder for the second component code is similar.

Let us denote the state of the encoder at time k by S_k, where $k = 0, 1, 2, \ldots, 2^M - 1$, M being the number of memory elements in the encoder. Following a derivation similar to the one in [31], we can show that, the log-likelihood of the information bits can be written as

$$
\begin{aligned}
\Lambda(d_k) &= \log \frac{\Pr[d_k = 1 \mid y^s, y^{1p}]}{\Pr[d_k = 0 \mid y^s, y^{1p}]} \\
&= \log \frac{\sum_m \sum_{m'} \gamma_1(y_k, m', m)\alpha_{k-1}(m')\beta_k(m)}{\sum_m \sum_{m'} \gamma_0(y_k, m', m)\alpha_{k-1}(m')\beta_k(m)}
\end{aligned}
$$

---

[3] The AWGN channel model is not essential, and an arbitrary channel model may also be used.

where $y_k = (y_k^s, y_k^{1p})$, $\alpha_k(m) = p(S_k = m, y_1, \ldots, y_k)$ and $\beta_k(m) = p(y_{k+1}, \ldots, y_N \mid S_k = m)$ are computed using the following forward and backward recursions respectively

$$\alpha_k(m) = \frac{\sum_{m'} \sum_{i=0}^{1} \gamma_i(y_k, m', m) \alpha_{k-1}(m')}{\sum_{m''} \sum_{m'} \sum_{i=0}^{1} \gamma_i(y_k, m', m'') \alpha_{k-1}(m')}$$

$$\beta_k(m) = \frac{\sum_{m'} \sum_{i=0}^{1} \gamma_i(y_{k+1}, m, m') \beta_{k+1}(m')}{\sum_{m''} \sum_{m'} \sum_{i=0}^{1} \gamma_i(y_{k+1}, m'', m') \beta_{k+1}(m')}$$

with the initial values of $\alpha_0(0) = 1$, $\alpha_0(m) = 0$ for $m \neq 0$, $\beta_N(0) = 1$ and $\beta_N(m) = 0$ for $m \neq 0$. For the MAP decoder of the second component code, the initial values are $\alpha_0(0) = 1$, $\alpha_0(m) = 0$ for $m \neq 0$, and $\beta_N(m) = \alpha_N(m)$. Finally, $\gamma_i(y^s, y_k^{1p}, m', m)$ is given by

$$\gamma_i(y^s, y_k^{1p}, m', m) = p(y_k^s \mid d_k = i, S_k = m, S_{k-1} = m') p(y_k^{1p} \mid d_k = i, S_k = m, S_{k-1} = m')$$

$$\Pr[d_k = i \mid S_k = m, S_{k-1} = m'] \Pr[S_k = m \mid S_{k-1} = m'].$$

The probabilities $p(y_k^s \mid d_k = i, S_k = m, S_{k-1} = m')$ and $p(y_k^{1p} \mid d_k = i, S_k = m, S_{k-1} = m')$ are directly dependent on the channel characteristics and the probability $\Pr[d_k = i \mid S_k = m, S_{k-1} = m']$ is either zero or one depending on whether the bit i is associated with the transition from state m' to state m or not. The last term, $\Pr[S_k = m \mid S_{k-1} = m']$ uses the a-priori likelihood information on the bit d_k. Assuming that $L(d_k)$ is the a-priori information, we can write

$$\Pr[S_k = m \mid S_{k-1} = m'] = \begin{cases} \frac{e^{L(d_k)}}{1+e^{L(d_k)}} & \text{if } \Pr[d_k = 1 \mid S_k = m, S_{k-1} = m'] = 1 \\ \frac{1}{1+e^{L(d_k)}} & \text{if } \Pr[d_k = 0 \mid S_k = m, S_{k-1} = m'] = 1 \end{cases}.$$

11.2.2 The Iterative Decoder Structure

In this section, we describe the use of the two MAP decoders for the component codes in order to decode the turbo code. First, let us write the log-likelihood of the information bit d_k as the sum of three different terms as

$$\Lambda(d_k) = \log \frac{\sum_m \sum_{m'} \gamma_1'(y_k, m', m) \alpha_{k-1}(m') \beta_k(m)}{\sum_m \sum_{m'} \gamma_0'(y_k, m', m) \alpha_{k-1}(m') \beta_k(m)} + L(d_k) + \log \frac{p(y_k^s \mid d_k = 1)}{p(y_k^s \mid d_k = 0)}$$

where $\gamma_i'(y_k^{1p}, m', m) = p(y_k^{1p} \mid d_k = i, S_k = m, S_{k-1} = m') \Pr[d_k = i \mid S_k = m, S_{k-1} = m']$.

In the above equation, the first term is the extrinsic information generated by the current decoder by using the code constraints, the second term is the a-priori information and the last term is the systematic likelihood information.

The iterative decoder works as follows. At each iteration step, one of the decoders takes the systematic information (directly from the observation of the systematic part) and the extrinsic log-likelihood information produced by the other decoder in the previous iteration step to compute its new extrinsic log-likelihood information. Then, this updated extrinsic information is fed into the other decoder for the next iteration step. The extrinsic information of both decoders are initialized to zero before the iterations start. The block diagram of the iterative decoding algorithm is presented in Figure 11.4.

Let us denote the systematic log-likelihood information of the input bit d_k by $L_s(d_k)$, the first extrinsic log-likelihood information by $L_{1e}(d_k)$ and the second one by $L_{2e}(d_k)$, $k = 1, 2, \ldots, N$. $L_s(d_k)$ is directly found from the systematic observation and $L_{1e}(d_k)$ and $L_{2e}(d_k)$ (at each iteration step) can be computed from the code constraints by using the MAP decoding algorithm. In fact, the extrinsic information can be computed by using any other decoding algorithm for systematic codes that accepts log-likelihood values of the information bits and produces updated (independent) log-likelihoods, such as the soft output Viterbi algorithm (SOVA) [32].

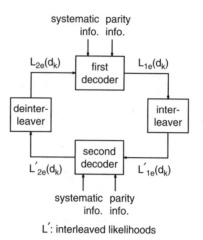

FIGURE 11.4   Iterative (turbo) decoding algorithm.

To summarize, at each iteration step, the log-likelihood of the information bit d_k is given by

$$\Lambda(d_k) = \log \frac{P(d_k = 1 \mid \text{observation})}{P(d_k = 0 \mid \text{observation})}$$
$$= L_s(d_k) + L_{1e}(d_k) + L_{2e}(d_k).$$

The a-priori information (coming from the other decoder) is approximately independent of the extrinsic information generated at the current iteration step. Therefore, further iterations do not deteriorate the performance, and the iterative decoding algorithm converges with high probability. In other words, after a number of iterations, the sum of the three likelihood values converges to the true log-likelihood of the kth information bit, which can be used to make the final decision.

11.2.3   Iterative Decoding of Serially Concatenated Convolutional Codes

Similar to parallel concatenated convolutional codes, the optimal decoding of the serially concatenated convolutional codes is almost impossible due to the interleaver used between the component codes. Therefore, to decode serially concatenated convolutional codes, a practically implementable, suboptimal, iterative decoding algorithm, utilizing soft output MAP component decoders, is used. The block diagram of the decoder is shown in Figure 11.5. At each iteration step, the inner decoder uses the noisy channel observations and the extrinsic log-likelihood ratio (LLR) information of its input block (the codeword generated by the outer encoder) calculated by the outer decoder in the previous iteration and then updates the extrinsic log-likelihood information of its input block using the inner code constraints. This updated extrinsic information, corresponding to the output of the outer code, is used by the outer decoder to

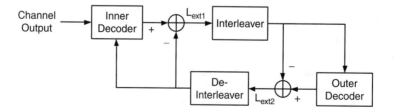

FIGURE 11.5   Iterative decoding of serially concatenated codes.

calculate the log-likelihood information of both the input and the output bits of the outer code based on the code constraints. The LLRs of the output bits of the outer code are sent to the inner decoder for use in the next iteration. After a number of iterations, the log-likelihood information of the input bits generated by the outer decoder is used to recover the transmitted information bits.

There is an important difference between the decoding algorithms of parallel and serial concatenated convolutional codes. For serial concatenation, although the inner decoder calculates only the LLR of its input block, the outer decoder calculates LLRs of both its input and output blocks. However, for parallel concatenation, only the LLRs of the original input block are calculated by both decoders, because in parallel concatenation, for both decoders the noisy channel outputs are available and the extrinsic information exchanged between the decoders are used as the a-priori information by the component decoders to perform the MAP decoding. However, for serial concatenation, the input of the inner code is the codeword generated by the outer code. Therefore, in order for the inner decoder to perform the MAP decoding, it requires the a-priori information, which are the LLRs of the coded bits of the outer code in this case. Likewise for the outer decoder, there is no channel output, so the LLR calculated by the inner decoder for its input is used by the outer decoder instead of the noisy channel outputs as in the parallel concatenation.

In terms of maximum-likelihood performance, serial concatenation is superior to parallel concatenation. In addition, the error floor, caused by the relatively low free distance of the turbo codes, is observed at lower bit error rates for serially concatenated codes.

11.2.4 Other Iterative Decoding Algorithms

The original iterative decoding algorithm for turbo codes [5] is complex. However, there are other simplified iterative decoding algorithms [33–36]. These algorithms provide an appreciable decrease in the complexity of the component decoders at the expense of some performance degradation (typically about 0.5 dB). Jung [37] presents a good comparison of various iterative decoding schemes for turbo codes. Other studies on the iterative decoding algorithm are reported in [38, 39]. Recently, it has also been observed that the original turbo decoding algorithm, among others, can be considered as a special case of a broader class of "belief propagation" algorithms for loopy Bayesian networks that are extensively studied in other fields, especially in artificial intelligence [40].

A general study of iterative decoding of block and convolutional codes is presented in [41]. Using log-likelihood algebra, the authors show that, any decoder that accepts soft inputs (including a-priori values) and delivers soft outputs that can be split into three terms: the soft channel input, the a-priori input, and the extrinsic value, can be used for decoding the component codes. The MAP decoding algorithm derived in [31] and used in [5, 30] and the SOVA algorithm developed in [32, 36] are in this category. Furthermore, the authors provide algorithms for soft decoding of systematic linear block codes, which makes it practical to use block codes (with certain restrictions due to increased complexity in decoding) as component codes in turbo code construction.

11.3 Performance of Turbo Codes over AWGN Channels

In this section, we present a set of results on the performance of turbo codes over AWGN channels. We consider the turbo code with 5/7 component convolutional codes with pseudo-random interleaver, for various block lengths (interleaver lengths). In Figure 11.6, the bit error rate performance is shown with respect to the number of decoding iterations for $N = 1000$, $R_c = 1/2$ turbo code obtained by puncturing half of the parity bits. While the bit error rate (BER) performance improves significantly during the first few iterations; after that, the performance gain is not very significant. Using this particular turbo code, at a BER of 10^{-5}, we can obtain an approximate coding gain of 7 dB over the uncoded system.

In Figure 11.7, the BER performance of $R_c = 1/2$ turbo code is shown for different interleaver lengths after 18 iterations. Larger interleavers improve the BER performance significantly, however the decoding

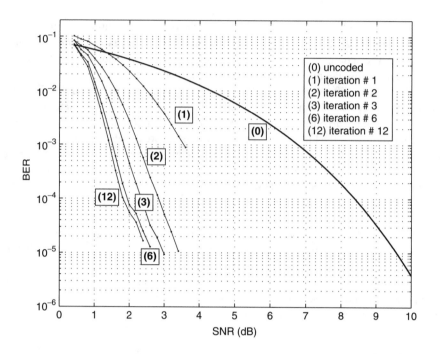

FIGURE 11.6   Performance of the iterative turbo decoder as a function of the number of iterations.

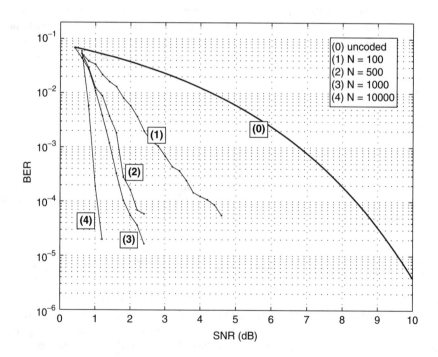

FIGURE 11.7   Performance of the turbo code with 5/7 component codes with different interleaver lengths.

delay increases as well. For example, by choosing $N = 10000$ we obtain a gain of 3 dB over the one using $N = 100$ at a BER of 10^{-4}, and the gain is even more for lower BER values. This improvement is due to the fact that, by using larger interleavers we effectively increase the probability of breaking down the input sequences that will result in low weight codewords, thus we have a larger interleaver gain. We also note that, the capacity for rate 1/2 transmission over an AWGN channel is at 0 dB and by using an interleaver of length 10000, we can obtain a bit error rate of 10^{-5} within 1.2 dB of the channel capacity. Simply by employing larger interleavers we can approach to the channel capacity further [5].

11.4 Recording Channels

Digital magnetic recording channels can be modelled as binary input inter-symbol interference (ISI) channels. In this chapter, we use two specific channel models. The first one is a simplistic partial response (PR) channel representing the recording channel with an equivalent discrete ISI channel with additive white Gaussian noise. This model can be used to represent an "ideal" recording channel. The second one is a Lorentzian channel model (assuming longitudinal recording) which is a more accurate approximation of the realistic magnetic recording channels.

11.4.1 Partial Response Channels

Partial response channels are nothing but ISI channels. For example in magnetic recording, important partial response channels include the PR4 channel (i.e., $(1 - D^2)$) and the EPR4 channel (i.e., $(1 + D - D^2 - D^3)$), where D is the delay operator. In other words, PR4 and EPR4 refer to the equivalent discrete channel models of $y_k = x_k - x_{k-2} + v_k$ and $y_k = x_k + x_{k-1} - x_{k-2} - x_{k-3} + v_k$ respectively, where y_k is the channel output, v_k is additive white Gaussian noise and x_k is the input to the ISI channel.

A PR channel can be considered as rate one nonbinary convolutional code (i.e., the channel outputs are not necessarily 0 or 1). Therefore, it can be modelled as a finite state machine and a trellis can be used to represent its input-output relationship. The optimal method (in terms of minimizing the bit error probability) to recover the information bits from the channel output is to use a MAP decoder (usually called the channel detector) matched to the trellis of the PR channel. This MAP decoder calculates the log-likelihood information of the channel inputs from which the information bits are recovered. Viterbi or soft output Viterbi algorithms can be used to recover the information bits as well.

We define the signal to noise ratio as SNR $= \frac{E_b}{N_0}$, where E_b is the energy per information bit and $N_0/2$ is the two sided additive white Gaussian noise power spectral density. Here $E_s = E_b \cdot R_c$, where R_c is the code rate and E_s is the average energy of the channel outputs.

11.4.2 Realistic Recording Channel Models

The block diagram of a more realistic recording channel model including an outer turbo code is shown in Figure 11.8. Instead of directly recording the uncoded data sequence, we first use an error correcting code, which is the turbo code for our case, to obtain the coded data sequence. This sequence is then interleaved using a pseudo-random interleaver, and may be precoded to obtain another bit sequence. The precoded bit sequence is fed into the write current driver which generates a two-level waveform called write current. The mapping from the binary (precoded) data sequence to the write current is done in such a way that a change in the write current corresponds to a "1", and no change corresponds to a "0". This is called NRZI (non-return-to-zero inverse) recording. Finally, the write head magnetizes the medium in one direction or the other depending on the polarity of the write current.

For longitudinal recording, in the readback process, the output voltage of the readback head corresponding to an isolated positive transition is well modelled as a Lorentzian pulse given by

$$h(t) = \frac{1}{1 + \left(\frac{2t}{T_{50}}\right)^2}$$

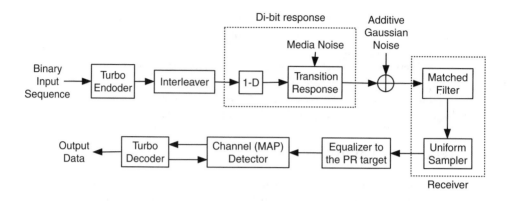

FIGURE 11.8   Block diagram of the magnetic recording system with actual write/read.

where T_{50} is the width of the Lorentzian pulse at 50% amplitude level. The normalized density of the system is defined as $D_n = \frac{T_{50}}{T}$, where T is the duration of each bit interval.

The noiseless readback signal is of the form

$$r(t) = \sum_{k=-\infty}^{\infty} b_k h(t - kT)$$

where $b_k = a_k - a_{k-1}$ is the transition sequence corresponding to the recording channel input a_k. The di-bit response shown in Figure 11.8 corresponds to impulse response of the channel and it is given as $h(t) - h(t - T)$.

The noise in a magnetic recording system is a combination of media noise, head noise and thermal noise generated in the preamplifier. The latter two components can be well modelled as additive Gaussian noise. On the other hand, media noise which may be the result of the stochastic fluctuations on the position of written transitions cannot be generally modelled as additive [42, 43] and it degrades the bit error rate performance significantly. Ignoring the media noise, we define the SNR as

$$\text{SNR} = \frac{E_i}{N_0}$$

where $E_i = \frac{\pi}{2 \times T_{50}}$ is the energy of the impulse response of the recording channel, that is, the derivative of the isolated transition response $h(t)$ and $N_0/2$ is the two sided additive white Gaussian noise power spectral density.

For this channel model the optimum detection consists of a filter matched to $p(t)$, symbol rate sampler and a maximum likelihood sequence detector [44]. In general, the overall noise is not Gaussian, therefore, the use of a matched filter is not optimal. Alternatively, one might use a low pass filter instead of the matched filter as done in most practical systems. After uniform sampling, the receiver output is usually equalized to an appropriate partial response target using a linear equalizer to reduce the computational complexity of the channel detector following the uniform sampler. The function of the channel detector is to compute the log-likelihood values of the transmitted bits for the outer error correction code decoder.

11.5   Turbo Codes for Recording Channels

Turbo codes are applied to digital magnetic recording successfully [45–52]. In particular the simplistic PR channel model assuming Gaussian noise and ideal equalization to the PR target are used [45–51], and large coding gains are obtained. Additionally in [52], it is shown that the performance improvement offered

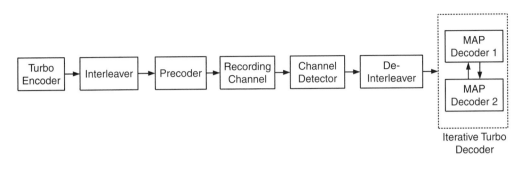

FIGURE 11.9 Block diagram of turbo coding without turbo equalization.

by the turbo coded system is preserved even if the ideal system is replaced by a more realistic Lorentzian channel model (without including the media noise).

We can classify the use of turbo codes for magnetic recording channels in three categories; turbo coding without turbo equalization, turbo coding with turbo equalization and convolutional coding with an interleaver (i.e., serial concatenation of a convolutional code with the ISI channel). These three schemes are detailed in the following subsections.

11.5.1 Turbo Coding without Turbo Equalization

In this scheme, whose block diagram is shown in Figure 11.9, the underlying error correcting code which is used to produce the channel coded bits is a parallel concatenated convolutional code. To obtain the bit sequence that will be stored in the magnetic medium, first the coded data sequence is generated by the turbo encoder. Then, the coded bits are interleaved using a pseudo-random interleaver, and may be precoded to obtain another bit sequence. In this case, the existence of the precoder is not crucial to obtain an interleaving gain, however different precoders will perform differently [53].

The outer decoder is an iterative turbo decoder which uses the log-likelihood values for the channel coded bits (i.e., the systematic and the transmitted parity bits) which are produced by the channel detector. Clearly, appropriate de-interleaving should be employed while passing the computed log-likelihood values to the turbo decoder. The soft information (i.e., the log-likelihoods) about the coded bits are — strictly speaking — correlated. However, since their statistics are not easy to characterize, the decoder assumes that they are the log-likelihoods of the observations from a BPSK transmission over an AWGN channel. To perform the turbo decoding, the variance of the additive Gaussian noise should be specified. Here, we assume that the noise variance is $\frac{N_0}{2}$ though better alternatives may exist. Fortunately, the iterative turbo decoding algorithm is very robust, and although there is not a good reason for assuming an AWGN channel with the specified noise variance, it works properly, that is, errors are corrected in the subsequent iteration steps as it will be illustrated later.

In this scheme, we do not allow the passage of information from the turbo decoder back to the channel detector. Therefore, this scheme is called turbo coding without turbo equalization.

11.5.2 Turbo Coding with Turbo Equalization

This scheme also uses a turbo code as the underlying error correcting code. The only difference is in the decoding algorithm. The extrinsic (new) information about the coded bits produced by the turbo decoder (using the code constraints) is uncorrelated with the original log-likelihoods computed by the channel detector. Therefore, the channel detector can make use of this new information to update the log-likelihoods of the turbo coded bits. By including the channel detector in the iterative decoding algorithm (together with the two MAP decoders), the performance of the decoding algorithm is improved.

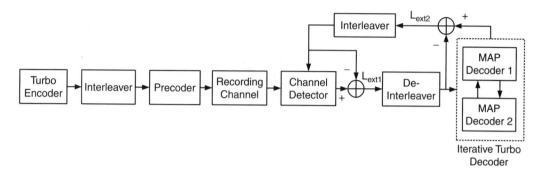

FIGURE 11.10   Block diagram of turbo coding with turbo equalization.

The process of feeding the extrinsic information of the turbo decoder back to the channel detector (and vice versa in the subsequent iteration steps) is named "turbo equalization" [54]. Figure 11.10 illustrates the passage of information between the channel detector and the turbo decoder to perform turbo equalization. Similar to the previous scheme, for turbo coding with turbo equalization, precoding is not essential to obtain an interleaver gain.

11.5.3   Convolutional Coding with an Interleaver

For turbo coding both with and without turbo equalization over recording channels, an outer turbo code is connected "serially" with the recording channel. Since the recording channel acts as a rate 1 inner code, the overall system can be viewed as a serial concatenation scheme [50]. A simpler system can be obtained by replacing the underlying turbo code with a convolutional code. This scheme is called the serial concatenation of a convolutional code with the partial response (ISI) channel [50]. The block diagram of this scheme including the decoder is shown in Figure 11.11.

The decoding algorithm for this scheme involves the exchange of information between the channel detector and a single MAP decoder for the convolutional code, which is similar to the turbo equalization. In the first iteration step, using the original channel observations, the channel detector computes the log-likelihood of its input by the MAP decoding algorithm and passes this information to the outer decoder. Then, the outer decoder computes its extrinsic information using the code constraints and passes it back to the channel detector. We note that, the outer decoder computes the extrinsic information for the parity bits as well as the information bits of the underlying convolutional code. The iterations are repeated several times to obtain the final likelihoods of the information bits transmitted and bit decisions are made.

Compared to the case of turbo decoding with turbo equalization, this algorithm is less complex due to the decrease in the number of the MAP decoders from three (two for the turbo decoder and one for the channel detector) to two (one for the outer decoder and one for the channel detector).

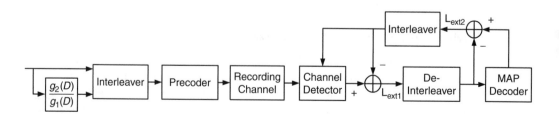

FIGURE 11.11   Block diagram of convolutional coding with an interleaver.

In [21], for serial concatenated codes it is shown that, for superior bit error rate performance, the inner encoder must be chosen as a recursive convolutional encoder. Therefore, unlike the previous two schemes, for the serial concatenation of a convolutional code with the PR channel, the use of a precoder is essential to obtain a large interleaver gain. Precoding makes the PR channel "recursive," thus the precoded channel has good distance properties [53]. As will be shown in the next section, the convolutional coding with an interleaver outperforms turbo coding when precoding is used.

11.6 Performance of Turbo Codes over Recording Channels

In this section, we present a set of results on the performance of turbo codes for magnetic recording channels by using both the simplistic PR channel and the more realistic Lorentzian channel models.

11.6.1 Over PR Channels

For the simulations we consider the $R_c = 4/5$, $N = 10000$, 5/7 turbo code over PR4 and EPR4 channels. After pseudo-random interleaving, the coded bits are precoded with $1/1 \oplus D^2$. Figure 11.12 and Figure 11.13 show the BER for PR4 and EPR4 channels respectively. For comparison purposes, the BER for the uncoded system is also included to the plots.

These results confirm that turbo coding both with and without turbo equalization introduces large coding gains. For instance at 10^{-5} probability of bit error, turbo coding without turbo equalization provides 6 dB and 5 dB coding gains over the uncoded system for PR4 and EPR4 channels respectively. With turbo equalization, the same code introduces an additional 0.5 dB coding gain compared to the system without turbo equalization. However, this additional gain comes at the expense of an increased complexity, because for turbo equalization channel detector is used at every iteration unlike the case without turbo equalization for which the channel detector is used only once.

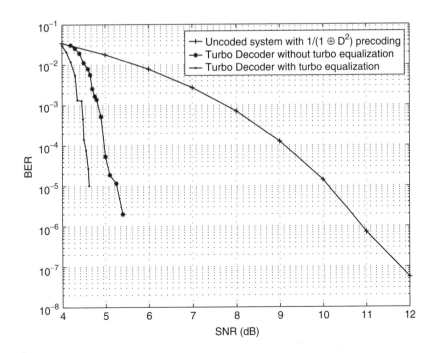

FIGURE 11.12 The BER of the 5/7 turbo code with $R_c = 4/5$ over PR4 channel, $N = 10000$.

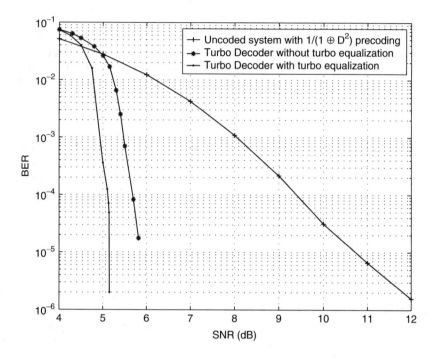

FIGURE 11.13   The BER of the 5/7 turbo code with $R_c = 4/5$ over the EPR4 channel, $N = 10000$.

11.6.2   Over Lorentzian Channels

In this section, we consider the 5/7 and 23/31 turbo codes with $R_c = 16/17$ over the Lorentzian channel with $D_n = 2.5$. In our simulations, we use the (appropriate) iterative decoding algorithm with 15 iterations and equalize the channel to an EPR4 target using a least mean squares (LMS) based linear equalizer with 21 taps.

In Figure 11.14, we present the performance of several codes for an interleaver length of $N = 10016$. We assume that $\frac{1}{1 \oplus D^2}$ precoder is used and there is no media noise. We observe that the high rate coding (turbo or convolutional with an interleaver) provides a significant coding gain of up to 4.5 dB over the uncoded system at 10^{-5} bit error probability. Therefore, the coding gain of turbo codes is mostly preserved when the simplistic PR channel model is replaced by the more realistic Lorentzian channel model. Furthermore, we observe that the convolutional code used together with an interleaver outperforms the parallel concatenated codes, which is in agreement with the observations made for the PR4 equalized ideal magnetic recording channel model of [50].

We believe that the reason why the convolutional code (together with an interleaver) performs better than the turbo code lies in the decoding algorithm. In the case of the convolutional code, the iterative suboptimal decoding algorithm requires the exchange of information between two MAP decoders, whereas in the parallel concatenation case, the exchange of information (when turbo equalization is employed) is between three MAP decoders which results in a "worse" decoder. Performance bounds based on the maximum likelihood decoding computed in [55] support our claim, since the simulation results for the convolutionally coded systems are very close to the bounds based on the optimal (maximum likelihood) decoding. On the other hand, for the turbo codes used in magnetic recording systems, the simulation results (based on the suboptimal iterative decoding algorithm) are worse than the bounds computed using maximum likelihood decoding.

In Figure 11.15, we present the performance of two different codes using different precoders (or, no precoder). We see that the performance of these schemes vary slightly. However the best choice of the precoding scheme is not clear, therefore for code design, one should take the various possibilities into

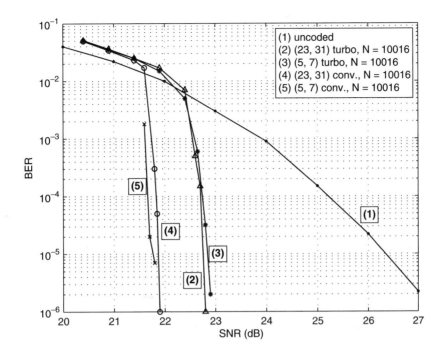

FIGURE 11.14 Performance of several codes and the uncoded system with $D_n = 2.5$ and $1/1 \oplus D^2$ precoder.

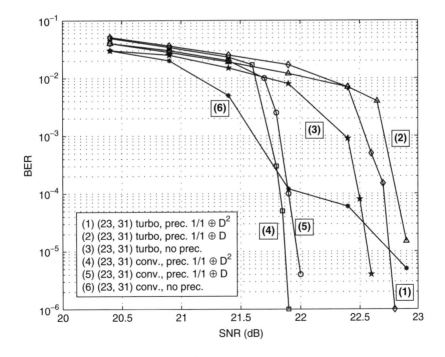

FIGURE 11.15 Performance of two different codes with different precoders with $D_n = 2.5$ and $N = 10016$.

account, as some schemes (including no precoding) may outperform the others. In our example, for the convolutional code the $\frac{1}{1 \oplus D^2}$ precoding is the best choice. On the other hand for the turbo code, no precoding is the best choice among the three different precoding schemes used. However as mentioned in [50], at high SNRs, the BER performance is better for both serial and parallel concatenation with precoding. Since we usually do not operate at very low SNRs especially for magnetic recording, we can conclude that precoding improves the performance of the recording channels.

For the simulations, when turbo equalization is performed, improvements of up to 1 dB are observed. It is interesting to note that for both parallel and serial concatenation, the improvement is more if the component code is simpler. This result agrees with the observation that the simple convolutional code outperforms the parallel concatenated code when concatenated serially with the channel [50].

In [51], for the uncoded system it is observed that, at high media noise levels the probability of error cannot be made smaller than a certain value no matter how large the SNR is. On the other hand, the convolutionally coded system is able to tolerate very high media noise levels. This property of turbo based codes is very attractive for use in media noise limited magnetic recording channels.

Although turbo codes provide very low bit error rates for AWGN and magnetic recording channels, they have a "burst" error problem. That is, when there is an error, it is likely that there are many errors within the decoded block [56]. Such errors pose challenging problems for various applications including magnetic recording systems, because the outer error correcting code (ECC) cannot correct the residual errors if they exceed its error correction capability. Therefore, it is standard in magnetic recording literature to study the error bursts and error event distributions to evaluate the performance of the system, see, for instance, [57]. In [58] various burst error identification techniques are introduced for turbo and LDPC coded systems.

11.7   Summary

In this chapter, we studied the parallel concatenated and serial concatenated convolutional codes, that is, turbo codes. First, use of these codes for additive white Gaussian channels is discussed. The encoding procedures and the iterative decoding algorithm based on the two soft output component decoders are presented. Then, the use of concatenated codes over recording channels are discussed. Partial response channels with additive white Gaussian noise are used as simple models to illustrate the main ideas. The concepts of turbo decoding without turbo equalization and with turbo equalization are explained in detail. As more accurate models, Lorentzian channels are also considered for longitudinal recording channels.

To illustrate the capabilities and limitations of the various turbo coding approaches for AWGN and recording channels, we presented a set of results, which verified that, the BER performance of turbo codes is excellent and they provide a large coding gain over the uncoded systems for both channels. Despite this appreciable coding gain, a negative side of turbo codes is the existence of error bursts even at low probability of error values.

References

[1] C.E. Shannon, A mathematical theory of communication, *Bell System Technical Journal*, pp. 1–10, January 1948.

[2] S. Lin and D.J. Costello, Jr., *Error Control Coding: Fundamentals and Applications*. Englewood Cliffs, NJ: Prentice Hall, 1983.

[3] A.M. Michelson and A.H. Levesque, *Error-Control Techniques for Digital Communications*, New York, NY: Wiley, 1985.

[4] G.D. Forney, Jr., *Concatenated Codes*. Cambridge, Massachusetts: M.I.T. Press, 1966.

[5] C. Berrou, A. Glavieux, and P. Thitimajshima, Near Shannon limit error-correcting coding and decoding: Turbo-codes, in *Proceedings of IEEE International Conference on Communications (ICC)*, pp. 1064–1070, 1993.

[6] S. Benedetto and G. Montorsi, Design of parallel concatenated convolutional codes, *IEEE Transactions on Communications,* pp. 591–600, May 1996.

[7] D. Divsalar and F. Pollara, On the design of turbo codes, TDA Progress Report 42-123, JPL, November 1995.

[8] S. Dolinar and D. Divsalar, Weight distributions for turbo codes using random and nonrandom permutations, TDA Progress Report 42-122, JPL, August 1995.

[9] T.M. Duman, *Interleavers for Serial and Parallel Concatenated (Turbo) Codes,* in *Wiley Encyclopedia of Telecommunications,* J.G. Proakis, Ed., Wiley, New York, December 2002, pp. 1141–1151.

[10] D. Divsalar and R.J. McEliece, Effective free distance of turbo codes, *Electronics Letters,* pp. 445–446, February 1996.

[11] L.C. Perez, J. Seghers, and D.J. Costello, Jr., A distance spectrum interpretation of turbo codes, *IEEE Transactions on Information Theory,* pp. 1698–1709, November 1996.

[12] S. Benedetto and G. Montorsi, Unveiling turbo codes: Some results on parallel concatenated coding schemes, *IEEE Transactions on Information Theory,* pp. 409–428, March 1996.

[13] D. Divsalar, S. Dolinar, F. Pollara, and R.J. McEliece, Transfer function bounds on the performance of turbo codes, TDA Progress Report 42-122, JPL, August 1995.

[14] T.M. Duman and M. Salehi, New performance bounds for turbo codes, *IEEE Transactions on Communications,* pp. 717–723, June 1998.

[15] I. Sason and S.S. (Shitz), Improved upper bounds on the ML decoding error probability of parallel and serial concatenated turbo codes via their ensemble distance spectrum, *IEEE Transactions on Information Theory,* vol. 46, pp. 24–47, January 2000.

[16] R.J. McEliece, E.R. Rodemich, and J. Cheng, The turbo decision algorithm, in *Proceedings of Allerton Conference on Communications, Control and Computing,* pp. 366–379, 1995.

[17] S. Benedetto and G. Montorsi, Performance of continuous and blockwise decoded turbo codes, *IEEE Communications Letters,* pp. 77–79, May 1997.

[18] D. Divsalar and F. Pollara, "Turbo codes for deep-space communications," TDA Progress Report 42-120, JPL, February 1995.

[19] W.J. Blackert, E.K. Hall, and S.G. Wilson, Turbo code termination and interleaver conditions, *Electronics Letters,* pp. 2082–2084, November 1995.

[20] O. Joerssen and H. Meyr, Terminating the trellis of turbo codes, *Electronics Letters,* pp. 1285–1286, August 1994.

[21] S. Benedetto, D. Divsalar, G. Montorsi, and F. Pollara, Serial concatenation of interleaved codes: Performance analysis, design, and iterative decoding, TDA Progress Report 42-126, JPL, August 1996.

[22] S. Benedetto and G. Montorsi, Average performance of parallel concatenated block codes, *Electronics Letters,* pp. 156–158, February 1995.

[23] D. Divsalar and F. Pollara, Multiple turbo codes for deep-space communications, TDA Progress Report 42-121, JPL, May 1995.

[24] S. Divsalar and F. Pollara, Hybrid concatenated codes and iterative decoding, TDA Progress Report 42-130, JPL, August 1997.

[25] R.G. Gallager, Low-density parity check codes, *IRE transactions on Information Theory,* vol. IT-8, pp. 21–28, January 1962.

[26] D.J.C. MacKay, Good error-correcting codes based on very sparse matrices, *IEEE Transactions on Information Theory,* vol. 45, pp. 399–431, March 1999.

[27] S.B. Wicker, *Error Control Systems for Digital Communication and Storage.* Upper Saddle River, N.J.: Prentice Hall, 1995.

[28] T. Richardson, A. Shokrollahi, and R. Urbanke, Design of capacity approaching irregular low-density parity check codes, *IEEE Transactions on Information Theory,* vol. 47, pp. 619–637, February 2001.

[29] Z. Wu, *Coding and Iterative Detection for Magnetic Recording Channels,* Norwell, Massachusetts: Kluwer Academic Publishers, 1999.

[30] P. Robertson, Illuminating the structure of code and decoder of parallel concatenated recursive systematic (turbo) codes, in *Proceedings of IEEE Global Communications Conference (GLOBECOM)*, pp. 1298–1303, 1994.

[31] L.R. Bahl, J. Cocke, F. Jelinek, and J. Raviv, Optimal decoding of linear codes for minimizing symbol error rate, *IEEE Transactions on Information Theory*, pp. 284–287, March 1974.

[32] J. Hagenauer and P. Hoeher, A Viterbi algorithm with soft-decision outputs and its applications, in *Proceedings of IEEE Global Communications Conference (GLOBECOM)*, pp. 47.1.1–47.1.7, 1989.

[33] C. Berrou, P. Adde, E. Angui, and S. Faudeil, A low complexity soft-output Viterbi decoding architecture, in *Proceedings of IEEE International Conference on Communications (ICC)*, pp. 737–740, 1994.

[34] P. Jung, Novel low complexity decoder for turbo codes, *Electronics Letters*, pp. 86–87, January 1995.

[35] S.S. Pietrobon and A.S. Barbulescu, Simplification of the modified Bahl decoding algorithm for systematic convolutional codes, in *Proceedings of IEEE International Symposium on Information Theory and Its Applications (ISITA)*, pp. 1073–1077, 1994.

[36] J. Hagenauer, P. Robertson, and L. Papke, Iterative (turbo) decoding of systematic convolutional codes with the MAP and SOVA algorithms, in *Proceedings of ITG Conference, vol. 130, Munich*, pp. 21–29, 1994.

[37] P. Jung, Comparison of turbo code decoders applied to short frame transmission systems, *IEEE Journal of Selected Areas in Communications*, pp. 530–537, April 1996.

[38] S. Benedetto, G. Montorsi, D. Divsalar, and F. Pollara, A soft-input soft-output maximum a posteriori (MAP) module to decode parallel and serial concatenated codes, TDA Progress Report 42-127, JPL, November 1996.

[39] S. Benedetto, D. Divsalar, G. Montorsi, and F. Pollara, Algorithm for continuous decoding of turbo codes, *Electronics Letters*, pp. 314–315, February 1996.

[40] R.J. McEliece, D.J.C. MacKay, and J. Cheng, Turbo decoding as an instance of Pearl's belief propagation algorithm, *IEEE Journal of Selected Areas in Communications*, pp. 140–152, February 1998.

[41] J. Hagenauer, E. Offer, and L. Papke, Iterative decoding of binary block and convolutional codes, *IEEE Transactions on Information Theory*, pp. 429–445, March 1996.

[42] J.G. Proakis, Equalization techniques for high density magnetic recording systems, *IEEE Signal Processing Magazine*, pp. 73–82, July 1998.

[43] H.N. Bertram, *Theory of Magnetic Recording*. Cambridge Press, London 1994.

[44] J.G. Proakis, *Digital Communications*. New York, NY: McGraw-Hill, Inc., 1995.

[45] C. Heegard, Turbo coding in magnetic recording in *Proceedings of the Information Theory Workshop, San Diego, CA*, pp. 18–19, 1998.

[46] W.E. Ryan, L.L. McPheters, and S. McLaughlin, Combined turbo coding and turbo equalization for PR4-equalized Lorentzian channels, in *Proceedings of Conference on Information Sciences and Systems*, pp. 489–494, March 1998.

[47] W. Pusch, D. Weinrichter, and M. Tafarner, Turbo codes matched to the $1 - D^2$ partial response channel, in *Proceedings of IEEE International Symposium on Information Theory (ISIT)*, p. 62, 1998.

[48] M.C. Reed and C.B. Schelegel, An iterative receiver for the partial response channel, in *Proceedings of IEEE International Symposium on Information Theory (ISIT)*, p. 63, 1998.

[49] W.E. Ryan, Performance of high rate turbo codes on a PR4 equalized magnetic recording channel, in *Proceedings of IEEE International Conference on Communications (ICC)*, pp. 947–951, June 1998.

[50] T. Souvignier, A. Friedman, M. Oberg, P.H. Siegel, R.E. Swanson, and J.K. Wolf, Turbo codes for PR4: Parallel versus serial concatenation, in *Proceedings of IEEE International Conference on Communications (ICC)*, pp. 1638–1642, June 1999.

[51] T.M. Duman and E. Kurtas, Comprehensive performance investigation of turbo codes over high density magnetic recording channels, in *Proceedings of IEEE Global Communications Conference (GLOBECOM)*, vol. 1b, pp. 744–748, December 1999.

[52] T. Souvignier and J.K. Wolf, Turbo decoding for partial response channels with colored noise, *IEEE Transactions on Magnetics*, pp. 2322–2324, September 1999.

[53] K.R. Narayanan, Effect of precoding on the convergence of turbo equalization for partial response channels, in *Proceedings of IEEE Global Communications Conference (GLOBECOM)*, vol. 3, pp. 1865–1871, December 2000.

[54] C. Douillard, M. Jezequel, C. Berrou, A. Picart, P. Didier, and A. Glaivieux, Iterative correction of intersymbol interference: Turbo-equalization, *European Transactions on Telecommunications*, vol. 6, pp. 507–511, September 1995.

[55] T.M. Duman and E. Kurtas, Performance bounds for high rate linear codes over partial response channels, *IEEE Transactions on Information Theory*, vol. 47, pp. 1201–1205, March 2001.

[56] M.N. Kaynak, Turbo and low density parity check codes for AWGN and partial response channels, Technical Report, Department of Electrical Engineering, Arizona State University, December 2002.

[57] T. Souvignier, Turbo decoding for partial response channels. Ph.D. Thesis, Department of Electrical Engineering, University of California, San Diego, 1999.

[58] M.N. Kaynak, T.M. Duman, and E.M. Kurtas, Burst error identification for turbo and LDPC coded systems, in *Proceedings of the 3rd International Symposium on Turbo Codes and Related Topics*, pp. 515–518, September 2003.

12

An Introduction to LDPC Codes

William E. Ryan
University of Arizona
Tucson, AZ

12.1 Introduction

Low-density parity-check (LDPC) codes are a class of linear block codes which provide near-capacity performance on a large collection of data transmission and storage channels while simultaneously admitting implementable decoders. LDPC codes were first proposed by Gallager in his 1960 doctoral dissertation [1] and were scarcely considered in the 35 years that followed. One notable exception is the important work of Tanner in 1981 [2] in which Tanner generalized LDPC codes and introduced a graphical representation of LDPC codes, now called Tanner graphs. The study of LDPC codes was resurrected in the mid-1990s with the work of MacKay, Luby, and others [3–5] who noticed, apparently independently of the work of Gallager, the advantages of linear block codes which possess sparse (low-density) parity-check matrices.

This tutorial chapter provides the foundations for the study and practice of LDPC codes. We will start with the fundamental representations of LDPC codes via parity-check matrices and Tanner graphs. Classification of LDPC ensembles via Tanner graph degree distributions will be introduced, but we will only superficially cover the design of LDPC codes with optimal degree distributions via constrained pseudo-random matrix construction. We will also review some of the other LDPC code construction techniques which have appeared in the literature. The encoding problem for such LDPC codes will be presented and certain special classes of LDPC codes which resolve the encoding problem will be introduced. Finally, the iterative message-passing decoding algorithm (and certain simplifications) which provides near-optimal performance will be presented.

12.2   Representations of LPDC Codes

12.2.1   Matrix Representation

Although LDPC codes can be generalized to nonbinary alphabets, we shall consider only binary LDPC codes for the sake of simplicity. Because LDPC codes form a class of linear block codes, they may be described as a certain k-dimensional subspace $\mathcal{C}$ of the vector space $\mathbb{F}_2^n$ of binary n-tuples over the binary field $\mathbb{F}_2$. Given this, we may find a basis $B = \{\mathbf{g}_0, \mathbf{g}_1, \ldots, \mathbf{g}_{k-1}\}$ which spans $\mathcal{C}$ so that each $\mathbf{c} \in \mathcal{C}$ may be written as $\mathbf{c} = u_0 \mathbf{g}_0 + u_1 \mathbf{g}_1 + \cdots + u_{k-1} \mathbf{g}_{k-1}$ for some $\{u_i\}$; more compactly, $\mathbf{c} = \mathbf{u}G$ where $\mathbf{u} = [u_0\, u_1 \ldots u_{k-1}]$ and G is the so-called $k \times n$ generator matrix whose rows are the vectors $\{\mathbf{g}_i\}$ (as is conventional in coding, all vectors are row vectors). The $(n-k)$-dimensional null space $\mathcal{C}^\perp$ of G comprises all vectors $\mathbf{x} \in \mathbb{F}_2^n$ for which $\mathbf{x}G^T = \mathbf{0}$ and is spanned by the basis $B^\perp = \{\mathbf{h}_0, \mathbf{h}_1, \ldots, \mathbf{h}_{n-k-1}\}$. Thus, for each $\mathbf{c} \in \mathcal{C}$, $\mathbf{c}\mathbf{h}_i^T = 0$ for all i or, more compactly, $\mathbf{c}H^T = \mathbf{0}$ where H is the so-called $(n-k) \times n$ parity-check matrix whose rows are the vectors $\{\mathbf{h}_i\}$, and is the generator matrix for the null space $\mathcal{C}^\perp$. The parity-check matrix H is so named because it performs $m = n - k$ separate parity checks on a received word.

A *low-density parity-check code* is a linear block code for which the parity-check matrix H has a low density of 1s. A *regular LDPC code* is a linear block code whose parity-check matrix H contains exactly w_c 1s in each column and exactly $w_r = w_c(n/m)$ 1s in each row, where $w_c \ll m$ (equivalently, $w_r \ll n$). The code rate $R = k/n$ is related to these parameters via $R = 1 - w_c/w_r$ (this assumes H is full rank). If H is low density, but the number of 1s in each column or row is not constant, then the code is an *irregular LDPC code*. It is easiest to see the sense in which an LDPC code is regular or irregular through its graphical representation.

12.2.2   Graphical Representation

Tanner considered LDPC codes (and a generalization) and showed how they may be represented effectively by a so-called bipartite graph, now call a Tanner graph [2]. The Tanner graph of an LDPC code is analogous to the trellis of a convolutional code in that it provides a complete representation of the code and it aids in the description of the decoding algorithm. A *bipartite graph* is a graph (nodes connected by edges) whose nodes may be separated into two types, and edges may only connect two nodes of different types. The two types of nodes in a Tanner graph are the *variable nodes* and the *check nodes* (which we shall call v-nodes and c-nodes, respectively).[1] The Tanner graph of a code is drawn according to the following rule: check node j is connected to variable node i whenever element h_{ji} in H is a 1. One may deduce from this that there are $m = n - k$ check nodes, one for each check equation, and n variable nodes, one for each code bit c_i. Further, the m rows of H specify the m c-node connections, and the n columns of H specify the n v-node connections.

Example 12.1

Consider a (10, 5) linear block code with $w_c = 2$ and $w_r = w_c(n/m) = 4$ with the following H matrix:

$$
H = \begin{bmatrix}
1 & 1 & 1 & 1 & 0 & 0 & 0 & 0 & 0 & 0 \\
1 & 0 & 0 & 0 & 1 & 1 & 1 & 0 & 0 & 0 \\
0 & 1 & 0 & 0 & 1 & 0 & 0 & 1 & 1 & 0 \\
0 & 0 & 1 & 0 & 0 & 1 & 0 & 1 & 0 & 1 \\
0 & 0 & 0 & 1 & 0 & 0 & 1 & 0 & 1 & 1
\end{bmatrix}
$$

---

[1] The nomenclature varies in the literature: variable nodes are also called bit or symbol nodes and check nodes are also called function nodes.

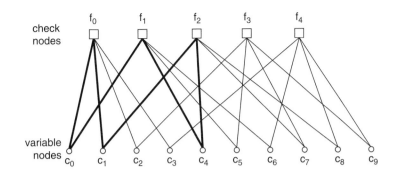

FIGURE 12.1 Tanner graph for example code.

The Tanner graph corresponding to H is depicted in Figure 12.1. Observe that v-nodes c_0, c_1, c_2, and c_3 are connected to c-node f_0 in accordance with the fact that, in the zeroth row of H, $h_{00} = h_{01} = h_{02} = h_{03} = 1$ (all others are zero). Observe that analogous situations hold for c-nodes f_1, f_2, f_3, and f_4 which corresponds to rows 1, 2, 3, and 4 of H, respectively. Note, as follows from the fact that $\mathbf{c}H^T = \mathbf{0}$, the bit values connected to the same check node must sum to zero. We may also proceed along columns to construct the Tanner graph. For example, note that v-node c_0 is connected to c-nodes f_0 and f_1 in accordance with the fact that, in the zeroth column of H, $h_{00} = h_{10} = 1$.

Note that the Tanner graph in this example is regular: each v-node has two edge connections and each c-node has four edge connections (i.e., the *degree* of each v-node is 2 and the degree of each c-node is 4). This is in accordance with the fact that $w_c = 2$ and $w_r = 4$. It is also clear from this example that $mw_r = nw_c$.

For irregular LDPC codes, the parameters w_c and w_r are functions of the column and row numbers and so such notation is not generally adopted in this case. Instead, it is usual in the literature (following [7]) to specify the v-node and c-node *degree distribution polynomials*, denoted by $\lambda(x)$ and $\rho(x)$, respectively. In the polynomial

$$\lambda(x) = \sum_{d=1}^{d_v} \lambda_d x^{d-1},$$

λ_d denotes the fraction of all edges connected to degree-d v-nodes and d_v denotes the maximum v-node degree. Similarly, in the polynomial

$$\rho(x) = \sum_{d=1}^{d_c} \rho_d x^{d-1},$$

ρ_d denotes the fraction of all edges connected to degree-d c-nodes and d_c denotes the maximum c-node degree. Note for the regular code above, for which $w_c = d_v = 2$ and $w_r = d_c = 4$, we have $\lambda(x) = x$ and $\rho(x) = x^3$.

A *cycle* (or *loop*) of length v in a Tanner graph is a path comprising v edges which closes back on itself. The Tanner graph in the above example possesses a length-6 cycle as exemplified by the six bold edges in the figure. The *girth* γ of a Tanner graph is the minimum cycle length of the graph. The shortest possible cycle in a bipartite graph is clearly a length-4 cycle, and such cycles manifest themselves in the H matrix as four 1s that lie on the corners of a submatrix of H. We are interested in cycles, particularly short cycles, because they degrade the performance of the iterative decoding algorithm used for LDPC codes. This fact will be made evident in the discussion of the iterative decoding algorithm.

12.3   LDPC Code Design Approaches

Clearly, the most obvious path to the construction of an LDPC code is via the construction of a low-density parity-check matrix with prescribed properties. A large number of design techniques exist in the literature, and we introduce some of the more prominent ones in this section, albeit at a superficial level. The design approaches target different design criteria, including efficient encoding and decoding, near-capacity performance, or low-error rate floors. (Like turbo codes, LPDC codes often suffer from low-error rate floors, owing both to poor distance spectra and weaknesses in the iterative decoding algorithm.)

12.3.1   Gallager Codes

The original LDPC codes due to Gallager [1] are regular LDPC codes with an H matrix of the form

$$H = \begin{bmatrix} H_1 \\ H_2 \\ \vdots \\ H_{w_c} \end{bmatrix}$$

where the submatrices H_d have the following structure. For any integers μ and w_r greater than 1, each submatrix H_d is $\mu \times \mu w_r$ with row weight w_r and column weight 1. The submatrix H_1 has the following specific form: for $i = 1, 2, \ldots, \mu$, the ith row contains all of its w_r 1s in columns $(i-1)w_r + 1$ to $i w_r$. The other submatrices are simply column permutations of H_1. It is evident that H is regular, has dimension $\mu w_c \times \mu w_r$, and has row and column weights w_r and w_c, respectively. The absence of length-4 cycles in H is not guaranteed, but they can be avoided via computer design of H. Gallager showed that the ensemble of such codes has excellent distance properties provided $w_c \geq 3$ and $w_r > w_c$. Further, such codes have low-complexity encoders since parity bits can be solved for as a function of the user bits via the parity-check matrix [1].

Gallager codes were generalized by Tanner in 1981 [2] and were studied for application to code-division multiple-access communication channel in [9]. Gallager codes were extended by MacKay and others [3, 4].

12.3.2   MacKay Codes

MacKay had independently discovered the benefits of designing binary codes with sparse H matrices and was the first to show the ability of these codes to perform near capacity limits [3, 4]. MacKay has archived on a web page [10] a large number of LPDC codes that he has designed for application to data communication and storage, most of which are regular. He provided in [4] algorithms to semi-randomly generate sparse H matrices. A few of these are listed below in order of increasing algorithm complexity (but not necessarily improved performance).

1. H is created by randomly generating weight-w_c columns and (as near as possible) uniform row weight.
2. H is created by randomly generating weight-w_c columns, while ensuring weight-w_r rows, and no two columns having overlap greater than one.
3. H is generated as in Step 2, plus short cycles are avoided.
4. H is generated as in Step 3, plus $H = [H_1 \ H_2]$ is constrained so that H_2 is invertible (or at least H is full rank).

One drawback of MacKay codes is that they lack sufficient structure to enable low-complexity encoding. Encoding is performed by putting H in the form $[P^T \ I]$ via Gauss-Jordan elimination, from which the generator matrix can be put in the systematic form $G = [I \ P]$. The problem with encoding via G is that the submatrix P is generally not sparse so that, for codes of length $n = 1000$ or more, encoding complexity is high. An efficient encoding technique employing only the H matrix was proposed in [6].

12.3.3 Irregular LDPC Codes

Richardson et al. [7] and Luby et al. [8] defined ensembles of irregular LDPC codes parameterized by the degree distribution polynomials $\lambda(x)$ and $\rho(x)$ and showed how to optimize these polynomials for a variety of channels. Optimality is in the sense that, assuming message-passing decoding (described below), a typical code in the ensemble was capable of reliable communication in worse channel conditions than codes outside the ensemble are. The worst-case channel condition is called the *decoding threshold* and the optimization of $\lambda(x)$ and $\rho(x)$ is found by a combination of a so-called *density evolution* algorithm and an optimization algorithm. Density evolution refers to the evolution of the probability density functions (pdfs) of the various quantities passed around the decoder's Tanner graph. The decoding threshold for a given $\lambda(x)$-$\rho(x)$ pair is determined by evaluating the pdfs of computed log-likelihood ratios (see the next section) of the code bits. The separate optimization algorithm optimizes over the $\lambda(x)$-$\rho(x)$ pairs.

Using this approach an irregular LDPC code has been designed whose simulated performance was within 0.045 dB of the capacity limit for a binary-input AWGN channel [11]. This code had length $n = 10^7$ and rate $R = 1/2$. It is generally true that designs via density evolution are best applied to codes whose rate is not too high ($R \lesssim 3/4$) and whose length is not too short ($n \gtrsim 5000$). The reason is that the density evolution design algorithm assumes $n \to \infty$ (hence, $m \to \infty$), and so $\lambda(x)$-$\rho(x)$ pairs which are optimal for very long codes, will not be optimal for medium-length and short codes. As discussed in [12–15], such $\lambda(x)$-$\rho(x)$ pairs applied to medium-length and short codes gives rise to a high error-rate floor.

Finally, we remark that, as for the MacKay codes, these irregular codes do not intrinsically lend themselves to efficient encoding. However, as mentioned above, Richardson and Urbanke [6] have proposed algorithms for achieving linear-time encoding for these codes.

12.3.4 Finite Geometry Codes

In [16, 17], regular LDPC codes are designed using techniques based on finite geometries [18]. The resulting LDPC codes fall into the cyclic and quasi-cyclic classes of block codes and lend themselves to simple encoder implementation via shift-register circuits. The cyclic finite geometry codes tend to have relatively large minimum distances, but the quasi-cyclic codes tend to have somewhat small minimum distances. Also, short LDPC codes (n on the order of 200 bits) designed using these techniques are generally better than short LDPC codes designed using pseudo-random H matrices.

The cyclic finite geometry codes have the drawback that the parity-check matrix used in decoding is $n \times n$ instead of $(n - k) \times n$. (It is possible to choose an $(n - k) \times n$ submatrix of the $n \times n$ matrix to decode, but the loss in performance is often non-negligible.) The $n \times n$ matrix is circulant, with its first row equal to a certain *incidence vector* [16]. Another drawback is that the values of w_r and w_c are relatively large which is undesirable since the complexity of the iterative message-passing decoder is proportional to these values. One final drawback is that there is no flexibility in the choice of length and rate, although this issue can be dealt with by code shortening and puncturing.

12.3.5 RA, IRA, and eIRA Codes

A type of code, called a *repeat-accumulate* (RA) code, which has the characteristics of both serial turbo codes and LDPC codes, was proposed in [20]. The encoder for an RA code is shown in Figure 12.2 where it is seen that user bits are repeated (2 or 3 times is typical), permuted, and then sent through an accumulator (differential encoder). These codes have been shown to be capable of operation near capacity limits, but they have the drawback that they are naturally low rate (rate 1/2 or lower).

The RA codes were generalized in such a way that some bits were repeated more than others yielding *irregular repeat-accumulate* (IRA) codes [21]. As shown in Figure 12.2, the IRA encoder comprises a low-density generator matrix, a permuter, and an accumulator. Such codes are capable of operation even closer to theoretical limits than RA codes, and they permit higher code rates. A drawback to IRA codes is that

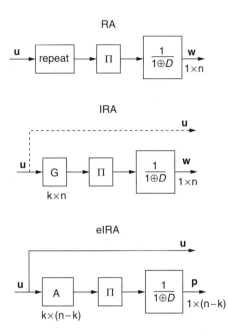

FIGURE 12.2 Encoders for the repeat-acccumulate (RA), the irregular RA (IRA) code, and the extended IRA code (eIRA).

they are nominally non-systematic codes, although they be put in a systematic form, but it is at the expense of greatly lowering the rate as depicted in Figure 12.2.

Yang and Ryan [13–15] have proposed a class of efficiently encodable irregular LDPC codes which are called extended IRA (eIRA) codes. (These codes were independently proposed in [22].) The eIRA encoder is shown in Figure 12.2. Note that the eIRA encoder is systematic and permits both low and high rates. Further, encoding can be efficiently performed directly from the H matrix which possesses an $m \times m$ submatrix which facilitates computation of the parity bits from the user bits [15, 22].

12.3.6 Array Codes

Fan has shown that a certain class of codes called array codes can be viewed as LDPC codes and thus can be decoded with a message passing algorithm [23, 24]. Subsequent to Fan's work, Eleftheriou and Ölçer [25] proposed a modified array code employing the following H matrix format

$$
H = \begin{bmatrix}
I & I & I & \cdots & I & I & \cdots & I \\
0 & I & \alpha & \cdots & \alpha^{j-2} & \alpha^{j-1} & \cdots & \alpha^{k-2} \\
0 & 0 & I & \cdots & \alpha^{2(j-3)} & \alpha^{2(j-2)} & \cdots & \alpha^{2(k-3)} \\
\vdots & \vdots & \vdots & \ddots & \vdots & \vdots & \cdots & \vdots \\
0 & 0 & \cdots & 0 & I & \alpha^{j-1} & \cdots & \alpha^{(j-1)(k-j)}
\end{bmatrix}
\tag{12.1}
$$

where k and j are two integers such that $k, j \leq p$ where p denotes a prime number. I is the $p \times p$ identity matrix, O is the $p \times p$ null matrix, and α is a $p \times p$ permutation matrix representing a single left- or right-cyclic shift. The upper triangular nature of H guarantees encoding linear in the codeword length (encoding is essentially the same as for eIRA codes).

These modified array codes have low error rate floors, and both low- and high-rate codes may be designed, although the high-rate designs perform better (relative to other design techniques). However, as is clear from the description of H above, only selected code lengths and rates are available.

12.3.7 Combinatorial LDPC Codes

In view of the fact that LDPC codes may be designed by constrained random construction of H matrices, it is not difficult to imagine that good LDPC codes may be designed via the application of combinatorial mathematics. That is, design constraints (such as no cycles of length 4) applied to an H matrix of size $(n - k) \times n$ may be cast as a problem in combinatorics. Several researchers have successfully approached this problem via such combinatorial objects as Steiner systems, Kirkman systems, and balanced incomplete block designs [16, 26–29].

12.4 Iterative Decoding Algorithms

12.4.1 Overview

In addition to introducing LDPC codes in his seminal work in 1960 [1], Gallager also provided a decoding algorithm that is typically near optimal. Since that time, other researchers have independently discovered that algorithm and related algorithms, albeit sometimes for different applications [4, 30]. The algorithm iteratively computes the distributions of variables in graph-based models and comes under different names, depending on the context. These names include: the sum-product algorithm (SPA), the belief propagation algorithm (BPA), and the message passing algorithm (MPA). The term "message passing" usually refers to all such iterative algorithms, including the SPA (BPA) and its approximations.

Much like optimal (maximum *a posteriori*, MAP) symbol-by-symbol decoding of trellis codes, we are interested in computing the *a posteriori* probability (APP) that a given bit in the transmitted codeword $\mathbf{c} = [c_0\ c_1 \ldots c_{n-1}]$ equals 1, given the received word $\mathbf{y} = [y_0\ y_1 \ldots y_{n-1}]$. Without loss of generality, let us focus on the decoding of bit c_i so that we are interested in computing the APP

$$\Pr(c_i = 1 \mid \mathbf{y})$$

or the APP ratio (also called the likelihood ratio, LR)

$$l(c_i) \triangleq \frac{\Pr(c_i = 0 \mid \mathbf{y})}{\Pr(c_i = 1 \mid \mathbf{y})}$$

Later we will extend this to the more numerically stable computation of the log-APP ratio, also called the log-likelihood ratio (LLR):

$$L(c_i) \triangleq \log \left[\frac{\Pr(c_i = 0 \mid \mathbf{y})}{\Pr(c_i = 1 \mid \mathbf{y})} \right]$$

where here and in the sequel the natural logarithm is assumed.

The MPA for the computation of $\Pr(c_i = 1 \mid \mathbf{y})$, $l(c_i)$, or $L(c_i)$ is an iterative algorithm which is based on the code's Tanner graph. Specifically, we imagine that the v-nodes represent processors of one type, c-nodes represent processors of another type, and the edges represent message paths. In one half-iteration, each v-node processes its input messages and passes its resulting output messages *up* to neighboring c-nodes (two nodes are said to be *neighbors* if they are connected by an edge). This is depicted in Figure 12.3 for the message $m_{\uparrow 02}$ from v-node c_0 to c-node f_2 (the subscripted arrow indicates the direction of the message, keeping in mind that our Tanner graph convention places c-nodes above v-nodes). The information passed concerns $\Pr(c_0 = b \mid \text{input messages})$, $b \in \{0, 1\}$, the ratio of such probabilities, or the logarithm of the ratio of such probabilities. Note in the figure that the information passed to c-node f_2 is all the information available to v-node c_0 from the channel and through its neighbors, excluding c-node f_2; that is, only *extrinsic information* is passed. Such extrinsic information $m_{\uparrow ij}$ is computed for each connected v-node/c-node pair c_i/f_j at each half-iteration.

In the other half-iteration, each c-node processes its input messages and passes its resulting output messages *down* to its neighboring v-nodes. This is depicted in Figure 12.4 for the message $m_{\downarrow 04}$ from

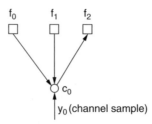

FIGURE 12.3   Subgraph of a Tanner graph corresponding to an H matrix whose zeroth column is $[1\,1\,1\,0\,0\ldots0]^T$. The arrows indicate message passing from node c_0 to node f_2.

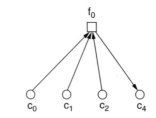

FIGURE 12.4   Subgraph of a Tanner graph corresponding to an H matrix whose zeroth row is $[1\,1\,1\,0\,1\,0\,0\ldots0]^T$. The arrows indicate message passing from node f_0 to node c_4.

c-node f_0 down to v-node c_4. The information passed concerns Pr(check equation f_0 is satisfied | input messages), the ratio of such probabilities, or the logarithm of the ratio of such probabilities. Note, as for the previous case, only extrinsic information is passed to v-node c_4. Such extrinsic information $m_{\downarrow ji}$ is computed for each connected c-node/v-node pair f_j/c_i at each half-iteration.

After a prescribed maximum number of iterations or after some stopping criterion has been met, the decoder computes the APP, the LR or the LLR from which decisions on the bits c_i are made. One example stopping criterion is to stop iterating when $\hat{c}H^T = \mathbf{0}$, where $\hat{c}$ is a tentatively decoded codeword.

The MPA assumes that the messages passed are statistically independent throughout the decoding process. When the y_i are independent, this independence assumption would hold true if the Tanner graph possessed no cycles. Further, the MPA would yield exact APPs (or LRs or LLRs, depending on the version of the algorithm) in this case [30]. However, for a graph of girth γ, the independence assumption is only true up to the $\gamma/2$th iteration, after which messages start to loop back on themselves in the graph's various cycles. Still, simulations have shown that the message passing algorithm is generally very effective provided length-4 cycles are avoided. Lin *et al.* [19] showed that some configurations of length-four cycles are not harmful. It was shown in [31] how the message-passing *schedule* described above and below (the so-called *flooding schedule*) may be modified to mitigate the negative effects of short cycles.

In the remainder of this section we present the "probability domain" version of the SPA (which computes APPs) and its log-domain version, the log-SPA (which computes LLRs), as well as certain approximations. Our treatment considers the special cases of the binary erasure channel (BEC), the binary symmetric channel (BSC), and the binary-input AWGN channel (BI-AWGNC).

12.4.2   Probability-Domain SPA Decoder

We start by introducing the following notation:

- $V_j = \{$v-nodes connected to c-node $f_j\}$.
- $V_j\backslash i = \{$v-nodes connected to c-node $f_j\}\backslash\{$v-node $c_i\}$.
- $C_i = \{$c-nodes connected to v-node $c_i\}$.

- $C_i \backslash j = \{\text{c-nodes connected to v-node } c_i\} \backslash \{\text{c-node } f_j\}$.
- $M_v(\sim i) = \{\text{messages from all v-nodes except node } c_i\}$.
- $M_c(\sim j) = \{\text{messages from all c-nodes except node } f_j\}$.
- $P_i = \Pr(c_i = 1 \mid y_i)$.
- $S_i = $ event that the check equations involving c_i are satisfied.
- $q_{ij}(b) = \Pr(c_i = b \mid S_i, y_i, M_c(\sim j))$, where $b \in \{0,1\}$. For the APP algorithm presently under consideration, $m_{\uparrow ij} = q_{ij}(b)$; for the LR algorithm, $m_{\uparrow ij} = q_{ij}(0)/q_{ij}(1)$; and for the LLR algorithm, $m_{\uparrow ij} = \log[q_{ij}(0)/q_{ij}(1)]$.
- $r_{ji}(b) = \Pr(\text{check equation } f_j \text{ is satisfied} \mid c_i = b, M_v(\sim i))$, where $b \in \{0,1\}$. For the APP algorithm presently under consideration, $m_{\downarrow ji} = r_{ji}(b)$; for the LR algorithm, $m_{\downarrow ji} = r_{ji}(0)/r_{ji}(1)$; and for the LLR algorithm, $m_{\downarrow ji} = \log[r_{ji}(0)/r_{ji}(1)]$.

Note that the messages $q_{ij}(b)$, while interpreted as probabilities here, are random variables (RVs) as they are functions of the RVs y_i and other messages which are themselves RVs. Similarly, by virtue of the message passing algorithm, the messages $r_{ji}(b)$ are RVs.

Consider now the form of $q_{ij}(0)$ which, given our new notation and the independence assumption, we may express as (see Figure 12.5)

$$
\begin{aligned}
q_{ij}(0) &= \Pr(c_i = 0 \mid y_i, S_i, M_c(\sim j)) \\
&= (1 - P_i) \Pr(S_i \mid c_i = 0, y_i, M_c(\sim j)) / \Pr(S_i) \\
&= K_{ij}(1 - P_i) \prod_{j' \in C_i \backslash j} r_{j'i}(0)
\end{aligned}
\tag{12.2}
$$

where we used Bayes' rule twice to obtain the second line and the independence assumption to obtain the third line. Similarly,

$$
q_{ij}(1) = K_{ij} P_i \prod_{j' \in C_i \backslash j} r_{j'i}(1).
\tag{12.3}
$$

The constants K_{ij} are chosen to ensure that $q_{ij}(0) + q_{ij}(1) = 1$.

To develop an expression for the $r_{ji}(b)$, we need the following result.

Result

(Gallager [1]) Consider a sequence of M independent binary digits a_i for which $\Pr(a_i = 1) = p_i$. Then the probability that $\{a_i\}_{i=1}^M$ contains an *even* number of 1s is

$$
\frac{1}{2} + \frac{1}{2} \prod_{l=i}^M (1 - 2p_i)
\tag{12.4}
$$

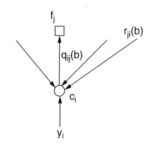

FIGURE 12.5 Illustration of message passing half-iteration for the computation of $q_{ij}(b)$.

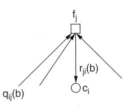

FIGURE 12.6   Illustration of message passing half-iteration for the computation of $r_{ji}(b)$.

Proof 12.1   Induction on M.   □

In view of this result, together with the correspondence $p_i \leftrightarrow q_{ij}(1)$, we have (see Figure 12.6)

$$r_{ji}(0) = \frac{1}{2} + \frac{1}{2} \prod_{i' \in V_j \setminus i} (1 - 2q_{i'j}(1)) \tag{12.5}$$

since, when $c_i = 0$, the bits $\{c_{i'} : i' \in V_j \setminus i\}$ must contain an even number of 1s in order for check equation f_j to be satisfied. Clearly,

$$r_{ji}(1) = 1 - r_{ji}(0) \tag{12.6}$$

The MPA for the computation of the APPs is initialized by setting $q_{ij}(b) = \Pr(c_i = b \mid y_i)$ for all i, j for which $h_{ij} = 1$ (i.e., $q_{ij}(1) = P_i$ and $q_{ij}(0) = 1 - P_i$). Here, y_i represents the channel symbol that was actually received (i.e., it is a not variable here). We consider the following special cases.

BEC. In this case, $y_i \in \{0, 1, E\}$ where E is the erasure symbol, and we define $\delta = \Pr(y_i = E \mid c_i = b)$ to be the erasure probability. Then it is easy to see that

$$\Pr(c_i = b \mid y_i) = \begin{cases} 1 & \text{when } y_i = b \\ 0 & \text{when } y_i = b^c \\ 1/2 & \text{when } y_i = E \end{cases}$$

where b^c represents the complement of b.

BSC. In this case, $y_i \in \{0, 1\}$ and we define $\varepsilon = \Pr(y_i = b^c \mid c_i = b)$ to be the error probability. Then it is obvious that

$$\Pr(c_i = b \mid y_i) = \begin{cases} 1 - \varepsilon & \text{when } y_i = b \\ \varepsilon & \text{when } y_i = b^c \end{cases}$$

BI-AWGNC. We first let $x_i = 1 - 2c_i$ be the ith transmitted binary value; note $x_i = +1(-1)$ when $c_i = 0(1)$. We shall use x_i and c_i interchangeably hereafter. Then the ith received sample is $y_i = x_i + n_i$ where the n_i are independent and normally distributed as $\eta(0, \sigma^2)$. Then it is easy to show that

$$\Pr(x_i = x \mid y_i) = [1 + \exp(-2yx/\sigma^2)]^{-1}$$

where $x \in \{\pm 1\}$.

12.4.2.1   Summary of the Probability-Domain SPA Decoder

1. For $i = 0, 1, \ldots, n - 1$, set $P_i = \Pr(c_i = 1 \mid y_i)$ where y_i is the ith received channel symbol. Then set $q_{ij}(0) = 1 - P_i$ and $q_{ij}(1) = P_i$ for all i, j for which $h_{ij} = 1$.
2. Update $\{r_{ji}(b)\}$ using Equation 12.5 and Equation 12.6.

3. Update $\{q_{ji}(b)\}$ using Equation 12.2 and Equation 12.3. Solve for the constants K_{ij}.

4. For $i = 0, 1, \ldots, n - 1$, compute

$$Q_i(0) = K_i(1 - P_i) \prod_{j \in C_i} r_{ji}(0) \qquad (12.7)$$

and

$$Q_i(1) = K_i P_i \prod_{j \in C_i} r_{ji}(1) \qquad (12.8)$$

where the constants K_i are chosen to ensure that $Q_i(0) + Q_i(1) = 1$.

5. For $i = 0, 1, \ldots, n - 1$, set

$$\hat{c}_i = \begin{cases} 1, & \text{if } Q_i(1) > Q_i(0) \\ 0, & \text{else} \end{cases}$$

If $\hat{c} H^T = \mathbf{0}$ or the number of iterations equals the maximum limit, stop; else, go to Step 2.

Remark 12.1 This algorithm has been presented for pedagogical clarity, but may be adjusted to optimize the number of computations. For example, Step 4 may be computed before Step 3 and Step 3 may be replaced with the simple division $q_{ij}(b) = K_{ij} Q_i(b)/r_{ji}(b)$. We note also that, for good codes, this algorithm is able to detect an uncorrected codeword with near-unity probability (Step 5), unlike turbo codes [4].

12.4.3 Log-Domain SPA Decoder

As with the probability-domain Viterbi and BCJR algorithms, the probability-domain SPA suffers because multiplications are involved (additions are less costly to implement) and many multiplications of probabilities are involved which could become numerically unstable. Thus, as with the Viterbi and BCJR algorithms, a log-domain version of the SPA is to be preferred. To do so, we first define the following LLRs:

$$L(c_i) = \log\left(\frac{\Pr(c_i = 0 \mid y_i)}{\Pr(c_i = 1 \mid y_i)}\right)$$

$$L(r_{ji}) = \log\left(\frac{r_{ji}(0)}{r_{ji}(1)}\right)$$

$$L(q_{ij}) = \log\left(\frac{q_{ij}(0)}{q_{ij}(1)}\right)$$

$$L(Q_i) = \log\left(\frac{Q_i(0)}{Q_i(1)}\right)$$

The initialization steps for the three channels under consideration thus become:

$$L(q_{ij}) = L(c_i) = \begin{cases} +\infty, \ y_i = 0 \\ -\infty, \ y_i = 1 \quad \text{(BEC)} \\ 0, \ y_i = E \end{cases} \qquad (12.9)$$

$$L(q_{ij}) = L(c_i) = (-1)^{y_i} \log\left(\frac{1 - \varepsilon}{\varepsilon}\right) \quad \text{(BSC)}$$

$$L(q_{ij}) = L(c_i) = 2y_i/\sigma^2 \quad \text{(BI-AWGNC)}$$

For Step 1, we first replace $r_{ji}(0)$ with $1 - r_{ji}(1)$ in Equation 12.6 and rearrange it to obtain

$$1 - 2r_{ji}(1) = \prod_{i' \in V_j \setminus i} (1 - 2q_{i'j}(1))$$

Now using the fact that $\tanh[\frac{1}{2}\log(p_0/p_1)] = p_0 - p_1 = 1 - 2p_1$, we may rewrite the equation above as

$$\tanh\left(\frac{1}{2}L(r_{ji})\right) = \prod_{i' \in V_j \setminus i} \tanh\left(\frac{1}{2}L(q_{i'j})\right) \tag{12.10}$$

The problem with these expressions is that we are still left with a product and the complex tanh function. We can remedy this as follows [1]. First, factor $L(q_{ij})$ into its sign and magnitude:

$$L(q_{ij}) = \alpha_{ij}\beta_{ij}$$
$$\alpha_{ij} = sign[L(q_{ij})]$$
$$\beta_{ij} = |L(q_{ij})|$$

so that Equation 12.10 may be rewritten as

$$\tanh\left(\frac{1}{2}L(r_{ji})\right) = \prod_{i' \in V_j \setminus i} \alpha_{i'j} \cdot \prod_{i' \in V_j \setminus i} \tanh\left(\frac{1}{2}\beta_{i'j}\right)$$

We then have

$$L(r_{ji}) = \prod_{i'} \alpha_{i'j} \cdot 2\tanh^{-1}\left(\prod_{i'} \tanh\left(\frac{1}{2}\beta_{i'j}\right)\right)$$

$$= \prod_{i'} \alpha_{i'j} \cdot 2\tanh^{-1}\log^{-1}\log\left(\prod_{i'} \tanh\left(\frac{1}{2}\beta_{i'j}\right)\right)$$

$$= \prod_{i'} \alpha_{i'j} \cdot 2\tanh^{-1}\log^{-1}\sum_{i'} \log\left(\tanh\left(\frac{1}{2}\beta_{i'j}\right)\right)$$

$$= \prod_{i' \in V_j \setminus i} \alpha_{i'j} \cdot \phi\left(\sum_{i' \in V_j \setminus i} \phi\left(\beta_{i'j}\right)\right) \tag{12.11}$$

where we have defined

$$\phi(x) = -\log[\tanh(x/2)] = \log\left(\frac{e^x + 1}{e^x - 1}\right)$$

and used the fact that $\phi^{-1}(x) = \phi(x)$ when $x > 0$. The function is fairly well behaved, as shown in Figure 12.7, and so may be implemented by a look-up table.

For Step 2, we simply divide Equation 12.2 by Equation 12.3 and take the logarithm of both sides to obtain

$$L(q_{ij}) = L(c_i) + \sum_{j' \in C_i \setminus j} L(r_{j'i}) \tag{12.12}$$

Step 3 is similarly modified so that

$$L(Q_i) = L(c_i) + \sum_{j \in C_i} L(r_{ji}) \tag{12.13}$$

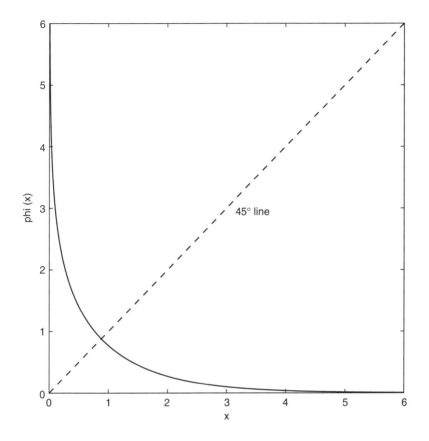

FIGURE 12.7 Plot of the $\phi(x)$ function.

12.4.3.1 Summary of the Log-Domain SPA Decoder

1. For $i = 0, 1, \ldots, n-1$, initialize $L(q_{ij})$ according to Equation 12.9 for all i, j for which $h_{ij} = 1$.
2. Update $\{L(r_{ji})\}$ using Equation 12.11.
3. Update $\{L(q_{ij})\}$ using Equation 12.12.
4. Update $\{L(Q_i)\}$ using Equation 12.13.
5. For $i = 0, 1, \ldots, n-1$, set

$$\hat{c}_i = \begin{cases} 1, & \text{if } L(Q_i) < 0 \\ 0, & \text{else} \end{cases}$$

If $\hat{c} H^T = \mathbf{0}$ or the number of iterations equals the maximum limit, stop; else, go to Step 2.

Remark 12.2 This algorithm can be simplified further for the BEC and BSC channels since the initial LLRs (see Equation 12.9) are ternary in the first case and binary in the second case. See the discussion of the min-sum decoder below.

12.4.4 Reduced Complexity Decoders

It should be clear from the above that the log-domain SPA algorithm has lower complexity and is more numerically stable than the probability-domain SPA algorithm. We now present decoders of lower complexity which often suffer only a little in terms of performance. The degradation is typically on the order of 0.5 dB, but is a function of the code and the channel as demonstrated in the example below.

12.4.4.1 The Min-Sum Decoder [32]

Consider the update Equation 12.11 for $L(r_{ji})$ in the log-domain decoder. Note from the shape of $\phi(x)$ that the term corresponding to the smallest β_{ij} in the summation dominates, so that

$$\phi\left(\sum_{i'}\phi(\beta_{i'j})\right) \simeq \phi\left(\phi\left(\min_{i'}\beta_{i'j}\right)\right)$$

$$= \min_{i' \in V_j \setminus i}\beta_{i'j}$$

Thus, the min-sum algorithm is simply the log-domain SPA with Step 1 replaced by

$$L(r_{ji}) = \prod_{i' \in V_j \setminus i}\alpha_{i'j} \cdot \min_{i' \in V_j \setminus i}\beta_{i'j}$$

It can also be shown that, in the BI-AWGNC case, the initialization $L(q_{ij}) = 2y_i/\sigma^2$ may be replaced by $L(q_{ij}) = y_i$ when the min-sum algorithm is employed. The advantage, of course, is that knowledge of the noise power σ^2 is unnecessary in this case.

12.4.4.2 The Min-Sum-Plus-Correction-Factor Decoder [34]

Note that we can write

$$r_{ji}(b) = \Pr\left(\sum_{i' \in V_j \setminus i} c_{i'} = b(\mathrm{mod}\ 2) \mid M_v(\tilde{i})\right)$$

so that $L(r_{ji})$ corresponds to the (conditional) LLR of a sum of binary RVs. Now consider the following general result.

Result

(Hagenauer et al. [33]) Consider two independent binary RVs a_1 and a_2 with probabilities $\Pr(a_i = b) = p_{ib}$, $b \in \{0, 1\}$, and LLR's $L_i = L(a_i) = \log(p_{i0}/p_{i1})$. The LLR of the binary sum $A_2 = a_1 \oplus a_2$, defined as

$$L(A_2) = \log\left[\frac{\Pr(A_2 = 0)}{\Pr(A_2 = 1)}\right]$$

is given by

$$L(A_2) = \log\left(\frac{1 + e^{L_1 + L_2}}{e^{L_1} + e^{L_2}}\right) \tag{12.14}$$

Proof 12.2

$$L(A_2) = \log\left(\frac{\Pr(a_1 \oplus a_2 = 0)}{\Pr(a_1 \oplus a_2 = 1)}\right)$$

$$= \log\left(\frac{p_{10}\,p_{20} + p_{11}\,p_{21}}{p_{10}\,p_{21} + p_{11}\,p_{20}}\right)$$

$$= \log\left(\frac{1 + \frac{p_{10}}{p_{11}}\frac{p_{20}}{p_{21}}}{\frac{p_{10}}{p_{11}} + \frac{p_{20}}{p_{21}}}\right)$$

$$= \log\left(\frac{1 + e^{L_1 + L_2}}{e^{L_1} + e^{L_2}}\right)$$

□

If more than two independent binary RVs are involved (as is the case for $r_{ji}(b)$), then the LLR of the sum of these RVs may be computed by repeated application of this result. For example, the LLR of $A_3 = a_1 \oplus a_2 \oplus a_3$ may be computed via $A_3 = A_2 \oplus a_3$ and

$$L(A_3) = \log\left(\frac{1 + e^{L(A_2)+L_3}}{e^{L(A_2)} + e^{L_3}}\right)$$

As a shorthand [33], we will write $L_1 \boxplus L_2$ to denote the computation of $L(A_2) = L(a_1 \oplus a_2)$ from L_1 and L_2; and $L_1 \boxplus L_2 \boxplus L_3$ to denote the computation of $L(A_3) = L(a_1 \oplus a_2 \oplus a_3)$ from L_1, L_2, and L_3; and so on for more variables.

We now define, for any pair of real numbers x, y,

$$\max{}^*(x,\ y) = \log(e^x + e^y) \tag{12.15}$$

which may be shown [35] to be

$$\max{}^*(x, y) = \max(x, y) + \log\left(1 + e^{-|x-y|}\right) \tag{12.16}$$

Observe from Equation 12.14 and Equation 12.15 that we may write

$$L_1 \boxplus L_2 = \max{}^*(0, L_1 + L_2) - \max{}^*(L_1, L_2) \tag{12.17}$$

so that

$$L_1 \boxplus L_2 = \max(0, L_1 + L_2) - \max(L_1, L_2) + s(L_1, L_2) \tag{12.18}$$

where $s(x, y)$ is a so-called *correction term* given by

$$s(x, y) = \log\left(1 + e^{-|x+y|}\right) - \log\left(1 + e^{-|x-y|}\right)$$

It can be shown [34] that

$$\max(0, L_1 + L_2) - \max(L_1, L_2) = \operatorname{sign}(L_1)\operatorname{sign}(L_2)\min(|L_1|, |L_2|)$$

so that

$$L_1 \boxplus L_2 = \operatorname{sign}(L_1)\operatorname{sign}(L_2)\min(|L_1|, |L_2|) + s(L_1, L_2) \tag{12.19}$$

which may be approximated as

$$L_1 \boxplus L_2 \simeq \operatorname{sign}(L_1)\operatorname{sign}(L_2)\min(|L_1|, |L_2|) \tag{12.20}$$

since $|s(x, y)| \leq 0.693$.

Returning to the computation of $L(r_{ji})$ which we said corresponds to the LLR of a sum of binary RVs, under the independence assumption, we may write

$$L(r_{ji}) = L(q_{1j}) \boxplus \ldots L(q_{i-1,j}) \boxplus L(q_{i+1,j}) \boxplus \ldots L(q_{nj})$$

This expression may be computed via repeated application of Result 2 together with Equation 12.18 (see [34] for an efficient way of doing this). Observe that, if the approximation Equation 12.20 is employed, we have the min-sum algorithm. At the cost of slightly greater complexity, performance can be enhanced by using a slightly tighter approximation, by substituting $\tilde{s}(x, y)$ for $s(x, y)$ in Equation 12.19 where [34]

$$\tilde{s}(x, y) = \begin{cases} c & \text{if } |x + y| < 2 \quad \text{and} \quad |x - y| > 2\,|x + y| \\ -c & \text{if } |x - y| < 2 \quad \text{and} \quad |x + y| > 2\,|x - y| \\ 0 & \text{otherwise} \end{cases}$$

and where c on the order of 0.5 is typical.

Example 12.2

We consider two regular Euclidean geometry (EG) LDPC codes and their performance with the three decoders discussed above: the (log-)SPA, the min sum, and the min sum with a correction factor (which we denote by min-sum-c, with c set to 0.5). The first code is a cyclic rate-0.82 (4095, 3367) EG LPDC code with minimum distance bound $d_{min} \geq 65$. Because the code is cyclic, it may be implemented using a shift-register circuit. The H matrix for this code is a 4095×4095 circulant matrix with row and column weight 64. The second code is a (generalized) quasi-cyclic rate-0.875 (8176, 7156) EG LDPC code. Because it is quasi-cyclic, encoding may be performed using several shift-register circuits. The H matrix for this code is 1022×8176 and has column weight 4 and row weight 32. It comprises eight 511×2044 circulant submatrices, each with column weight 2 and row weight 8. These codes are being considered for CCSDS standardization for application to satellite communications [37] (see also [16, 17, 19]).

The performance of these codes for the three decoders on a BI-AWGNC is presented in Figure 12.8 and Figure 12.9. We make the following observations (all measurements are with respect to a BER of 10^{-5}).

- The length-4095 code is 1.6 dB away from the rate-0.82 BI-AWGNC capacity limit. The length-8176 code is closer to its capacity limit, only 0.9 dB away. Regular LDPC codes of these lengths might be designed which are a bit closer to their respective capacity limits, but one would have to resort to irregular LDPC codes to realize substantial gains. Of course, an irregular LDPC code would in general require a more complex encoder.

- For the length-4095 code, the loss relative to the SPA decoder suffered by the min-sum decoder is 1.1 dB and the loss suffered by the min-sum-c decoder is 0.5 dB. For the length-8176 code, these losses are 0.3 and 0.01 dB, respectively. We attribute the large losses for the length-4095 code to the fact that its decoder relies on an H matrix with weight-64 rows. Thus, a minimum among 64 small nonnegative numbers is taken at each check node in the code's Tanner graph, so that a value near zero is usually produced and passed to a neighboring variable node.

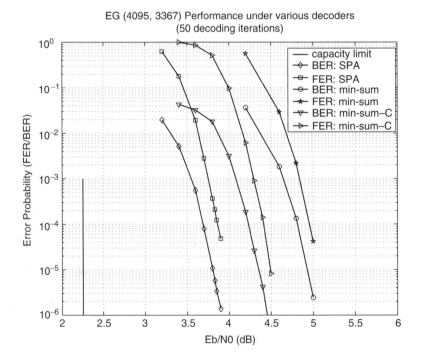

FIGURE 12.8 Performance of a cyclic EG(4095, 3367) LDPC code on a binary-input AWGN channel and three decoding algorithms.

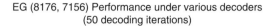

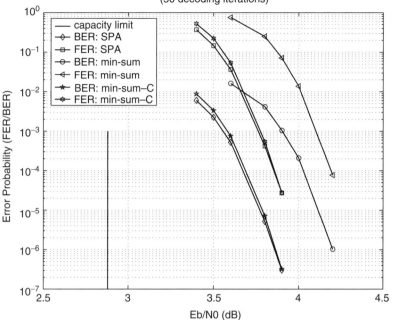

FIGURE 12.9 Performance of a quasi-cyclic EG(8176, 7156) LDPC code on a binary-input AWGN channel and three decoding algorithms.

12.5 Concluding Remarks

Low-density parity-check codes are being studied for a large variety of applications, much as turbo codes, trellis codes, and other codes were when they were first introduced to the coding community. As indicated above, LDPC codes are capable of near-capacity performance while admitting an implementable decoder. Among the advantages LDPC codes have over turbo codes are: (1) They allow a parallelizable decoder; (2) They are more amenable to high code rates; (3) They generally possess a lower error-rate floor (for the same length and rate); (4) They possess superior performance in bursts (due to interference, fading, and so on), (5) They require no interleavers in the encoder and decoder; and (6) A single LDPC code can be universally good over a collection of channels [36]. Among their disadvantages are: (1) Most LDPC codes have somewhat complex encoders, (2) The connectivity among the decoder component processors can be large and unwieldy; and (3) Turbo codes can often perform better when the code length is short. It is easy to find in the literature many of the applications being explored for LPDC codes, including application to deep space and satellite communications, wireless (single and multi-antenna) communications, magnetic storage, and internet packet transmission.

Acknowledgments

The author would like to thank the authors of [19] and [31] for kindly providing preprints. He would also like to thank Yifei Zhang of the University of Arizona for producing Figure 12.7 and Figure 12.8.

References

[1] R. Gallager, Low-density parity-check codes, *IRE Trans. Inf. Theory*, pp. 21–28, January 1962.

[2] R.M. Tanner, A recursive approach to low complexity codes, *IEEE Trans. Inf. Theory*, pp. 533–547, September 1981.

[3] D. MacKay and R. Neal, Good codes based on very sparse matrices, in *Cryptography and Coding, 5th IMA Conf.*, C. Boyd, Ed., *Lecture Notes in Computer Science*, pp. 100–111, Springer, Berlin, Germany, 1995.

[4] D. MacKay, Good error correcting codes based on very sparse matrices, *IEEE Trans. Inf. Theory*, pp. 399–431, March 1999.

[5] N. Alon and M. Luby, A linear time erasure-resilient code with nearly optimal recovery, *IEEE Trans. Inf. Theory*, pp. 1732–1736, November 1996.

[6] T.J. Richardson and R. Urbanke, Efficient encoding of low-density parity-check codes, *IEEE Trans. Inf. Theory*, vol. 47, pp. 638–656, February 2001.

[7] T. Richardson, A. Shokrollahi, and R. Urbanke, Design of capacity-approaching irregular low-density parity-check codes, *IEEE Trans. Inf. Theory*, vol. 47, pp. 619–637, February 2001.

[8] M. Luby, M. Mitzenmacher, M. Shokrollahi, and D. Spielman, Improved low-density parity check codes using irregular graphs, *IEEE Trans. Inf. Theory*, pp. 585–598, February 2001.

[9] V. Sorokine, F.R. Kschischang, and S. Pasupathy, Gallager codes for CDMA applications: Part I, *IEEE Trans. Commn.*, pp. 1660–1668, October 2000 and Gallager codes for CDMA applications: Part II, *IEEE Trans. Commn.*, pp. 1818–1828, November 2000.

[10] D.J.C. Mackay, http://wol.ra.phy. cam.ac.uk/mackay.

[11] S.Y. Chung, G.D. Forney, T.J. Richardson, and R. Urbanke, On the design of low-density parity-check codes within 0.0045 dB of the Shannon limit, *IEEE Commn. Lett.*, vol. 5, pp. 58–60, February 2001.

[12] M. Chiani and A. Ventura, Design and performance evaluation of some high-rate irregular low-density parity-check codes, *Proc. 20001 IEEE GlobeCom Conf.*, pp. 990–994, November 2001.

[13] M. Yang, Y. Li, and W.E. Ryan, Design of efficiently-encodable moderate-length high-rate irregular LDPC codes, *Proc. 40th Annual Allerton Conference on Communication, Control, and Computing, Champaign, IL.*, pp. 1415–1424, October 2002.

[14] M. Yang and W.E. Ryan, Lowering the error rate floors of moderate-length high-rate LDPC codes, *Proc. 2003 Int. Symp. on Inf. Theory*, June–July, 2003.

[15] M. Yang, W.E. Ryan, and Y. Li, Design of efficiently encodable moderate-length high-rate LDPC codes, *IEEE Trans. Commn.*, pp. 564–571, April 2004.

[16] Y. Kou, S. Lin, and M. Fossorier, Low-density parity-check codes based on finite geometries: a rediscovery and new results, *IEEE Trans. Inf. Theory*, vol. 47, pp. 2711–2736, November 2001.

[17] R. Lucas, M. Fossorier, Y. Kou, and S. Lin, Iterative decoding of one-step majority-logic decodable codes based on belief propagation, *IEEE Trans. Commn.*, pp. 931–937, June 2000.

[18] S. Lin and D. Costello, *Error-Control Coding: Fundamentals and Applications*, Prentice Hall, New York, 1983.

[19] H. Tang, J. Xu, S. Lin, and K. Abdel-Ghaffar, Codes on finite geometries, submitted to *IEEE Trans. Inf. Theory*, 2002, in review.

[20] D. Divsalar, H. Jin, and R. McEliece, Coding theorems for turbo-like codes, *Proc. 36th Annual Allerton Conf. on Commn., Control, and Computing*, pp. 201–210, September 1998.

[21] H. Jin, A. Khandekar, and R. McEliece, Irregular repeat-accumulate codes, *Proc. 2nd. Int. Symp. on Turbo Codes and Related Topics*, Brest, France, pp. 1–8, September 4, 2000.

[22] R. Narayanaswami, Coded Modulation with Low-Density Parity-Check Codes, M.S. thesis, Texas A&M University, 2001, chap. 7.

[23] J. Fan, *Constrained Coding and Soft Iterative Decoding for Storage*, Ph.D. dissertation, Stanford University, December 1999. (See also Fan's Kluwer monograph by the same title.)

[24] J. Fan, Array codes as low-density parity-check codes, *2nd Int. Symp. on Turbo Codes and Related Topics*, Brest, France, pp. 543–546, September 2000.

[25] E. Eleftheriou and S. Ölçer, Low-density parity-check codes for digital subscriber lines, *Proc. 2002 Int. Conf. on Commn.*, pp. 1752–1757, April–May, 2002.

[26] D. MacKay and M. Davey, Evaluation of Gallager codes for short block length and high rate applications, in *Codes, Systems, and Graphical Models: Volume 123 of IMA Volumes in Mathematics and its Applications*, pp. 113–130, Spring-Verlag, New York, 2000.

[27] B. Vasic, Structured iteratively decodable codes based on Steiner systems and their application to magnetic recording, *Proc. 2001 IEEE GlobeCom Conf.*, pp. 2954–2958, November 2001.

[28] B. Vasic, Combinatorial constructions of low-density parity-check codes for iterative decoding, *Proc. 2002 IEEE Int. Symp. Inf. Theory*, p. 312, June/July 2002.

[29] S. Johnson and S. Weller, Construction of low-density parity-check codes from Kirkman triple systems, *Proc. 2001 IEEE GlobeCom Conf.*, pp. 970–974, November 2001.

[30] J. Pearl, *Probabilistic Reasoning in Intelligent Systems*, Morgan Kaufmann, San Mateo, CA, 1988.

[31] H. Xiao and A. Banihashemi, Message-passing schedules for decoding LDPC codes, submitted to *IEEE Trans. Commn.*, May 2003, in revision.

[32] N. Wiberg, Codes and Decoding on General Graphs, Ph.D. dissertation, U. Linköping, Sweden, 1996.

[33] J. Hagenauer, E. Offer, and L. Papke, Iterative decoding of binary block and convolutional codes, *IEEE Trans. Inf Theory*, pp. 429–445, March 1996.

[34] X-Y. Hu, E. Eleftheriou, D-M. Arnold, and A. Dholakia, Efficient implementation of the sum-product algorithm for decoding LDPC codes, *Proc. 2001 IEEE GlobeCom Conf.*, pp. 1036–1036E, November 2001.

[35] A. Viterbi, An intuitive justification and a simplified implementation of the MAP decoder for convolutional codes, *IEEE JSAC*, pp. 260–264, February 1998.

[36] C. Jones, A. Matache, T. Tian, J. Villasenor, and R. Wesel, The universality of LDPC codes on wireless channels, *Proc. 2003 IEEE Milcom conf.*, pp. 440–445 October 2003.

[37] W. Fong, White paper for LDPC codes for CCSDS channel Blue Book, NASA GSFC White Paper submitted to the Panel 1B CCSDS Standards Committee, October 2002.

13

Concatenated Single-Parity Check Codes for High-Density Digital Recording Systems

Jing Li
Lehigh University
Bethlehem, PA

Krishna R. Narayanan
Texas A&M University
College Park, TX

Erozan M. Kurtas
Seagate Technology
Pittsburgh, PA

Travis R. Oenning
IBM Corporation
Rochester, MI

13.1 Introduction

The breakthrough of turbo codes [1] and low-density parity check codes [11, 37] has revolutionized the coding research with several new concepts, among which code concatenation and iterative decoding are being actively exploited both for wireless communications and future digital magnetic recording systems. After being precoded, filtered and equalized to some simple partial response (PR) target, the magnetic recording channel appears much like an intersymbol interference (ISI) channel to an outer code and, hence, many of the techniques used in concatenated coding can be adopted. In particular, the observation that an ISI channel can be effectively viewed as a rate-1 convolutional code leads to the natural format of a serial concatenated system where the ISI channel is considered as the inner code and the error correction code (ECC) as the outer code. With reasonable complexity, iterative decoding and equalization (IDE), or turbo equalization, can be used to obtain good performance gains.

It has been shown by many that turbo codes (based on punctured recursive systematic convolutional codes) and LDPC codes can provide 4-5 dB of coding gain over uncoded systems at bit error rates (BER) of around 10^{-5} or 10^{-6} [2–5], [6–10]. Since magnetic recording applications require BERs in the order of 10^{-15} and since turbo and LDPC codes (as well as many other codes) would have already hit an error floor well before they reach this point, significant coding gains cannot be guaranteed. An effective remedy is to wrap a t-error correcting Reed-Solomon error correction code (RS-ECC) on top of these codes to clear up the residue errors. In this set-up, it is important that the output of the LDPC or turbo decoder (or other ECC codes) do not contain more than t byte errors that may cause the RS-ECC decoder to fail.

Due to the high decoding complexity of turbo codes, current research focuses on lower complexity solutions that are easily implementable in hardware. It has been recognized that very simple (almost useless) component codes can result in an overall powerful code when properly concatenated using random interleavers. Of particular interest is single-parity check codes due to their intrinsic high rates and the availability of a simple and optimal soft decoder. Researchers have applied single-parity check codes directly onto recording channels [32] as well as using them to construct a variety of well-performing codes including LDPC codes, block turbo codes (BTC) (e.g., [14, 20, 29]), multiple-branch turbo codes [17, 18], and concatenated tree (CT) codes [42].

In this work, we propose and study concatenated SPC codes for use in PR recording channels. Two constructions are considered: the first in the form of turbo codes where two branches of SPC codes are concatenated using a random interleaver, and the other in the form of product codes where arrays of SPC codewords are lined up in a multidimensional fashion. We denote the former as turbo/SPC scheme and the latter TPC/SPC scheme, that is, single parity check turbo product codes[1] [7, 19].

We undertake a comprehensive study of the properties of concatenated SPC codes and their applicability to PR recording channels, and highlight new results of this application. The fact that SPC codes are intrinsically very weak codes and that turbo/SPC and TPC/SPC codes are generally worse than LDPC codes on additive white Gaussian noise (AWGN) channels tend to indicate their inferiority on PR channels also. However, our studies show that, when several codewords are combined to form a larger effective block size and when the PR channel is properly precoded, concatenated SPC codes are capable of achieving large coding gains just as LDPC codes, but with less complexity[2].

Theoretical analysis is first conducted to explain the well-known spectral thinning phenomenon and to quantify the interleaving gain. Next, we compute the thresholds for iterative decoding of concatenated SPC codes using density evolution (DE) [21–24] to cast insight into the asymptotic performance limit of such schemes. Finally, we study the distribution of errors at the output of the decoder (i.e., at the input to the RS-ECC decoder) and show that concatenated SPC codes have an error distribution more favorable than that of LDPC codes in the presence of an outer RS-ECC code.

The paper proceeds as follows. Section 13.2 presents the PR system model, followed by a brief introduction to the turbo/SPC and TPC/SPC concatenated schemes. Section 13.3 conducts distance spectrum analysis and Section 13.4 computes the iterative thresholds. Section 13.5 presents the simulation results, which include bit error rate and bit/byte error statistics, and compares them with that of (random) LDPC codes. Finally, Section 13.6 concludes the paper with a discussion of future work in this area.

---

[1] Single parity check turbo product codes are also termed as array codes in [20] and hyper codes in [29].

[2] Randomly constructed LDPC codes typically have quadratic encoding complexity in the length N of the code ($O(N^2)$). It has been shown that several greedy algorithms can be applied to triangulate matrices (preprocessing) to reduce encoding complexity, but the complexity of preprocessing may be as much as $O(N^{3/2})$ [16]. Further, with the exception of a few LDPC codes that have cyclic or quasi-cyclic structures (mostly from combinatorial or geometric designs, see for examples [15]), large memory is generally required (for storage of generator and/or parity check matrices), which is a big concern in hardware implementation.

13.2   System Model

13.2.1   System Model

The digital recording channel is modeled as a L-tap ISI channel, where the channel impulse response is assumed to be a partial response polynomial with AWGN:

$$r_k = \sum\nolimits_{i=0}^{L-1} h_i x_{k-i} + n_k \tag{13.1}$$

where x_i, y_i, h_i and n_i are transmitted symbols, received symbols, channel response and AWGN, respectively. Specifically, we focus on the PR4 channel with channel response polynomial $H(D) = 1 - D^2$ and the EPR4 channel with $H(D) = (1 - D^2)(1 + D)$.

The overall system model is presented in Figure 13.1. Conforming to the set-up of the current and immediate future recording systems, we use a Reed-Solomon code as the outer wrap (referred to as RS-ECC) to clear up the residue errors. The data is first encoded using an RS-ECC and then the concatenated SPC code (either the turbo/SPC scheme or the TPC/SPC scheme) which we refer to as the outer code. The precoded ISI channel is treated as the inner code. The random interleaver between the outer and inner code works to break the correlation among neighboring bits, to eliminate error bursts, and (in conjunction with the precoder) to improve the overall distance spectrum by mapping low-weight error events to high-weight ones (the so-called spectrum thinning phenomenon). We call this interleaver the "channel interleaver" to differentiate it with the "code interleaver" in the outer code (i.e., the concatenated SPC code). The outer code, the inner code and the channel interleaver together form a serial concatenated system, where turbo equalization can be exploited to iterate soft information between the outer decoder and the inner decoder/equalizer, and then feed the hard decision decoding to the RS-ECC code.

13.2.2   Concatenated SPC Codes

Magnetic recording systems require a high code rate R since for recording systems code rate loss (in dB) is in the order of $10 \log_{10}(R^2)$ rather than $10 \log_{10}(R)$ as in an AWGN channel [27]. Hence, we consider only high-rate codes, that is, 2-branch turbo/SPC codes or 2-dimensional TPC/SPC codes.

Turbo/SPC codes: The turbo/SPC code comprises two parallel branches of $(t + 1, t)$ SPC codes and a random interleaver (Figure 13.2(a)). It should be noted that the interleaver between the parallel branches is of size $K = Pt$, that is, P blocks of data bits are taken and interleaved together before being encoded

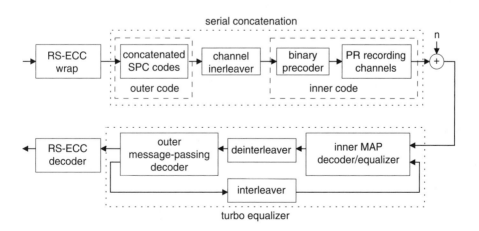

FIGURE 13.1   System model of concatenated SPC codes on (precoded) PR channels.

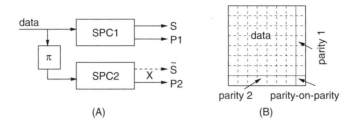

FIGURE 13.2   (a) Structure of a turbo/SPC code. (b) Structure of a 2-dimensional TPC/SPC code.

by the second branch. As we will show later, such combination of P blocks is essential in achieving the interleaving gain. Since the (interleaved) systematic bits are not transmitted in the second branch, the turbo/SPC code we consider is of parameters $(N, K, R) = (P(t + 2), Pt, t/(t + 2))$.

TPC/SPC codes: Another way of concatenating SPC codes is to take the form of product codes. A product code is formed by multidimensional arrays of codewords from linear block codes, where the codeword length, the user data block size, and the minimum distance of the overall code are the product of those of the component codes, respectively [12–14]. When the decoder takes an iterative (i.e. turbo) approach as is typically the case, product codes are also called turbo product codes (TPC) or block turbo codes. We consider 2-dimensional TPC/SPC codes whose code structure is illustrated in Figure 13.2(b). It is worth mentioning that, for the same reason that we combine SPC codewords in the turbo/SPC scheme, several blocks of TPC/SPC codewords will also be combined to form a larger effective block size before passing to the channel interleaver (see Figure 13.1). The resulting TPC/SPC scheme has parameters $(N, K, R) = (Q(t + 1)^2, Qt^2, t^2/(t + 1)^2)$ [7, 9, 19] and it may also appear in the paper as a $Q(t + 1, t)^2$ TPC/SPC code.

It is interesting to note that the two schemes, although different in lengths and rates, bear structural similarities. Depending on whether it has "parity-on-parity" bits or not, a TPC/SPC code can be either viewed as a serial or parallel concatenation where a linear block interleaver is used between the component codes. It is also worth noting that, from the graph-based point of view, both concatenated SPC schemes can be viewed as special types of LDPC codes where each SPC codeword satisfies a check. The equivalent LDPC code for the turbo/SPC scheme has a variable-node degree profile $\lambda(x) = \frac{1}{t+1} + \frac{t}{t+1}x$ and a check-node degree profile $\rho(x) = x^t$, that for the TPC/SPC code has a variable node degree profile $\lambda(x) = x$ and a check node degree profile $\rho(x) = x^t$, where the coefficient of the term x^j denotes the portion of edges connecting to nodes of degree $j + 1$ [22]. The encoding of these codes involves adding a single parity check bit in each SPC code, and hence is extremely simple and a dense generator matrix need not be explicitly stored as for random LDPC codes.

13.2.3   Iterative Decoding and Equalization

Since an overall maximum likelihood (ML) decoding and equalization of the coded PR system is prohibitively complex, the practical yet effective way, is to use turbo equalization to iterate soft extrinsic information (in the form of log-likelihood ratio or LLR) between the inner decoder/equalizer and the outer decoder. From the coding theory, we know that the inner code needs to be recursive in order for a serial concatenated system to achieve interleaving gain. The PR channel alone can be viewed as a rate-1 binary input real-valued output nonrecursive convolutional code. When a binary precoder is properly placed before the PR channel, the combination of the precoder and the channel becomes a recursive code that can be jointly equalized/decoded using maximum a posteriori probability (APP) decoding or the BCJR algorithm [25]. As long as the memory size of the precoder is not larger than that of the PR channel, no additional decoding complexity is introduced by the precoder.

The decoder of the outer concatenated SPC code performs the soft-in soft-out (SISO) message-passing decoding based on component SPC codes. We use l to denote the number of local iterations within the outer decoder (before LLRs are passed to the inner decoder for the next round of global iteration).

TABLE 13.1   Decoding Complexity in Terms of Number of Operations per Coded Bit per Iteration

Operations	Turbo/SPC	TPC/SPC	LDPC	BCJR
Addition	$d\left(3 - \frac{1}{t+1}\right)$	$d\left(3 - \frac{1}{t+1}\right)$	$4s$	$15 \cdot 2^m + 9$
Min/max				$5 \cdot 2^m - 2$
Table lookup	$2d$	$2d$	$2s$	$5 \cdot 2^m - 2$

d-number of branches of the turbo/SPC code or dimensionality of the TPC/SPC code; t-parameter of the component SPC code $(t + 1, t)$; s-average column weight of the LDPC code; m-memory size of the convolutional code.

For a $(t + 1, t)$ SPC code with $a_1 \oplus a_2 \oplus \cdots \oplus a_{t+1} = 0$, where $\oplus$ denotes the binary addition, the soft extrinsic information of bit a_i can be obtained from all other bits in the SPC codes as

$$L_e(a_i) = 2 \tanh^{-1}\left(\prod_{j=1, \, j \neq i}^{t+1} \tanh \frac{L_a(a_j)}{2} \right) \quad i = 1, 2, \ldots, t + 1 \tag{13.2}$$

where L_a stands for the a priori LLR value and L_e the extrinsic LLR value, and $\tanh(\cdot)$ is the hyper tangent function. This is known as the tanh rule decoding of SPC codes.

Since turbo/SPC and TPC/SPC schemes can be viewed as special types of LDPC codes, the same LDPC decoder can be used where all checks are decoded and updated simultaneously. However, below we discuss a slightly different approach, where checks are grouped into two groups (corresponding to upper/lower branches or to row/column component codes) and updated alternatively. This "serial" update is expected to converge a little faster than the "parallel" update.

The turbo/SPC decoder operates much like that of the turbo (convolutional) code. Each branch-decoder takes the extrinsic LLRs from the other branch, adds it to that from the equivalent channel (i.e., the inner code) to form the a priori LLR, and uses Equation 13.2 to compute new extrinsic LLRs to be passed to the other branch for use in the subsequent decoding iterations. After l (local) iterations, the extrinsic LLRs from both branches are added together and passed to the inner code. The TPC/SPC code can be iteratively decoded in a similar manner. Detailed discussion and pseudo-code of the message-passing TPC/SPC decoder can be found in [7, 9].

Table 13.1 compares the complexity of turbo/SPC, TPC/SPC, conventional LDPC (for comparison) and log-domain BCJR decoders (for the channel decoder/equalizer). It is assumed that multiplications are converted to additions in the log-domain [28], and that $\log(\tanh(\frac{x}{2}))$ and its reverse function $2 \tanh^{-1}(e^x)$ are implemented through table lookups. We see that the decoding algorithms for the turbo/SPC and TPC/SPC code require about 2/3 the complexity and about 1/3 the storage space of that for a regular column-weight-3 LDPC code in each decoding iteration.

13.3   Analysis of Distance Spectrum

Despite their similarities in decoding algorithms, turbo/SPC and TPC/SPC codes have very different distance spectra from random LDPC codes. For example, the minimum distance of a randomly-constructed regular LDPC code of column weight ≥ 3 will, with high probability, increase linearly with block length N (for large N). However, the minimum distance of a 2-dimensional TPC/SPC code is fixed to $d_m = 4$ regardless of block lengths and that of turbo/SPC codes in the worst-case scenario is $d_m = 2$. This partially explains why they perform noticeably worse than conventional LDPC codes on AWGN channels. However, on (precoded) ISI channels where iterative decoding and equalization is deployed, these codes perform on par with LDPC codes due to the interleaving gain [7].

Before proceeding to discussion, we first note the results of [30, 31] which state that, for a serially concatenated system with a recursive inner code, there exists an SNR threshold γ such that for any

$E_b/N_o \geq \gamma$, the asymptotic word error rate (WER) is upper bounded by:

$$P_w^{UB} = O\left(N^{-\left\lfloor \frac{d_m^{(o)}-1}{2} \right\rfloor}\right) \tag{13.3}$$

where N is the interleaver size and $d_m^{(o)}$ the minimum distance of the outer code. This suggests that (i) the outer code needs to have $d_m \geq 3$ in order to achieve the interleaving gain, and (ii) interleaving gain is attainable when $d_m^{(o)} \geq 3$ (and the inner code is recursive). While these serve as a quick and general guideline, caution needs to be exercised in extrapolating the result. Specifically, we will discuss two interesting results with the concatenated SPC system[3]. First, although a TPC/SPC system has $d_m^{(o)} = 4 \geq 3$, interleaving gain is not obtainable unless multiple TPC/SPC codewords are combined before passing to the channel interleaver (i.e., $Q \geq 2$). Second, although the ensemble of turbo/SPC systems has $d_m^{(o)} = 2 < 3$ (worst case), interleaving gain still exists so long as multiple SPC codewords are grouped in each branch ($P \geq 2$). Hence, while the concatenated SPC codes we consider in this work are not "good" codes by themselves[4], the combination of these codes with a precoded ISI channel become "good" codes due to spectral thinning.

13.3.1   Distance Properties of TPC/SPC Coded PR Systems

For ease of proposition, we use the precoded PR4 channel with channel response $H(D) = \frac{1-D^2}{1 \oplus D^2}$ as an example to quantify the interleaving gain of the TPC/SPC system. The result can be generalized to any ISI channel.

We first introduce some notations that will be used here and in the analysis of turbo/SPC systems. Let $A_{w,h}$ denote the input-output weight enumerator (IOWE) of a binary code, which enumerates the number of codewords with input Hamming weight w and output Hamming weight h, and A_h the output weight enumerator (OWE) where $A_h = \sum_w A_{w,h}$. Similar notations, A_{w,d_E} and A_{d_E}, are used with the ISI channel, where d_E stands for the (output) Euclidean distance.

We consider an TPC/SPC system with parameters $(N, K, R) = (Q(t+1)^2, Qt^2, t^2/(t+1)^2)$, where each "codeword" of (effective) length N is composed of Q TPC/SPC codewords of length $(t+1)^2$ each. To argue that TPC/SPC codes are capable of interleaving gain on precoded PR channels, we need to show that the average OWE of the TPC/SPC system, $A_{d_E}^{TPC/SPC}$, decreases with interleaver size for small d_E, thus providing a reduction in error rate. Using the ideas in [30, 32, 33] and assuming a uniform interleaver, we have the average OWE given by:

$$A_{d_E}^{TPS/SPC} = \sum_{\substack{l=4 \\ l \text{ even}}}^{N} \frac{A_l^o \times A_{l,d_E}^i}{\binom{N}{l}} \tag{13.4}$$

The sum of the series starts with $l = 4$ because TPC/SPC code has $d_m = 4$ and only even terms are considered because all codewords of the TPC/SPC code have even weights. Since a precoded PR channel is in general a nonregular convolutional code, the all-zeros sequence cannot be treated as the reference codeword. Since it is computationally prohibitive to perform an exact analysis of the full compound of error events pertaining to the inner code (i.e., $A_{l,d_E}^{(i)}$) [35], we take a similar approach developed in [7, 32], and assume that the input to the inner code are independent and identically distributed (i.i.d.) sequences of $\{0, 1\}^N$. It then follows that the equivalent trellis corresponding to odd/even bits of the precoded PR4 channel takes the form as in Figure 13.3.

---

[3]We use "system" to denote the concatenation of an outer code with the (precoded) ISI channel, and "scheme" or "code" to denote the outer code only.

[4]A "good" code as defined in [37] is a code that possesses a SNR threshold such that when the channel is better than this threshold, the code can achieve arbitrarily small error probability as the block size goes to infinity.

FIGURE 13.3   The equivalent trellis for even/odd bits of a precoded PR4 channel with response $H(D) = \frac{1-D^2}{1 \oplus D^2}$. Left: trellis; right: state diagram.

Following similar derivations as in [33], the average error enumerating function with a uniform interleaver is computed as

$$T(X, Y) = \frac{X^2 Y^8}{1 - \frac{1}{2}(1 + Y^{16})}$$

$$= X^2 Y^8 \left[\left(1 + \frac{1}{2} + \frac{1}{2^2} + \cdots \right) + Y^{16} \left(\frac{1}{2} + \frac{2}{2^2} + \frac{3}{2^3} + \cdots \right) + O(Y^{32}) \right]$$

$$= X^2 Y^8 [2 + 2Y^{16} + O(Y^{32})] \tag{13.5}$$

where the exponent of X is the Hamming distance between the error sequence and the i.i.d. reference sequence at the input, and the exponent of Y is the corresponding squared Euclidean distance at the output. The fractional terms in the branch weight enumerator such as $1/2(1 + Y^{16})$ (Figure 13.3) are a direct consequence of the assumption that the input corresponding to that branch can be a 0 or 1 with equal probability [7, 37].

Several things can be observed from the transfer function (Equation 13.5). First, the i.i.d. input error sequence (i.e., without an outer TPC/SPC code) always has input weight 2 for the precoded PR4 channel, since every term in Equation 13.5 corresponds to X^2. Second, any input error sequence of the form $1 + D^{2j}$ results in an error event, and $j = 1$ results in the minimum Euclidean distance (among all such error events) which is 8. Third, every finite weight codeword is the concatenation of k weight-2 input error events for some integer k [7]. For large value of N, let $T_N(X^{2k}, Y)$ denote the truncated weight enumerator truncated to length N, where each error event results from the joint effect of k input error sequences each of weight 2. Hence,

$$T_N(X^{2k}, Y) \propto \binom{N}{k} X^{2k} Y^{8k} [2 + 2Y^{16} + O(Y^{32})]^k \tag{13.6}$$

since there are approximately $\binom{N}{k}$ ways to arrange k error events in a block of length N. For the least nonzero l in the TPC/SPC system, namely $l = 4$ (i.e., $k = 2$ in Equation 13.6), we have $A^i_{l=4, d_E=4} \approx 4\binom{N}{2}$, and $A^o_{l=4} \approx Q[\sqrt{\frac{N/Q}{2}}]^2$, (there are $[(\sqrt{\frac{N/Q}{2}})]^2$ ways in which we can arrange a block of weight 4 within a single TPC/SPC block and there are Q blocks in a codeword of length N.) Substituting them into (Equation 13.4) and using the approximation $\binom{N}{n} \approx N^n/n!$ for large N, we have

$$A^c_{d_E=4} \propto \frac{\frac{N^2}{Q} \cdot N^2}{N^4} \propto Q^{-1} \tag{13.7}$$

Clearly, Equation 13.7 states that the reduction in word error rate is proportional to the number of blocks Q of the TPC/SPC that form a codeword, rather than N as what would be expected from Equation 13.3. This implies that a single TPC/SPC codeword ($Q = 1$), however long it is, does not get additional gain due to the length of the code. This is important for finite-length block sizes, since the achievable interleaving gain is limited to the number of blocks of the outer TPC/SPC codewords that are combined and interleaved. Although we have only discussed the error event corresponding to the least nonzero l (i.e., $l = 4$), it can be shown that for other values of small l, similar arguments hold.

To handle a general ISI channel, it is convenient to consider the precoder separately from the channel. That is, we treat the concatenation of the TPC/SPC and the precoder as a code whose codewords are passed through the ISI channel. Since the interleaving gain is dependent only on the recursive nature of the inner code, an interleaving gain will result regardless of the type of ISI channel. This idea will be further explained in the analysis of turbo/SPC systems.

13.3.2   Distance Properties of Turbo/SPC Coded PR Systems

In this subsection, we show that although the minimum distance of the ensemble turbo/SPC codes (with random interleavers[5]) is only 2, an interleaving gain still exists. Here we take a different approach from what we did with the TPC/SPC system, namely, we separate the precoder from the ISI channel, and argue that the combination of the binary precoder (i.e., recursive inner code) and the turbo/SPC code (outer code) results in an interleaving gain, irrespective of what ISI channel follows. This approach obviates the trouble of handling nonregular inner code, and the all-zeros sequence can therefore be used as the reference (for the serial concatenation of the turbo/SPC code and the precoder). For simplicity, we take the precoder $1/(1 \oplus D)$ as an example.

First, it is easy to show, as can also be inferred from Equation 13.3, that outer codewords of weight 3 or more will lead to an interleaver gain. Hence we focus the investigation on weight-2 outer codewords only, and show that the number of these codewords vanishes as P increases, where P is the number of SPC blocks combined in each branch.

Let $A_{w,h}^{(j)}$, $j = 1, 2$, denote the IOWE of the j_{th} SPC branch code that is parallelly concatenated in the outer code. The IOWE of the turbo/SPC codewords, $A_{w,h}^{(o)}$, averaged over the code ensemble is given by:

$$A_{w,h}^{(o)} = \sum_{h_1} \frac{A_{w,h_1}^{(1)} A_{w,h-h_1}^{(2)}}{\binom{K}{w}} \tag{13.8}$$

where $K = Pt$ is the input sequence length.

In each branch where P blocks of $(t + 1, t)$ SPC codewords are combined, the IOWE function is given by (assuming even parity check):

$$A^{SPC}(w, h) = \left(1 + \binom{t}{1} wh^2 + \binom{t}{2} w^2 h^2 + \binom{t}{3} w^3 h^4 + \cdots + \binom{t}{t} w^t h^{2\lceil t/2 \rceil}\right)^P$$

$$= \left(\sum_{j=0}^{t} \binom{t}{j} w^j h^{2\lceil j/2 \rceil}\right)^P \tag{13.9}$$

where the coefficient of the term $w^u h^v$ denotes the number of codewords with input weight u and output weight v. From the coefficients of Equation 13.9, we can obtain the IOWEs of the first SPC branch code: $A_{u,v}^{(1)} = A_{u,v}^{SPC}$. For the second SPC branch, since only parity bits are transmitted, $A_{u,v}^{(2)} = A_{u,v+u}^{(1)}$.

With a little computation, it is easy to show that the number of weight-2 outer codewords is given by

$$A_{h=2}^{(o)} = \sum_{w} A_{w,h=2}^{(o)} = P \binom{t}{2} \left(\frac{P\binom{t}{2}}{\binom{Pt}{2}}\right) = O(t^2) \tag{13.10}$$

where the last equation assumes a large P (i.e., a large block size). Equation 13.10 indicates that the number of weight-2 outer codewords is a function of a single parameter t, which is related only to the rate of SPC codes and not the block length. Now consider the serial concatenation of the outer codewords with the

---

[5]If S-random interleavers are used such that bits within S distance are mapped to at least S distance apart, then the outer codewords are guaranteed to have a minimum distance of at least 3 as long as $S \geq t$.

inner recursive precoder. The ensemble average OWE of the turbo/SPC system, $A_h^{\text{turbo/SPC}}$, can thus be computed as:

$$A_h^{\text{turbo/SPC}} = \sum_{h'} A_{h'}^{(o)} \frac{A_{h',h}^{(i)}}{\binom{N}{h'}} = \sum_{h'} \sum_w A_{w,h'}^{(o)} \frac{A_{h',h}^{1/(1 \oplus D)}}{\binom{N}{h'}} \tag{13.11}$$

where the IOWE of the $1/(1 \oplus D)$ precoder is given by [31, 40]:

$$A_{w,h}^{1/(1 \oplus D)} = \binom{N-h}{\lfloor w/2 \rfloor} \binom{h-1}{\lceil w/2 \rceil - 1} \tag{13.12}$$

Substituting Equation 13.10 and Equation 13.12 in Equation 13.11, we get the number of weight-s codewords (for small-s) produced by weight-2 outer codewords (i.e., $h' = 2$), denoted as $A_{h=s}^{\text{turbo/SPC:2}}$, is given as

$$A_{h=s}^{\text{turbo/SPC:2}} = \frac{(t-1)^2}{2} \frac{N-s}{\binom{N}{2}} = O(t P^{-1}) \tag{13.13}$$

where $N = P(t+2)$ is the length of the overall codeword (or the channel interleaver size). This indicates that the number of small weight s codewords of the overall system due to weight-2 outer codewords (which are caused by weight-2 input data sequences) vanishes as P increases. When the input weight is greater than 2, the outer codeword always has weight greater than 2 and, hence, an interleaving gain can be guaranteed. Hence, an interleaving gain exists for turbo/SPC systems and it is proportional to P.

It is also worth noting that the system model we considered above, namely, the combination of an outer concatenated SPC code and an inner $1/(1 \oplus D)$ code, essentially forms a product accumulate code, which is shown to be a class of high-performance, low-complexity, high-rate "good" codes [38, 40]. Hence, depending on different view-stands, the coded ISI systems we discuss in this work can either be viewed as concatenated SPC codes on precoded ISI channels or product accumulate codes on non-precoded ISI channels.

13.4 Thresholds Analysis using Density Evolution

13.4.1 Introduction to Density Evolution and Gaussian Approximation

Distance spectrum analysis shows that both turbo/SPC and TPC/SPC systems possess good distance spectra and that interleaving gain is achievable for both systems. However, since the analysis assumes a maximum likelihood decoder which differs from the practical iterative decoder, it would be more convincing to account for the suboptimality of iterative decoding in the analysis. Such analysis is possible using the recently developed technique of density evolution [21–24].

Introduced for the analysis of LDPC codes, density evolution unveiled an SNR threshold effect for LDPC codes in that error rate goes to zero when the channel is better than this threshold and that the error rate is bounded away from zero otherwise [21–24]. This threshold clearly marks the performance limit we can expect with the existing suboptimal decoder. Interestingly, the same threshold effect also presents in turbo/SPC and TPC/SPC systems and, hence, the same DE treatment can be used for capacity analysis. However, certain modifications are required.

Due to the space limitation, we go through the critical points in the application of density evolution to turbo/SPC and TPC/SPC systems. Detailed discussion on density evolution and its application to a variety of systems can be found in [21–24] [7].

13.4.2 Problem Formulation

Consider a unified architecture where the precoded PR channel is modeled as an inner rate-1 recursive convolutional code, and the outer code is either a turbo/SPC or TPC/SPC code. A turbo equalizer iterates extrinsic LLR information, denoted as $L_i^{(q)}(a_j)$ and $L_o^{(q)}(a_j)$, between the inner and outer decoders, where subscript i and o denote the quantities associated with the inner and outer codes, respectively. Assuming infinite length and perfect interleaving, the LLR messages are approximated as i.i.d. random variables. During the q_{th} iteration, the outer message-passing decoder generates extrinsic information on the j_{th} coded bit a_j, denoted by $L_o^{(q)}(a_j)$, and passes it to the inner decoder. The inner MAP decoder then uses this extrinsic information (treat as a priori) with the received signal and generates extrinsic information, $L_i^{(q+1)}(a_j)$.

The idea in density evolution is to examine the probability density function (pdf) of $L_o^{(q)}(a_j)$ during the q_{th} iteration, denoted by $f_{L_o^{(q)}}(x \mid a_j)$. For infinite lengths and perfect interleaving, these random variables are i.i.d. Hence, we drop the dependence on j. If the sign of $(\text{sign}(a) \cdot L_o^{(q)}(a))$ is positive, then the decoding algorithm has converged to the correct codeword. The probability that $\text{sign}(a) \cdot L_o^{(q)}(a) < 0$ is

$$\Pr(\text{sign}(a) \cdot L_o^{(q)}(a) < 0)$$

$$= \Pr(a = +1) \int_{-\infty}^{0} f_{L_o^{(q)}}(x \mid a = +1)\, dx + \Pr(a = -1) \int_{0}^{\infty} f_{L_o^{(q)}}(x \mid a = -1)\, dx \quad (13.14)$$

The key is to find the critical SNR value, or the threshold of the system, such that

$$\gamma = \inf_{SNR} \left\{ SNR : \lim_{q \to \infty} \lim_{N \to \infty} \Pr\left(\text{sign}(a) \cdot L_o^{(q)}(a) < 0\right) \to 0 \right\} \quad (13.15)$$

where N denotes the block size.

Since it is quite difficult to analytically evaluate $f_{L_o^{(q)}}(x)$ for all q, simplification can be made by approximating $f_{L_o^{(q)}}(x)$ to be Gaussian (or Gaussian mixture). This is what is used by Wiberg et al. [34], Chung et al. [22], El Gamal et al. [36], and many others to analyze concatenated codes. Further, Richardson and Urbanke have shown that, for binary input, output symmetric channels, a consistency condition is preserved under density evolution for all messages, such that the pdf's satisfy the condition $f_{L_o^{(q)}}(x) = f_{L_o^{(q)}}(-x) \cdot e^x$ [21]. Imposing this constraint to the approximate Gaussian densities at every step leads to $(\sigma_o^{(q)})^2 = 2m_o^{(q)}$, that is, the variance of the message density equals twice the mean. Under i.i.d. and Gaussian assumptions, the mean of the messages, $m_o^{(q)}$ then serves as the sufficient statistics of the message density. The problem thus reduces to:

$$\gamma = \inf_{SNR} \left\{ SNR : \lim_{q \to \infty} \lim_{N \to \infty} \text{sign}(a) m_o^{(q)}(x \mid a) \to \infty \right\} \quad (13.16)$$

To solve the problem formulated in Equation 13.16, we need to examine the message flow within the outer decoder, the inner decoder as well as between the two. In general, we need to evaluate $m_o^{(q)}$ as a function of $m_i^{(q)}$ and vice versa.

13.4.3   Message Flow Within the Inner MAP Decoder

Since it is not straight-forward to derive $m_i^{(q+1)}$ as a function of $m_o^{(q)}$ for the inner MAP decoder, Monte Carlo simulation is used to determine a relationship between $m_i^{(q+1)}$ and $m_o^{(q)}$. This process is denoted as

$$m_i^{(q)} = \mathcal{F}(m_o^{(q-1)}) \quad (13.17)$$

where the mean of the message $m_i^{(q)}$ is evaluated at the output of the inner MAP decoder given the input a priori information is independent and Gaussian with mean $\pm m_o^{(q-1)}$ and variance $2m_o^{(q-1)}$.[6] Detailed discussion of Monte Carlo simulation technique for computing γ_i can be found, for example, in [26].

---

[6] Again, due to the nonlinearity of the ISI channel, a sequence of i.i.d. bits are used as the transmitted data.

13.4.4 Message Flow Within the Outer Code

Below we discuss how to compute $m_o^{(q)}$ as a function of $m_i^{(q)}$ for outer concatenated SPC codes. Since both turbo/SPC and TPC/SPC codes can be viewed as special types of LDPC codes, we start with LDPC codes to pinpoint the key steps, and then move onto turbo/SPC and TPC/SPC codes.

Irregular LDPC codes: Both turbo equalization and LDPC decoding are iterative processes. Let us use superscript (q, l) to denote quantities during the qth (global) iteration of turbo equalization and lth (local) iteration within the LDPC decoder. For irregular LDPC codes with bit-node and check-node degree profiles $\lambda(x) = \sum_k \lambda_k x^{k-1}$ and $\rho(x) = \sum_j \rho_j x^{j-1}$, the code rate is given by $R = 1 - \frac{\sum_k \lambda_k / k}{\sum_j \rho_j / j}$. Message flow on the code graph is a two-way procedure that composes of bit updates and check updates, which correspond to summation in the real domain and the so-called check-sum operation or *tanh* rule, respectively. [9, 22, 23]. After L local iterations of message exchange, the message passed over to the inner MAP decoder is the LLR of the bit in the Lth iteration after subtracting $L_i^{(q)}$ which was obtained from the inner code and was used as a priori information.

Under the Gaussian assumption, we are interested in tracking the means of $L_b^{(q,l)}$ and $L_c^{(q,l)}$, denoted as $m_b^{(q,l)}$ and $m_c^{(q,l)}$, where subscripts b and c refer to quantities pertaining to bit-nodes and check-nodes, respectively. To handle the irregularity of LDPC codes, we further introduce notations $m_{b,k}^{(q,l)}$ and $m_{c,j}^{(q,l)}$ to denoted message mean associated with bit-nodes of degree k and check-nodes of degree j, respectively. Treating extrinsic information as independent, the means of the extrinsic information at each local iteration l are shown to be [22]

$$\text{bit} - \text{to} - \text{check}: \quad m_{b,k}^{(q,l)} = m_i^{(q)} + (k-1) \cdot m_c^{(q,l-1)} \tag{13.18}$$

$$\text{check} - \text{to} - \text{bit}: \quad m_{c,j}^{(q,l)} = \psi^{-1}\left(1 - \left[1 - \sum_k \lambda_k \psi\left(m_{b,k}^{(q,l)}\right)\right]^{j-1}\right) \tag{13.19}$$

$$m_c^{(q,l)} = \sum_j \rho_j m_{c,j}^{(q,l)} \tag{13.20}$$

where $\psi(x)$ is the expected value of $1 - \tanh(\frac{u}{2})$ where u follows a Gaussian distribution with mean x and variance $2x$. Specifically, $\psi(x)$ is given by:

$$\psi(x) = \begin{cases} 1 - \frac{1}{\sqrt{4\pi x}} \int_{-\infty}^{\infty} \tanh\left(\frac{u}{2}\right) e^{-\frac{(u-x)^2}{4x}} du, & x > 0 \\ 1, & x = 0 \end{cases} \tag{13.21}$$

$\psi(x)$ is continuous and monotonically decreasing on $[0, \infty)$ with $\psi(0) = 1$ and $\psi(\infty) = 0$. The initial condition is $m_b^{(q,0)} = m_c^{(q,0)} = 0$. When x is large (corresponding to low error probability), $\psi(x)$ is shown to be proportional to the error probability [22]. The above derivation is essentially an extension of Chung et al.'s work [22] to the case of turbo equalization. For detailed discussions, readers are directed to [21–24] and the references therein.

After L LDPC decoding iterations, the messages passed from the outer LDPC code to the inner MAP decoder/equalizer follows a mixed Gaussian distribution where λ_k' fraction of bits follow a Gaussian distribution with mean value

$$m_{o,k}^{(q)} = k m_{b,k}^{(q,L)}, \quad k = 1, 2, \ldots \tag{13.22}$$

and variance $2m_{o,k}^{(q)}$. Here λ_k' denotes the percentage of bits that have degree k, and is given by $\lambda_k' = (\lambda_k / k)/(\sum_k \lambda_k / k)$. Hence, we can describe what is passed from the outer decoder to the inner decoder in the qth turbo equalization as $m_o^{(q)} \sim \{<\lambda_k', m_{o,k}^{(q)}>, k = 1, 2, \ldots\}$. This will in turn be used by the inner decoder to generate $m_i^{(q+1)}$.

After q global iterations between the outer and inner decoder (where each iteration contains L local iterations within the LDPC decoder), the threshold can be evaluated as:

$$\gamma_{LDPC} = \inf_{SNR} \left\{ SNR : \lim_{q \to \infty} \cdot m_c^{(q,L)} \to \infty \right\} \tag{13.23}$$

It is instructive to note that the value of L has a slight impact on the asymptotic threshold, but a quite noticeable impact for finite-length finite-complexity performance. Specifically, it has been shown in [7] that for a given complexity constraint, an optimal value of L can be computed using density evolution for the concatenated system to reach the best performance.

Turbo/SPC codes: Using the degree profiles $\lambda(x) = \frac{1}{t+1} + \frac{t}{t+1}x$ and $\rho(x) = x^t$, the message flow within a Turbo/SPC decoder can be tracked following exactly the same steps as we described above. An alternative procedure stems naturally from the decoding algorithm where checks corresponding to different branches take turns to update. As expected, such a serial scheduling expedites the convergence and improves the performance for finite-length systems, but has little impact on the asymptotic thresholds.

TPC/SPC codes: Although TPC/SPC codes can be viewed as a special type of LDPC codes, the above DE procedure cannot be applied directly. This is because the underlying assumption of a cycle-free code graph does not hold for TPC/SPC codes. Since any rectangular error pattern (or their combination) in the 2-dimensional TPC/SPC bit-array results in a loop in the code graph [7], there always exist cycles of length $4(k + 1)$ irrespective of block sizes (k can be any positive integer). Hence, messages being passed in the code graph are not always independent (loop-free operation), and adjustment needs to be made before it is applicable to TPC/SPC codes.

Notice that when the number of local iterations within the TPC/SPC code is restricted to be small, the density evolution method would have operated on cycle-free subgraphs of TPC/SPC codes. In other words, the messages exchanged along each step are statistically independent as long as the cycles have not "closed." Here, we restrict the number of local iterations within TPC/SPC codes to be one row update and one column update, since any more updates in either direction will either pass information to its source or pass duplicate information to the same node, which are unacceptable. On the other side, due to the perfect random interleaver, infinite number of turbo iterations can be performed between the inner and outer decoders if the messages within the outer TPC/SPC code are reset to zero in every new turbo iteration.

For an exact threshold, the density evolution procedure should, in addition to avoiding looping messages, also ensure completeness in the sense that every bit should have fully utilized all the messages (through dependencies) from all the checks. Unfortunately, one row update followed by one column update is not sufficient to exploit all the information contained in all bits and checks [7]. Hence, the resulting threshold is a lower bound[7].

13.4.5 Thresholds

The lower bound on the threshold for TPC/SPC codes, and the threshold for turbo/SPC codes are plotted in Figure 13.4 for precoded PR4 and EPR4 channels. For comparison, we also evaluate LDPC codes on nonprecoded PR channels[8] We consider regular LDPC codes with column weight 3, since regular LDPC codes are shown to be slightly advantageous over irregular ones at short block sizes and high rates such as what will be used in data storage applications [39]. It can be seen that the thresholds of concatenated SPC systems (and their lower bound) are within a few tenths a dB from that of LDPC systems. This indicates they have comparable performance asymptotically.

---

[7]By lower bound, we mean that the exact thresholds of TPC/SPC system should be better than this. Put another way, for a given dB, the achievable code rate (bandwidth efficiency) could be higher, or equivalently, for a given rate, the required SNR could be smaller.

[8]It has been shown that randomly constructed LDPC codes perform better on ISI channels that are not precoded. Hence, the comparison represents the best in both cases and is thus fair.

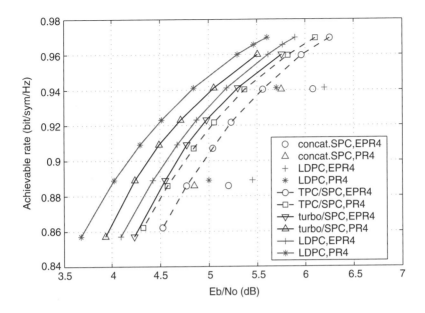

FIGURE 13.4 Iterative thresholds computed using density evolution with a Gaussian approximation. (Solid lines: bounds of LDPC or turbo/SPC systems; dashed lines: lower bounds of TPC/SPC systems.)

Also presented in Figure 13.4 are the corresponding simulation results of a length-4K block evaluated at a BER of 10^{-5}. It is interesting to observe that for practical block sizes concatenated SPC systems may actually (slightly) outperform LDPC systems. Considering the small block size used in the simulation, the $0.5 \sim 1$ dB gap between performance points and threshold curves indicates a good agreement between simulation and analysis.

13.5 Simulation Results

To be applicable to high-density recording systems, the concatenated SPC codes we consider have high rates of 0.89 and 0.94 which are formed from (17,16) and (33,32) TPC codes, respectively. Several codewords are combined and interleaved together to form an effective data block size of (around) 4K bits, the size of a block in a hard-disk drive. At such high code rates and short block sizes, the two concatenated SPC schemes differ very little in lengths, rates and performance. Hence, we do not differentiate them in discussing the simulation results. For comparison, also presented are the results of regular LDPC codes with column weight 3 and similar rates and lengths. In all the simulations presented, 2 iterations are performed within the concatenated SPC decoders and 4 iterations within the LDPC decoder. This makes the decoding complexity of LDPC codes slightly higher than that of concatenated SPC codes, but they are the most efficient scheduling schemes we have found in my experiments.

Bit error rate: Figure 13.5 shows the performance of LDPC codes and concatenated SPC codes over PR4 and EPR4 channels. Gains of 4.4 to 5 dB over uncoded partial response maximum likelihood (PRML) systems are obtained for concatenated SPC codes at a BER of 10^{-5}, which are comparable to those of LDPC codes. All concatenated SPC systems use a binary precoder with polynomial $1/(1 \oplus D^2)$ which is the best-performing precoder for PR4/EPR4 channels, as shown in Figure 13.6. The ISI channels are not precoded in LDPC systems, since as discussed in [26] nonprecoding leads to a better performance than otherwise. This is also confirmed by our simulations. Hence, the comparison is fair as it represents the best cases for both systems.

Error bursts: Although both concatenated SPC and LDPC codes can offer significant coding gains at a BER level of 10^{-7}, it is unclear whether and when they will have an error floor. Therefore, the conventional use of RS-ECC is still necessary to reduce the BER to 10^{-15} as is targeted for recording systems. The RS-ECC

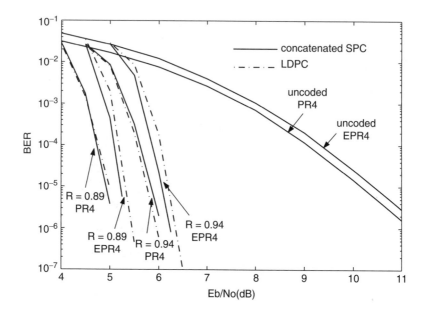

FIGURE 13.5 BER performance of concatenated SPC codes with comparison to LDPC codes on PR recording channels. (Code rate: 0.89 and 0.94; channel model: PR4 and EPR4; evaluated after 10 turbo iterations.

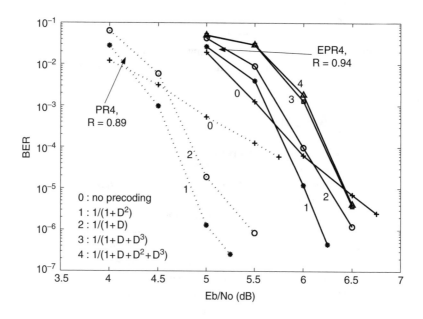

FIGURE 13.6 Effect of binary precoding on the bit error rate of concatenated SPC codes on PR4 and EPR4 channels.

code works on the byte level, capable of correcting up to t byte errors in each data block of size 4K bits or 512 bytes (t is usually around 10 to 20). Hence, the maximum number of uncorrected errors left in each block after the turbo decoding/equalization of concatenated SPC codes and LDPC codes has to be small to guarantee the proper functioning of the RS-ECC code. In other words, block error statistics is crucial to the overall system performance. Unfortunately, this has been largely neglected in most of the previous work.

Figure 13.7 and Figure 13.8 plot the histograms of the number of bit/byte errors for LDPC codes and TPC/SPC codes on EPR4 channels, respectively. The effective block size is 4K and the code rate is 0.94.

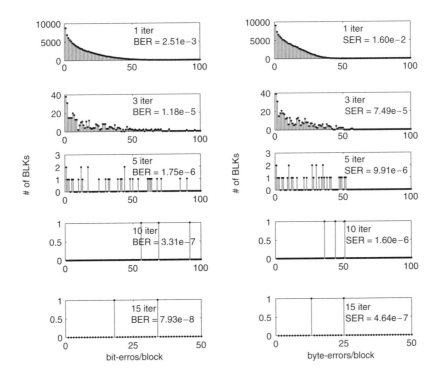

FIGURE 13.7   Statistics of error bursts for LDPC codes on EPR4 channels. (Code rate 0.94, block size $K = 4k$ bits, nonprecoded EPR4 channel, 6.5 dB SNR; X-axis: number of bit/byte errors in each block, Y-axis: number of erroneous blocks; one byte consists of eight consecutive bits; statistics are collected after 1, 3, 5, 10, and 15 iterations over more than 100,000 blocks.)

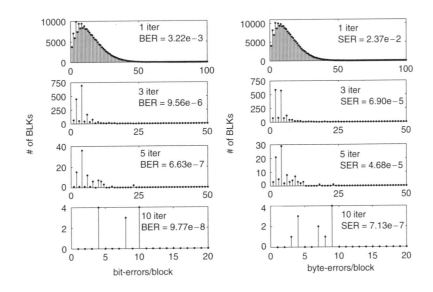

FIGURE 13.8   Statistics of error bursts for concatenated SPC codes on EPR4 channels. (Code rate 0.94, block size $K = 4k$ bits, EPR4 channel with precoder $1/(1 \oplus D^2)$, 6.5 dB SNR; X-axis: number of bit/byte errors in each block, Y-axis: number of erroneous blocks; one byte consists of eight consecutive bits; statistics are collected after 1, 3, 5, and 10 iterations over more than 165,000 blocks.)

The left column in each figure plots bit error histograms and the right byte error histograms, where a byte composes of eight consecutive bits. The statistics are collected over more than $100,000$ blocks of data size 4K bits. At an SNR of 6.5 dB and after the 10_{th} iteration (big loop), the maximum number of symbol errors observed in a single block is less than 10 for concatenated SPC codes (which would be corrected by the RS-ECC code), but around 50 for LDPC codes. If further iterations are allowed, error bursts in LDPC codes are alleviated. Nevertheless a block containing 25 symbol errors is observed after 15 turbo iterations and this may still cause the RS-ECC code to fail. Unless a more powerful RS-ECC is employed, LDPC codes are prone to cause block failure, where all data in that block are presumed lost. It should be noted that although what we have observed suggests that concatenated SPC codes may be more compatible to magnetic recording systems than LDPC codes, the statistics are nonetheless insufficient. Due to the very long simulation time in software, hardware tests over millions or billions of blocks are needed before a convincing argument can be made.

13.6   Conclusion

This paper investigates the potential of applying concatenated single-parity check codes on PR magnetic recording channels, with LDPC codes as a comparison study. Two ways of concatenation are studied, one in the form of parallel turbo codes and the other product codes. While they have very different structures, analysis and simulations show that the two schemes exhibit similar properties in terms of interleaving gain, iterative thresholds and finite-length performances (especially at high rates). Below summarizes the main results of this study.

1. Despite their relatively small minimum distances on AWGN channels, both concatenating schemes are capable of achieving interleaving gain on precoded ISI channels. The key is to have several codeword blocks combined and interleaved together.

2. Density evolution is an effective tool in the analysis of iterative decoding processes by accounting for both the code structure (i.e, codeword space and the mapping of data space to codeword space) and the iterative nature of the decoding algorithms. Thresholds (or their lower bounds) computed using density evolution with a Gaussian approximation indicate these codes have comparable performance asymptotically, and finite-length simulations agree with the analysis.

3. Simulations on high-rate and short-length concatenated SPC systems demonstrate considerable coding gains over uncoded PRML systems. In particular, gains of more than 4.4 dB are observed for length 4K and rate > 0.88 codes on PR4 and EPR4 channels, revealing a performance similar to or slightly better than that of random LDPC codes.

4. In addition to the slightly smaller complexity than that of random LDPC codes, concatenated SPC codes are linear time encodable. Further, they do not require large storage for parity check and generator matrices. The interleaving pattern needs be stored, but algebraic interleavers which can be pseudo-randomly generated "on the fly" can be used to save space in hardware implementation [40, 41].

5. In contrast to LDPC codes whose large error bursts within a block tend to exceed the capacity of RS-ECC wraps, our experiments indicate that the number of errors in each block is typically small with concatenated SPC codes (although there may be more blocks in error). Such a spread error distribution is not only desirable, but also crucial to ensure an overall low error probability in the recording system. However, extensive experiments on massive data need to be conducted before a firm conclusion can be reached.

To summarize, our study indicates that concatenated SPC codes can be a promising candidate for future high-density magnetic recording systems. However, further experiments need to be carried out on more realistic channel models, like the Lorentzian channel model and (eventually) the real data set collected in the lab.

There are many other interesting problems regarding future digital data recording systems. For example, how to efficiently incorporate the run-length limit (RLL) constraint into the SISO decoding/equalization scheme without affecting much of the overall code rate, performance and complexity? Interleaving is necessary for coded PR systems that use turbo equalizations. How to implement interleaving and deinterleaving schemes that are both space-efficient and time-efficient but still exhibiting good randomness? Most of all, how to achieve a good trade-off among the many competing factors like performance, complexity, delay and cost in a practical setting? These are a few of the many interesting problems that await to be addressed.

References

[1] C. Berrou, A. Glavieux, and P. Thitimasjshima, Near Shannon limit error-correcting coding and decoding: turbo-codes (1), *Proc. IEEE Intl. Conf. Commun.*, pp. 1064–1070, Geneva, Switzerland, May 1993.

[2] T.M. Duman and E. Kurtas, Performance of turbo codes over EPR4 equalized high density magnetic recording channels, *Proc. Global Telecom. Conf.* 1999.

[3] T. Souvignier, A. Friedmann, M. Oberg, P. Siegel, R. Swanson, and J. Wolf, Turbo decoding for PR4: parallel vs. serial concatenation, *Proc. Intl. Conf. Comm.*, pp. 1638–1642, June 1999.

[4] W. Ryan, L. McPheters, and S. McLaughlin, Combined turbo coding and turbo equalization for PR4-equalized Lorentzian channels, *Proc. Conf. Intl. Sci. and Sys.*, 1998.

[5] M. Oberg and P. H. Siegel, Performance analysis of turbo-equalized dicode partial-response channel, *Proc. of 36th Annual Allerton Confer. Commun., Control and Computing*, pp. 230–239, September 1998.

[6] J.L. Fan, A. Friedmann, E. Kurtas, and S. McLaughlin, Low density parity check codes for magnetic recording, *37-th Allerton Conf.*, 1999.

[7] J. Li, K.R. Narayanan, E. Kurtas, and C.N. Georghiades, On the performance of high-rate TPC/SPC and LDPC codes over partial response channels, *IEEE Trans. Commun.*, vol. 50, no. 5, pp. 723–734, May 2002.

[8] T.R. Oenning and J. Moon, Low density parity check coding for magnetic recording channels with media noise, *Proc. Intl. Conf. Commun.*, pp. 2189–2193, June 2001.

[9] J. Li, E. Kurtas, K.R. Narayanan, and C.N. Georghiades, On the performance of turbo product codes and LDPC codes over partial response channels, *Proc. Intl. Conf. Commun.*, Helsinki, Finland, June 2001.

[10] H. Song, R.M. Todd, and J.R. Cruz, Performance of low-density parity-check codes on magnetic recording channels, *Proc. 2nd Intl. Symp. on Turbo Codes and Related Topics*, France, September 2000.

[11] R.G. Gallager, Low-density parity check codes, *IRE Trans. Inform. Theory*, pp. 21–28, 1962.

[12] P. Elias, Error-free coding, *IRE Trans. Inform. Theory*, vol. IT-4, pp. 29–37, September 1954.

[13] J. Hagenauer, E. Offer, and L. Papke, Iterative decoding of binary block and convolutional codes, *IEEE Trans. Inform. Theory*, vol. 42, no. 2, March 1996.

[14] R.M. Pyndiah, Near-optimum decoding of product codes: block turbo codes, *IEEE Trans. Commun.*, vol. 46, no. 8, August 1998.

[15] Y. Kou, S. Lin, and M. Fossorier, LDPC codes based on finite geometries, a rediscovery and more, submitted to *IEEE Trans. Inform. Theory*, 1999.

[16] T.J. Richardson and R.L. Urbanke, Efficient encoding of low-density parity-check codes, *IEEE Trans. Inform. Theory*, vol. 47, no. 2, pp. 638–656, February 2001.

[17] L. Ping, S. Chan, and K.L. Yeung, Iterative decoding of multidimensional concatenated single parity check codes, *Proc. IEEE Intl. Conf. Commun.*, 1998.

[18] T.R. Oenning and J. Moon, A low-density generator matrix interpretation of parallel concatenated single bit parity codes, *IEEE Trans. Magn.*, pp. 737–741, March 2001.

[19] T. Souvignier, C. Argon, S.W. McLaughlin, and K. Thamvichai, Turbo product codes for partial response channels, *Proc. Intl. Conf. Commun.*, vol. 7, pp. 2184–2188, 2001.

[20] M. Blaum, P.G. Farrell, and H.C.A. van Tilborg, *Array Codes, Handbook of Coding Theory*, V.S. Pless, W.C. Huffman, and R.A. Brualdi, Eds., North-Holland, Amsterdam, pp. 1855–1909, November 1998.

[21] T.J. Richardson, A. Shokrollahi, and R. Urbanke, Design of capacity-approaching irregular low-density parity-check codes, *IEEE Trans. Inform.*, vol. 47, pp. 619–637, February 2001.

[22] S.-Y. Chung, R. Urbanke, and T.J. Richardson, Analysis of sum-product decoding of low-density parity-check codes using a Gaussian approximation, *IEEE Trans. Inform. Theory*, vol. 47, pp. 657–670, February 2001.

[23] J. Li, E. Kurtas, K.R. Narayanan, and C.N. Georghiades, Thresholds for iterative equalization of partial response channels using density evolution, *Proc. Intl. Symp. on Inform. Theory*, Washington D.C., June 2001.

[24] M.G. Luby, M. Mitzenmacher, M.A. Shokrollahi, and D.A. Spielman, Analysis of low density codes and improved designs using irregular graphs, available at http://www.icsi.berkeley.edu/~luby/.

[25] L.R. Bahl, I. Cocke, F. Jelinek, and J. Raviv, Optimal decoding of linear codes for minimizing symbol error rate, *IEEE Trans. Inform. Theory*, pp. 284–287, March 1974.

[26] K.R. Narayanan, Effect of precoding on the convergence of turbo equalization for partial response channels, *IEEE J. Select. Area. Commun.*, vol 19, pp. 686–698, April 2001.

[27] K. Immink, Coding techniques for the noisy magnetic recording channel: a state-of-the-art report, *IEEE Trans. Commun.*, pp 413–419, May 1989.

[28] P. Robertson, E. Villebrun, and P. Hoeher, A comparison of optimal and sub-optimal MAP decoding algorithms operating in the log domain, *Proc. Intl. Conf. Commun.*, vol. 2, pp. 1009–1013, June 1995.

[29] P.-P. Sauvé, A. Hunt, S. Crozier, and P. Guinand, Hyper-codes: high-performance, low-complexity codes, *Proc. 2nd Intl. Symp. on Turbo Codes and Related Topics*, France, September 2000.

[30] S. Benedetto, D. Divsalar, G. Montorsi, and F. Pollara, Serial concatenation of interleaved codes: Design and performance analysis, *IEEE Trans. Inform. Theory*, vol. 44, pp. 909–926, May 1998.

[31] D. Divsalar, H. Jin, and R. J. McEliece, Coding theorems for "turbo-like" codes, *Proc. Allerton Conf. Commun. and Control*, pp. 201–210, September 1998.

[32] M. Öberg and P. H. Siegel, Parity check codes for partial response channels, *Proc. Global Telecom. Conf.*, vol. 1b, pp. 717–722, December 1999.

[33] L.L. McPheters, S.W. McLaughlin, and K.R. Narayanan, Precoded PRML, serial concatenation, and iterative (turbo) decoding for digital magnetic recording, *IEEE Trans. Magn.*, vol. 35, no. 5, September 1999.

[34] N. Wiberg, Codes and Decoding on General Graphs, Doctoral dissertation, Linköping University, 1996.

[35] J. Li, K.R. Narayanan, and C.N. Georghiades, An efficient algorithm to compute the Euclidean distance spectrum of a general inter-symbol interference channel and its applications, *IEEE Trans. Commun.* preprint (in press).

[36] H. El Gamal, A.R. Hammons, Jr., and E. Geraniotis, Analyzing the turbo decoder using the Gaussian approximation, *Proc. Intl. Symp. Inform. Theory*, 2000.

[37] D.J.C. MacKay, Good error-correcting codes based on very sparse matrices, *IEEE Trans. Info. Theory*, vol. 45, no. 2, March 1999.

[38] J. Li, K.R. Narayanan, and C.N. Georghiades, A class of linear-complexity, soft-decodable, high-rate, 'good' codes: construction, properties and performance, *Proc. IEEE Intl. Symp. Inform. Theory*, pp. 122–122, Washington D.C., June 2001.

[39] D.J.C. MacKay and M.C. Davey, Evaluation of Gallager codes for short block length and high rate applications, available at http://www.keck.ucsf.edu/~mackay/seagate.ps.gz.

[40] J. Li, K.R. Narayanan, and C.N. Georghiades, Product accumulate codes: a class of capacity-approaching, low complexity codes, *IEEE Trans. Inform. Theory*, January 2004, in preprint (in press).

[41] G.C. Clark, Jr. and J.B. Cain, *Error-Correction Coding for Digital Communications*, Plenum Press, New York, 1981.

[42] L. Ping and K.Y. Wu, Concatenated tree codes: a low-complexity, high-performance approach, *IEEE Trans. Inform. Theory*, vol. 47, pp. 791–799, February 2001.

14

Structured Low-Density Parity-Check Codes

Bane Vasic
University of Arizona
Tucson, AZ

Erozan M. Kurtas
Seagate Technology
Pittsburgh, PA

Alexander Kuznetsov
Seagate Technology
Pittsburgh, PA

Olgica Milenkovic
University of Colorado
Boulder, CO

14.1 Introduction

Iterative coding techniques that improve the reliability of input-constrained, intersymbol interference (ISI) channels have recently attracted considerable attention for data storage applications. Inspired by the success of turbo codes [6], several authors have considered iterative decoding architectures for coding schemes comprised of a concatenation of an outer block, convolutional or turbo encoder with a rate one code representing the channel. Such an architecture is equivalent to a serial concatenation of codes [4], with the inner code being the ISI channel. Application of this concatenated scheme in magnetic and optical recording systems is considered in [23, 24].

Theory and practice of the soft iterative detection were facilitated by using the concept of codes on graphs. As shown in the chapter on message passing algorithms, a graph of a code representing the ISI channel is trivial. Therefore, we will focus only on the design and graphical description of the outer code. The prime examples of codes on graphs are low-density parity check codes (LDPC). One of the key results in codes on graphs comes from Frey and Kschischang [36, 45] who observed that iterative decoding algorithms developed for these codes are instances of probability propagation algorithms operating on a graphical model of the code. The belief propagation algorithms and graphical models have been developed in the expert systems literature by Pearl [65] and Lauritzen and Spiegelhalter [45]. MacKay and Neal [48, 58], and McEliece et al. [59] showed that Gallager's algorithm [31] for decoding low-density parity-check codes proposed in the early 1960s is essentially an instance of Pearl's algorithm. Extensive simulation results of MacKay and Neal showed that Gallager codes could perform nearly as well as earlier developed turbo codes [6]. The same authors also observed that turbo decoding is an instance of "belief" propagation and provided a description of Pearl's algorithm, and made explicit its connection to the basic turbo decoding algorithm described in [6]. The origins of the algorithm can be found in the work of Battail [3], Hartmann and Rudolph [35], Gallager [32] and Tanner [87].

Application of the LDPC codes and the MPA for their decoding in magnetic and optical recording systems is considered in [51, 52, 82, 83]. In fact, hard iterative decoding of the LDPC codes and their application in storage systems was considered earlier in [46, 47, 86]. A detailed asymptotic analysis of the minimum distance of the LDPC codes and the BER achieved by low complexity decoders can be found in [109].

More recently, Wiberg et al. [107] showed that graphs introduced by Tanner [87] more than two decades ago to describe a generalization of Gallager codes provide a natural setting in which to describe and study iterative soft-decision decoding techniques, in the same way as the code trellis [29] is an appropriate model for describing and studying conventional maximum likelihood soft-decision decoding using Viterbi's algorithm. Forney [27] generalized Wilberg's results and explained connections of various two-way propagation algorithms with coding theory. Frey and Kschischang [30, 36, 45] showed that various graphical models, such as Markov random fields, Tanner graphs, and Bayesian networks all support the basic probability propagation algorithm in factor graphs, similarly as a trellis diagram supports Bahl, Cocke, Jelinek, Raviv's (BCJR) algorithm [3]. Frey and Kschischang also derived a general distributed marginalization algorithm for functions described by factor graphs. From this general algorithm, Pearl's belief propagation algorithm as well as its instances: turbo decoding and message-passing can be easily derived as special cases of probability propagation in a graphical model of the code. A good tutorial on iterative decoding of block and convolutional codes is due to Hagenauer, Offer and Papke [33].

The theory of codes on graphs has not only improved the error performance, but it has also opened new research avenues for investigating alternative suboptimal decoding schemes. It seems that almost any proposed concatenated coding configuration has good performance provided that the used codewords are long and iterative decoding is employed. The iterative decoding algorithms employed in the current research literature are suboptimal, although simulations have demonstrated their performance to be near optimal (e.g., near maximum-likelihood). Although suboptimal, these decoders still have very high complexity and are incapable of operating in the high data rate regimes. The high complexity of the proposed schemes is a direct consequence of the fact that for random codes a large amount of information is necessary to specify positions of the nonzero elements in a parity-check matrix. In this chapter we will introduce well-structured LDPC codes, a concept opposed to the prevalent practice of using random code constructions. Our main focus will be on low-complexity coding schemes and structured LDPC codes: their construction and performance in ISI magnetic recording channels.

In the past few years several random low-density parity-check (LDPC) codes have been designed with performances very close to the Shannon limit [80], see for example, Richardson, Shokrollahi, and Urbanke [75], MacKay [59], and Luby, Mitzenmacher, Shokrollahi, and Spielman [57]. At the same time, significant progress has been made in designing structured LDPC codes. Examples of structured LDPC codes include Kou, Lin and Fossorier's [43] finite geometry codes, Tanner, Sridhara and Fuja's [88] codes constructed from groups, Rosenthal and Vontobel's codes on regular graphs [77], and Johnson and Weller's [37] Steiner system codes. MacKay and Davey [60] also used Steiner systems (a subclass of BIBDs) to construct Gallager codes. Rosenthal and Vontobel [77] constructed some short high-girth codes using a technique by Margulis [62] based on $k-$regular Caley graphs of $SL_2(GF(q))$, the special linear group, and $PGL_2(GF(q))$, the projective general linear group of dimension two over $GF(q)$, the finite field with q elements. Jon-Lark Kim et. al [39] gave another explicit construction of LDPC codes using Lazebnik and Ustimenko's [49] method based on regular graphs. In a series of articles Vasic [91–93, 95, 96, 100], and Vasic, Kurtas and Kuznetsov [94, 97, 98, 100], and [103] introduced several new classes of combinatorially constructed regular LDPC codes, and analyzed their performance in longitudinal and perpendicular recording systems. Sankaranarayanan, Vasic, and Kurtas [79] also showed how to construct irregular codes with a desired degree distribution starting from a combinatorial design. In what follows, we will present a short overview of some of these code constructions.

Although iterative decoding of LDPC codes is based on the concept of codes on graphs, a LDPC code itself can be substantially simplified if it is based on combinatorial objects known as *designs*. In fact, the construction considered in this chapter is purely combinatorial, and is based on *balanced incomplete block designs* (BIBD) [8]. Such combinatorial objects were extensively studied in connection with a large number of problems in applied mathematics and communication theory. Combinatorial designs and codes are very closely connected combinatorial entities, since one can be used to construct the other. For example, codewords of fixed weight in many codes, including the Golay code and the class of quadratic residue (QR) codes, support designs (see, [61, 85]).

As will be shown below, the parity-check matrix of the combinatorially constructed codes can be defined as the incidence matrix of a 2-$(v, c, 1)$ BIBD [17, 55], where v represents the number of parity bits, and c represents the column weight of the parity-check matrix. We are interested in very high rate and relatively short (less than 5000 bits) codes. High rates are necessary to control the equalization loss that is unavoidable in *partial response* (PR) channels [14, 38, 41], while short block lengths are required to maintain compatibility with existing data formats and enable simpler system architecture [91].

The first construction presented uses difference families such as the Bose [10, 11], Netto [66] and Buratti [13] difference families over the group Z_v, while the second construction is related to rectangular integer lattices. The class of codes based on difference families offers the best tradeoff between code rate and code length, but does not produce many high-rate, short-length codes, especially for large column weights of the parity-check matrix. The second construction gives a much larger family of codes, at the expense of the code rate. A third class of structured codes to be discussed in this chapter is the class of codes on projective planes, such as codes on projective and affine geometries, codes on Hermitian unitals and codes constructed from oval designs.

The encoding complexity of combinatorially constructed codes is very low and determined either by the size of the cyclic difference family upon which a block design is based, or by the "vertical" dimension of the lattice for the case of lattice constructions.

We start by introducing BIBDs in Section 14.2 and describing their relation to bipartite graphs and parity-check matrices of regular Gallager codes. In Section 14.3 we introduce several constructions for 2-$(v, 3, 1)$ systems (so called Steiner triple systems), based on cyclic difference families. We present three constructions of cyclic difference families that result in regular codes with column weight $c = 3, 4$ and 5, and give a list of known infinite families. We also give an overview of finite geometry codes in Section 14.4, and a construction based on integer lattices in Section 14.5. The last section of this chapter, Section 14.6, focuses on the performance evaluation of combinatorially constructed LDPC codes over PR channels. The first attempt to apply iterative decoding for ISI channels is due to Douillard et al. [22]. Douillard and his co-authors presented an iterative receiver structure, the so-called "turbo-equalizer," capable of combating ISI due to multipath effects in Gaussian and Rayleigh transmission channels. Soft-output decisions from the channel detector and from the convolutional decoder were used in an iterative fashion to generate bit estimates. Motivated largely by the potential applications to magnetic recording systems, several authors have explored turbo coding methods for some $(1 - D)(1 + D)^N$ channels. Heegard [36] and Pusch, et al. [71] illustrated the design and iterative decoding process of turbo codes for the $(1 - D)$ channel, using codes of rates 1/2 and lower. Ryan, et al. [72, 73] demonstrated that by using a parallel-concatenated turbo code as an outer code, punctured to achieve rates 4/5, 8/9, and 16/17 it is possible to reduce the bit error rate relative to previously known high-rate trellis-coding techniques on a precoded $1 - D$ or $1 - D^2$ channel. Souvignier et al. [84] and McPheters et al. [65] considered serial concatenated systems with an outer code that is a high-rate convolutional code, rather than a turbo code. They found that these convolutional codes perform as well as turbo codes. Additionally, they showed that removal of the channel precoder improves the performance of the turbo-coded system at low SNR, and degrades the performance of the convolutionally-coded system. Oberg and Siegel [67] found the error rates of precoded and nonprecoded serial concatenated systems with high rate block codes. Recently, the performance of LDPC codes in magnetic recording systems was analyzed by Fan, Kurtas, Friedmann and McLauglin [27] and by Dholakia and Eleftheriou [21]. LDPC codes combined with noise prediction are discussed in [74] and [20]. Ryan, McLaughlin et al. [72] combined an iterative decoding scheme with a maximum run length constrained code. Vasic and Pedagani [104] introduced a coding scheme for which the modulation code is completely removed and a channel constraint is imposed by structured (deliberate) error insertion. The scheme uses the power of a LDPC code to correct both random and structured errors. Here, we also should note that the simple turbo product codes with single parity checks (TPC/SPC) can be considered as a subclass of the structured LDPC codes. The TPC with single and multiple parities in columns and rows are considered in [51, 52].

Lattice based LDPC codes in partial response channels were analyzed in [94], and iteratively decodable codes based on Netto CDFs were introduced and studied in [95]. The application of these codes in PR channels assuming decoding with the min-sum algorithm and limited number of iterations was also discussed in [94]. Affine geometry codes that do not contain so-called Pasch configurations and hence have increased minimum distance were introduced in [96]. In [96] LDPC codes based on mutually orthogonal Latin rectangles were constructed and analyzed with respect to their performance in longitudinal magnetic recording channels. The performance of LDPC codes based on Kirkman systems and low-density generator matrix codes in perpendicular magnetic recording were extensively discussed in [98] and [79].

14.2   Combinatorial Designs and their Bipartite Graphs

A *balanced incomplete block design* (BIBD) with parameters (v, c, l) is an ordered pair (V, B), where V is a v-element set and B is a collection of b c-subsets of V, called blocks, such that each element of V is contained in exactly r blocks and any 2-subset of V is contained in exactly l blocks. Notice that $c \cdot b = R \cdot v$, so that r is uniquely determined by the remaining parameters of the design. We consider designs for which every block contains exactly c points, and every point is contained in exactly r blocks. The notation BIBD(v, c, l) is used for a BIBD on v points, block size c, and index l. A BIBD with block size $c = 3$ is called *a Steiner triple system*. A design is called *resolvable* if there exists a nontrivial partition of its block set B into disjoint subsets each of which partitions the set V. Each of these subsets is referred to as *a parallel class*. Resolvable Steiner triple systems with index $l = 1$ are called *Kirkman systems*. These combinatorial objects have been intensively studied in the combinatorial literature, and some construction methods for them are described in [17, 18, 54]. A BIBD is called symmetric if $b = v$ and $r = c$. A symmetric BIBD(v, c, l) with $c \geq 3$ is equivalent to a finite projective plane [55]. In addition to resolvable designs, one can also use l-configurations for the design of LDPC codes. An *l-configuration* is a combinatorial structure comprised of v points and b blocks such that: (a) each block contains c points; (b) each point is incident with r blocks; (c) two different points are contained in *at most l* blocks.

We define the point-block incidence matrix of a (V, B) design as a $v \times b$ matrix $A = (a_{ij})$, in which $a_{ij} = 1$ if the ith element of V occurs in the jth block of B, and $a_{ij} = 0$ otherwise. If one thinks of points of the design as parity-check equations and of blocks of the design as bits of a linear block code, then it is possible to define the parity-check matrix of a LDPC code as the block-point incidence matrix of the design. Since in a BIBD each block contains the same number of points c, and every point is contained in the same number r of blocks, A defines a parity-check matrix H of a regular LDPC (Gallager) code [32].

For example, the collection $B = \{B_1, B_2, \ldots, B_{12}\}$ of blocks $B_1 = \{1, 2, 3\}$, $B_2 = \{1, 5, 8\}$, $B_3 = \{1, 4, 7\}$, $B_4 = \{1, 6, 9\}$, $B_5 = \{4, 8, 9\}$, $B_6 = \{3, 4, 6\}$, $B_7 = \{2, 6, 8\}$, $B_8 = \{2, 4, 5\}$, $B_9 = \{5, 6, 7\}$, $B_{10} = \{2, 7, 9\}$, $B_{11} = \{3, 5, 9\}$, $B_{12} = \{3, 7, 8\}$ is a resolvable BIBD(9, 3, 1) system or Steiner system with $v = 9$, and $b = 12$. The resolvability classes are $\{B_1, B_5, B_9\}$, $\{B_2, B_6, B_{10}\}$, $\{B_3, B_7, B_{11}\}$ and $\{B_4, B_8, B_{12}\}$. The point-block incidence matrix is of the form:

$$
A = H = \begin{bmatrix}
1 & 1 & 1 & 1 & 0 & 0 & 0 & 0 & 0 & 0 & 0 & 0 \\
1 & 0 & 0 & 0 & 0 & 0 & 1 & 1 & 0 & 1 & 0 & 0 \\
1 & 0 & 0 & 0 & 0 & 1 & 0 & 0 & 0 & 0 & 1 & 1 \\
0 & 0 & 1 & 0 & 1 & 1 & 0 & 1 & 0 & 0 & 0 & 0 \\
0 & 1 & 0 & 0 & 0 & 0 & 0 & 1 & 1 & 0 & 1 & 0 \\
0 & 0 & 0 & 1 & 0 & 1 & 1 & 0 & 1 & 0 & 0 & 0 \\
0 & 0 & 1 & 0 & 0 & 0 & 0 & 0 & 1 & 1 & 0 & 1 \\
0 & 1 & 0 & 0 & 1 & 0 & 1 & 0 & 0 & 0 & 0 & 1 \\
0 & 0 & 0 & 1 & 1 & 0 & 0 & 0 & 0 & 1 & 1 & 0
\end{bmatrix}
\tag{14.1}
$$

The rate of the code is $R = (b - rank(H))/b$. In general, the rank of H is hard to find, but the following simple formula can be used to bound the rate of a LDPC code based on the parameters of a 2-(v, c, l)

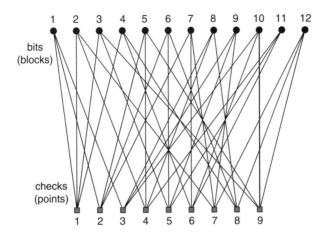

FIGURE 14.1 The bipartite graph representation of the Steiner (9,3,1) system.

design:

$$R \geq \frac{l \cdot x - v}{l \cdot x} \quad \text{where} \quad x = \frac{v(v-1)}{c(c-1)} \tag{14.2}$$

A more precise characterization of the rank (and "p-rank") of the incidence matrix of 2-designs was given by Hamada in [34]. It should be noticed that the above bound is generally loose. For example, for the case of codes constructed from projective planes, the bound is trivially equal to zero, while the actual rate of the codes is quite high (see, [43]). Therefore, for a given code rate, many BIBD codes will have a larger dimension than predicted by the above bound. The construction of maximum rate BIBD codes with $c = 2$ is trivial end reduces to finding K_v, the complete graph [55].

The parity check matrix of a linear block code can be represented as a bipartite graph [63, 87, 107]. The first vertex subset (B) contains bits or variables, while the second subset is comprised of parity-check equations (V). An edge between a bit and an equation exists if the bit is involved in the check. For example, the bipartite graph representation of the Steiner (9,3,1) system whose incidence matrix is given in Equation 14.1 is shown in Figure 14.1.

In order to be properly protected, each bit should be involved in as many equations as possible. On the other hand, iterative decoding algorithms require that the bipartite graph does not contain short cycles [45]. In other words, the *girth* of the graph (i.e., the length of the shortest cycle) must be large. Additionally, in order to allow for efficient iterative decoding, the out-degree of check nodes must be limited too. By using the incidence matrix of a design to define a code, one can observe that the constraint $l = 1$ imposed for BIBDs automatically implies that there are no cycles of length four in the bipartite graph of the code.

14.3 LDPC Codes on Difference Families

Let V be a finite additive Abelian group of order v. Then t c-element subsets of V, $B_i = \{b_{i,1}, \ldots, b_{i,c}\}$, $1 \leq i \leq t$, form a (v, c, l) *difference family* (DF) if every nonzero element of V can be represented in exactly l ways as a difference of two elements within the same member of a family. In other words, every nonzero element of V occurs l times among the differences $b_{i,m} - b_{i,n}$, $1 \leq i \leq t$, $1 \leq m, n \leq c$. The sets B_i are called base blocks. If V is isomorphic to Z_v, the additive group of integers modulo v, then the corresponding (v, c, l) DF is called a *cyclic difference family* (CDF).

For example the blocks $B_1 = \{3, 13, 15\}$, $B_2 = \{9, 8, 14\}$, $B_3 = \{27, 24, 11\}$, $B_4 = \{19, 10, 2\}$, and $B_5 = \{26, 30, 6\}$ are the base block of a (31,3,1) CDF of a group $V = Z_{31}$ since the nonzero elements of the difference arrays

$$D^{(1)} = \begin{bmatrix} 0 & 10 & 12 \\ 21 & 0 & 2 \\ 19 & 29 & 0 \end{bmatrix} \quad D^{(2)} = \begin{bmatrix} 0 & 30 & 5 \\ 1 & 0 & 6 \\ 26 & 25 & 0 \end{bmatrix} \quad D^{(3)} = \begin{bmatrix} 0 & 28 & 15 \\ 3 & 0 & 18 \\ 16 & 13 & 0 \end{bmatrix} \quad D^{(4)} = \begin{bmatrix} 0 & 22 & 14 \\ 9 & 0 & 23 \\ 17 & 8 & 0 \end{bmatrix} \quad D^{(5)} = \begin{bmatrix} 0 & 4 & 11 \\ 27 & 0 & 7 \\ 20 & 24 & 0 \end{bmatrix}$$

formed as $D_{i,j}^{(k)} = b_{2,i} - b_{2,j}, 1 \le k \le 5$) are all different.

If G is a group that acts on a set X, then the set $\{gx : g \in G\}, x \in X$, is called the orbit of x. For the case that $G = V$ and $X = B$, where B is the set of all base blocks of a CDF, a BIBD can be defined as the union of orbits of B. If the number of base blocks is s, the number of blocks in BIBD is $b = sv$. For a given constraint (v, c, l), the CDF construction maximizes the code rate, because for a given v the number of blocks is maximized. The parity check matrix of a (v, c, l) CDF LDPC code can be written in the form

$$H = [H_1 H_2, \ldots, H_t] \tag{14.3}$$

where each submatrix is of dimension $v \times v$ and each of the base blocks $B_i = \{b_{i,1}, \ldots, b_{i,c}\}, 1 \le i \le t$, specifies positions of nonzero elements in the first column of H_i. The CDF codes have a quasi-cyclic structure similar to Townsend and Weldon's [90] self-orthogonal quasi-cyclic codes and Weldon's difference set codes [105].

The blocks $B_1 = \{3, 13, 15\}$, $B_2 = \{9, 8, 14\}$, $B_3 = \{27, 24, 11\}$, $B_4 = \{19, 10, 2\}$, and $B_5 = \{26, 30, 6\}$ are the base block of a $(31,3,1)$ CDF of the group $V = Z_{13}$. The orbits are given in Table 14.1.

A bound for d_{min} of Gallager codes with column weight $c = 3$ was first derived in [60]. Another lower bound for d_{min} is due to Tanner [88], and can be applied to an arbitrary linear code with matrix of parity checks H, represented by a bipartite graph. Using purely combinatorial arguments, lower bounds for the minimum distance were derived in [93]. A general, nontrivial lower bound on d_{min} for codes based on BIBDs with block size c can easily be obtained by using the fact that these codes are majority-logic decodable. A code is one-step majority-logic decodable if for every bit there exists a set of L parity-check equations that are orthogonal on that bit. In this context, the orthogonality condition imposes the requirement that each of the check equations include the bit under consideration, and that no other bit is checked more than once by any of the equations. If a code is one-step majority-logic decodable, then the minimum distance of the code is at least $L + 1$. From the described construction of LDPC codes based on BIBDs, it follows that $L = c$ and $d_{min} \ge c + 1$.

As explained in the previous section, once a CDF is known, it is straightforward to construct a BIBD. Constructing a CDF is a complex problem and it is solved only for certain values of v, c and l. One of the first constructions of a difference set is due to Bose [10, 11] and Singer [81]. Most constructions of CDFs

TABLE 14.1   The Orbits of Base Blocks in a $(31,3,1)$ BIBD

B_1 orbit	B_2 orbit	B_3 orbit	B_4 orbit	B_5 orbit	B_1 orbit	B_2 orbit	B_3 orbit	B_4 orbit	B_5 orbit
3 13 15	9 8 14	27 24 11	19 10 2	26 30	6 19 29	0 25 24 30	12 9 27	4 26 18	11 15 22
4 14 16	10 9 15	28 25 12	20 11 3	27 0	7 20 30	1 26 25	0 13 10 28	5 27 19	12 16 23
5 15 17	11 10 16	29 26 13	21 12 4	28 1	8 21 0	2 27 26	1 14 11 29	6 28 20	13 17 24
6 16 8	12 11 17	30 27 4	22 13 5	29 2	9 22 1	3 28 27	2 15 12 30	7 29 21	14 18 25
7 17 19	13 12 18	0 28 15	23 14 6	30 3	10 23 2	4 29 28	3 16 13 0	8 30 22	15 19 26
8 18 20	14 13 19	1 29 16	24 15 7	0 4	11 24 3	5 30 29	4 17 14 1	9 0 23	16 20 27
9 19 21	15 14 20	2 30 17	25 16 8	1 5	12 25 4	6 0 30	5 18 15 2	10 1 24	17 21 28
10 20 22	16 15 21	3 0 18	26 17 9	2 6	13 26 5	7 1 0	6 19 16 3	11 2 25	18 22 29
11 21 23	17 16 22	4 1 19	27 18 10	3 7	14 27 6	8 2 1	7 20 17 4	12 3 26	19 23 30
12 22 24	18 17 23	5 2 20	28 19 11	4 8	15 28 7	9 3 2	8 21 18 5	13 4 27	20 24 0
13 23 25	19 18 24	6 3 21	29 20 12	5 9	16 29 8	10 4 3	9 22 19 6	14 5 28	21 25 1
14 24 26	20 19 25	7 4 22	30 21 13	6 10	17 30 9	11 5 4	10 23 20 7	15 6 29	22 26 2
15 25 27	21 20 26	8 5 23	0 22 14	7 11	18 0 10	12 6 5	11 24 21 8	16 7 30	23 27 3
16 26 28	22 21 27	9 6 24	1 23 15	8 12	19 1 11	13 7 6	12 25 22 9	17 8 0	24 28 4
17 27 29	23 22 28	10 7 25	2 24 16	9 13	20 2 12	14 8 7	13 26 23 10	18 9 1	25 29 5
18 28 30	24 23 29	11 8 26	3 25 17	10 14	21				

that followed Bose's work are based on the same idea of using finite fields. For example, if one defines the set of integers S as $\{i : 0 \leq i \leq q^2 - 1, \omega^i + \omega \in GF(q)\}$, then S consists of q elements and represents a cyclic difference set modulo $q^2 - 1$, with $l = 1$.

The first construction, considered in [103], is due to Netto [13, 16]. It applies for $c = 3$, and v a power of a prime such that $v \equiv 1 \mod 6$. Let ω be a primitive element of the field [64]. If $v = 6t + 1, t \geq 1$, for $d \mid v - 1$ let Ψ^d be the group of d−th powers in $GF(v)$. Let $\omega^i \Psi^d$ be a coset of d−th powers of the field. Then the set $\{\omega^i \Psi^{2t} \mid 1 \leq i \leq t\}$ defines a Steiner triple system difference family [66, 108] with parameters $(6t + 1, 3, 1)$. The base blocks of this family are typically given in the form $\{0, \omega^i(\omega^{2t} - 1), \omega^i(\omega^{4t} - 1)\}$ or less frequently in the form $\{\omega^i, \omega^{i+2t}, \omega^{i+4t}\}$. Similarly, for $v \equiv 7 \mod 12$ one can show that the set $\{\omega^{2i} \Psi^{2t} \mid 1 \leq i \leq t\}$ defines the base blocks of a so-called Netto triple systems [66].

The second construction is due to Burratti, and is applicable for $c = 4$ and $c = 5$ [13]. For $c = 4$, Burratti's method gives CDFs with v points, provided that v is a prime of the form $v \equiv 1 \mod 12$. The CDF is a set of the form $\{\omega^{6i} B : 1 \leq i \leq t\}$, where base blocks have the form $B = \{0, 1, b, b^2\}$, where again w is a primitive element of $GF(v)$. The numbers $b \in GF(v)$ for several values of v are given in [13]. Similarly, for $c = 5$, the CDF is given by $\{\omega^{10i} B : 1 \leq i \leq t\}$, where $B = \{0, 1, b, b^2, b^3\}$, and $b \in GF(20t + 1)$.

The third construction, also due to Bose [11], is based on a mixed difference system. The sets $\{0.1, 0.2, 0.3\}$, $\{1.i, (2 \cdot u).i, 0.(i + 1)\}$, $\{2.i, (2 \cdot u - 1).i, 0.(i + 1)\}, \ldots, \{u.i, (u + 1).i, 0.(i + 1)\}, 1 \leq i \leq t$, where the elements are all taken $\mod(2 \cdot u + 1)$ and the suffices are taken modulo three, form a mixed difference system. The notation 0.1, $1.i$ etc. used above means, for example, that the symbols 1 and i appearing after the decimal point are indices of 0 and 1, respectively. The mixed difference system uses several copies of the original point set that can be distinguished based on the second index as described above. For more information about this construction, see [1]. It can be shown that a Steiner triple system (where $t = 3$) of order $6 \cdot u + 3$ exists for all $u \geq 0$. In this case $v = 3 \cdot (2 \cdot u + 1), c = 3, l = 1$ are the parameters of BIBD design. Some other interesting constructions based on Latin squares can be found in [17, 18].

14.4 Codes on Projective Planes

The above constructions give a very small set of design with parameters of practical interests. However, so called *infinite families* [16] give an infinite range of block sizes. Infinite families of BIBDs include finite projective geometries, finite Euclidean (affine) geometries, unitals, ovals, Denniston designs as well as certain geometric equivalents of 2-designs (see [16, 62]). For certain choices of the parameters involved, they reduce to finite Euclidean and finite projective geometries. In the following paragraph, we will introduce some basic definitions and concepts regarding the above listed geometric entities. For more details on codes on finite geometries, the reader is referred to [61] and [62].

The parity check matrix of a projective or an affine geometry code is defined as a line-point incidence matrix of the geometry in which points are nonzero m-tuples (vectors) with elements in $GF(p^s)$, where $s > 0$, and p is a prime. The points and lines have the following set of properties: (a) every line consists of the same number of points; (b) any two points are connected by one line only; (c) two lines intersect in at most one point; and (d) each point is an intersection of a fixed number of lines. A finite projective geometry $PG(m, p^s)$ is constructed by using $(m+1)$-tuples of elements x_i from $GF(p^s)$, not all simultaneously equal to zero, called points. Two $(m+1)$-tuples $\mathbf{x} = (x_0, x_1, \ldots, x_m)$ and $\mathbf{y} = (y_0, y_1, \ldots, y_m)$ represent the same point if $\mathbf{x} = \mu \mathbf{x}$, for some nonzero $\mu \in GF(p^s)$. Hence, each point can be represented in $p^s - 1$ ways. All such representations of a point are referred to as its equivalence class. It is straightforward to see that the number of points in $PG(m, p^s)$ is $v = (p^{(m+1)s} - 1)/(p^s - 1)$. The equivalence classes, or points, can be represented by $[\alpha^i] = \{\alpha^i, \beta\alpha^i, \beta^2\alpha^i, \ldots, \beta^{p^s-2}\alpha^i\}$, where $0 \leq i \leq v$ and $\beta = \alpha^v$. Let $[\alpha^i]$ and $[\alpha^j]$ be two distinct points in $PG(m, p^s)$. Then the line passing through them consists of points of the form $[\lambda_1\alpha^i + \lambda_2\alpha^j]$, where $\lambda_1, \lambda_2 \in GF(p^s)$ are not both equal to zero. Since $[\lambda_1\alpha^i + \lambda_2\alpha^j]$ and $[\beta^k\lambda_1\alpha^i + \beta^k\lambda_2\alpha^j]$ are the same point, each line in $PG(m, p^s)$ consists of $p^s + 1$ points. The number of lines intersecting at a given point is $(p^{ms} - 1)/(p^s - 1)$, and the number of lines in projective geometry is

$$b = \left(p^{s(m+1)} - 1\right) \big/ \left(p^{sm} - 1\right)/(p^{2s} - 1)/(p^s - 1) \tag{14.4}$$

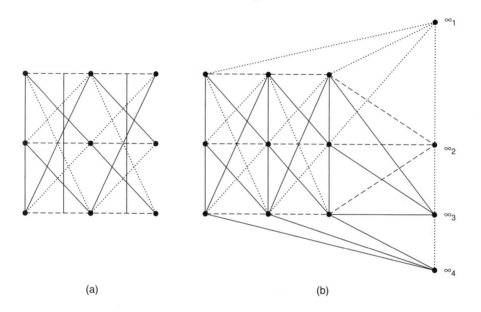

(a)   (b)

FIGURE 14.2   Obtaining a projective plane from an affine plane by adding points: (a) affine plane; (b) projective plane.

The main difference between projective and affine geometries is that projective planes do not contain parallel lines. Affine planes may be considered as special case of projective planes, since they can be obtained by removing the points belonging to the line at infinity (say $x_0 = 0$). Conversely, a projective plane can be obtained from an affine plane by adding an extra point to each line (a point at infinity), and connecting these points by a line. Figure 14.2 illustrates this idea for a geometry with nine points. Each line in the projective geometry contains one more point, and the points at infinity $\infty_1, \infty_2, \infty_3, \infty_4$ create a line so that no parallel classes exist.

In the parity check matrix each column has weight $p^s + 1$ and each row has weight $(p^{ms} - 1)/(p^s - 1)$. As an illustration of the above discussion, the parity-check matrix of a $PG(2,3)$ LDPC code is given below.

$$
H = \begin{bmatrix}
1 & 1 & 1 & 1 & 0 & 0 & 0 & 0 & 0 & 0 & 0 & 0 & 0 \\
1 & 0 & 0 & 0 & 1 & 1 & 1 & 0 & 0 & 0 & 0 & 0 & 0 \\
1 & 0 & 0 & 0 & 0 & 0 & 0 & 1 & 1 & 1 & 0 & 0 & 0 \\
1 & 0 & 0 & 0 & 0 & 0 & 0 & 0 & 0 & 0 & 1 & 1 & 1 \\
0 & 1 & 0 & 0 & 1 & 0 & 0 & 1 & 0 & 0 & 1 & 0 & 0 \\
0 & 1 & 0 & 0 & 0 & 1 & 0 & 0 & 1 & 0 & 0 & 1 & 0 \\
0 & 1 & 0 & 0 & 0 & 0 & 1 & 0 & 0 & 1 & 0 & 0 & 1 \\
0 & 0 & 1 & 0 & 1 & 0 & 0 & 0 & 0 & 1 & 0 & 1 & 0 \\
0 & 0 & 1 & 0 & 0 & 1 & 0 & 1 & 0 & 0 & 0 & 0 & 1 \\
0 & 0 & 1 & 0 & 0 & 0 & 1 & 0 & 1 & 0 & 1 & 0 & 0 \\
0 & 0 & 0 & 1 & 1 & 0 & 0 & 0 & 1 & 0 & 0 & 0 & 1 \\
0 & 0 & 0 & 1 & 0 & 1 & 0 & 0 & 0 & 1 & 1 & 0 & 0 \\
0 & 0 & 0 & 1 & 0 & 0 & 1 & 1 & 0 & 0 & 0 & 1 & 0
\end{bmatrix}
\tag{14.5}
$$

Finite geometry LDPC codes were first considered by Kou, Lin and Fossorier [43], for the case $p = 2$.

Since in a finite geometry two lines cannot be incident with the same pair of points, the corresponding parity checks are orthogonal. One advantage of these codes is that the code parameters, such as the minimum distance and the girth of a bipartite graph, are easily controllable (see [43]). These features are the result of the highly regular structures of these codes.

The second class of finite geometry LDPC can be obtained from *algebraic curves* in a projective plane. In a projective plane, an algebraic curve is a collection of points that satisfy a fixed homogeneous algebraic equation of order n, for example, $f(x_0, x_1, x_2) = 0$. An algebraic curve is irreducible if $f(x_0, x_1, x_2)$ is irreducible over the ground field $GF(q)$. The curve meets a line at most in n points. A *conic* is an algebraic curve of order two defined by an equation of the form:

$$f(x_0, x_1, x_2) = ax_0^2 + bx_1^2 + cx_2^2 + fx_1x_2 + gx_2x_0 + hx_0x_2 = 0$$

A conic is irreducible if $f(x_0, x_1, x_2)$ is irreducible over the ground field $GF(q)$. For $k > m$, a $\{k; m\}$-arc in $PG(2, q)$ is a set of k points such that no $m + 1$ points lie on a line. A c-arc in $PG(2, q)$ is a set of c points such that no three points lie on the same line. A c-arc is complete if it is not properly contained in any $(c + 1)$-arc. A line of the plane is said to be a secant, a tangent or an exterior line with respect to the oval, if the number of common points of the line with the oval is 2,1, and 0, respectively. For a given value of q, $(q + 1)$-arcs of $PG(2, q$ odd) are called ovals, and $(q + 1)$-arcs of $PG(2, q$ even) together with a nucleus point (a point for which every line incident to it is a tangent of the oval) are called hyperovals. In $PG(2, 2^m)$, an oval design is an incidence structure with points comprised from the lines exterior to the oval and blocks formed from points not on the oval. A block contains a point if and only if the corresponding exterior point lies on the exterior line. It is a resolvable 2-$(s(s - 1)/2, s/2, 1)$ Steiner design where $s = 2^m$. The rank of the incidence matrix of an oval design is $3^m - 2^m$. In the section containing the simulation results, we will present the performance of a code constructed from the nondegenerate conic $x_0x_2 = x_1^2$. Figure 14.3 shows an oval and its corresponding lines. Codes on ovals were first described in [106].

Unitals or Hermitian arcs are defined as follows. In $PG(2, q)$, with q a perfect square, a Hermitian arc is a $\{q\sqrt{q} + 1, \sqrt{q} + 1\}$-arc. The arc is constructed from an algebraic curve of order $\sqrt{q} + 1$ given by the equation:

$$x_0^{\sqrt{q}+1} + x_1^{\sqrt{q}+1} + x_2^{\sqrt{q}+1} = 0 \tag{14.6}$$

The arc intersects any line of the plane in one or $\sqrt{q} + 1$ points. A code based on a unital can be obtained as the arc constructed from an algebraic curve of order $\sqrt{q} + 1$, where the unital is a $\{q\sqrt{q} + 1, \sqrt{q} + 1, 1\}$ Steiner system. For q a power of 2, the rank of the incidence matrix is $\{q\sqrt{q}$, and for q a power of an odd prime, the rank of the incidence matrix is $(q - \sqrt{q} + 1)\sqrt{q}$. Such designs are treated in a great detail by Assmus and Key and in great detail in [2], but so far they have not been used for the construction of LDPC

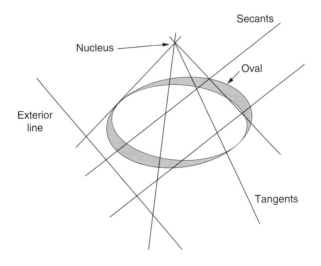

FIGURE 14.3 Visualization of an oval.

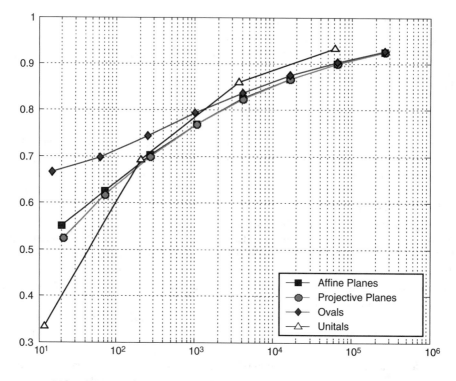

FIGURE 14.4　Achievable length-rate pairs for LDPC codes on projective planes.

codes. Figure 14.4 shows achievable length-rate pairs for LDPC codes on projective planes. Notice that in [43] it was shown that affine and projective geometry codes can be modified in different ways so as to produce a large set of code parameters and to achieve a good trade-off between length and code rate.

14.5　Lattice Construction of LDPC Codes

In this section we address the problem of constructing LDPC codes of large block lengths. The number of parity bits is $m \times c$, where m is a prime, and the blocks are defined as lines of different slopes connecting points of an $m \times c$ integer lattice. The number of blocks is equal to m^2. Integer lattices define l-congurations with index $l = 1$. Here, each 2-tuple is contained in at most $l = 1$ blocks [92]. The goal of the lattice-based construction is to trade the code rate and number of blocks for the simplicity of the construction and for the flexibility of choosing the design parameters. Additionally, as shown in [99], 1-congurations greatly simplify the construction of codes of large girth.

14.5.1　Codes on a Rectangular Subset of an Integer Lattice

Consider a rectangular subset L of the integer lattice, defined by

$$L = \{(x, y) : 0 \le x \le c - 1, 0 \le y \le m - 1, \}$$

where $m \ge c$ is a prime. Let $\lambda : L \to V$ be an one-to-one mapping from the set L to the point set V. An example of such mapping is a simple linear mapping $\lambda(x, y) = m \cdot x + y + 1$. The numbers $\lambda(x, y)$ are referred to as point labels.

A line is a set of c points specified by its *slope* $s, 0 \le s \le m - 1$ and starting point *starting at the point* $(0, a)$, contains the points $\{(x, a + sx \bmod m) : 0 \le x \le c - 1\}$, where $0 \le a \le m - 1$. Figure 14.5 depicts a rectangular subset of the integer lattice with, $m = 5$ and $c = 3$. In the same figure, four lines with slopes $s = 0$, $s = 1$, $s = 2$, and $s = 3$ are shown. Notice that in the configuration lines of infinite slope

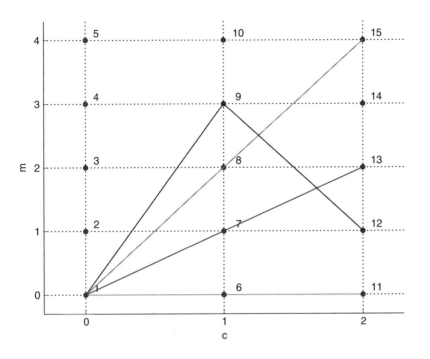

FIGURE 14.5 An example of the rectangular grid for $m = 5$ and $c = 3$.

are not included. In [94] it was shown that the set of blocks B containing all m c-element sets of points in V obtained by taking labels of points along the lines with slopes $s, 0 \leq s \leq m - 1$ forms a resolvable 1-configuration. Each point in the configuration occurs in exactly m blocks. The lines with slopes $s = 0$, $s = 1, s = 2, s = 3$ and $s = 4$, are $\{\{1, 6, 11\}, \{2, 7, 12\}, \{3, 8, 13\}, \{4, 9, 14\}, \{5, 10, 15\}\}, \{\{1, 7, 13\},\{2, 8, 14\}, \{3, 9, 15\}, \{4, 10, 11\}, \{5, 6, 12\}\}, \{\{1, 8, 15\},\{2, 9, 11\}, \{3, 10, 12\}, \{4, 6, 13\}, \{5, 7, 14\}\}, \{\{1, 9, 12\},\{2, 10, 13\}, \{3, 6, 14\}, \{4, 7, 15\}, \{5, 8, 11\}\}$, and $\{\{1, 10, 14\}, \{2, 6, 15\}, \{3, 7, 11\}, \{4, 8, 12\}, \{5, 9, 13\}\}$. The corresponding parity-check matrix is given by:

$$H = \begin{bmatrix} 1\ 0\ 0\ 0\ 0 & 1\ 0\ 0\ 0\ 0 & 1\ 0\ 0\ 0\ 0 & 1\ 0\ 0\ 0\ 0 & 1\ 0\ 0\ 0\ 0 \\ 0\ 1\ 0\ 0\ 0 & 0\ 1\ 0\ 0\ 0 & 0\ 1\ 0\ 0\ 0 & 0\ 1\ 0\ 0\ 0 & 0\ 1\ 0\ 0\ 0 \\ 0\ 0\ 1\ 0\ 0 & 0\ 0\ 1\ 0\ 0 & 0\ 0\ 1\ 0\ 0 & 0\ 0\ 1\ 0\ 0 & 0\ 0\ 1\ 0\ 0 \\ 0\ 0\ 0\ 1\ 0 & 0\ 0\ 0\ 1\ 0 & 0\ 0\ 0\ 1\ 0 & 0\ 0\ 0\ 1\ 0 & 0\ 0\ 0\ 1\ 0 \\ 0\ 0\ 0\ 0\ 1 & 0\ 0\ 0\ 0\ 1 & 0\ 0\ 0\ 0\ 1 & 0\ 0\ 0\ 0\ 1 & 0\ 0\ 0\ 0\ 1 \\ 1\ 0\ 0\ 0\ 0 & 0\ 0\ 0\ 0\ 1 & 0\ 0\ 0\ 1\ 0 & 0\ 0\ 1\ 0\ 0 & 0\ 1\ 0\ 0\ 0 \\ 0\ 1\ 0\ 0\ 0 & 1\ 0\ 0\ 0\ 0 & 0\ 0\ 0\ 0\ 1 & 0\ 0\ 0\ 1\ 0 & 0\ 0\ 1\ 0\ 0 \\ 0\ 0\ 1\ 0\ 0 & 0\ 1\ 0\ 0\ 0 & 1\ 0\ 0\ 0\ 0 & 0\ 0\ 0\ 0\ 1 & 0\ 0\ 0\ 1\ 0 \\ 0\ 0\ 0\ 1\ 0 & 0\ 0\ 1\ 0\ 0 & 0\ 1\ 0\ 0\ 0 & 1\ 0\ 0\ 0\ 0 & 0\ 0\ 0\ 0\ 1 \\ 0\ 0\ 0\ 0\ 1 & 0\ 0\ 0\ 1\ 0 & 0\ 0\ 1\ 0\ 0 & 0\ 1\ 0\ 0\ 0 & 1\ 0\ 0\ 0\ 0 \\ 1\ 0\ 0\ 0\ 0 & 0\ 0\ 0\ 1\ 0 & 0\ 1\ 0\ 0\ 0 & 0\ 0\ 0\ 0\ 1 & 0\ 0\ 1\ 0\ 0 \\ 0\ 1\ 0\ 0\ 0 & 0\ 0\ 0\ 0\ 1 & 0\ 0\ 1\ 0\ 0 & 1\ 0\ 0\ 0\ 0 & 0\ 0\ 0\ 1\ 0 \\ 0\ 0\ 1\ 0\ 0 & 1\ 0\ 0\ 0\ 0 & 0\ 0\ 0\ 1\ 0 & 0\ 1\ 0\ 0\ 0 & 0\ 0\ 0\ 0\ 1 \\ 0\ 0\ 0\ 1\ 0 & 0\ 1\ 0\ 0\ 0 & 0\ 0\ 0\ 0\ 1 & 0\ 0\ 1\ 0\ 0 & 1\ 0\ 0\ 0\ 0 \\ 0\ 0\ 0\ 0\ 1 & 0\ 0\ 1\ 0\ 0 & 1\ 0\ 0\ 0\ 0 & 0\ 0\ 0\ 1\ 0 & 0\ 1\ 0\ 0\ 0 \end{bmatrix} \qquad (14.7)$$

In general, the parity-check matrix of a lattice code can be written in the form:

$$H = \begin{bmatrix} I & I & I & \cdots & I \\ I & P_{2,2} & P_{2,3} & \cdots & P_{2,m-1} \\ I & P_{3,2} & P_{3,3} & \cdots & P_{3,m-1} \\ \cdots & \cdots & \cdots & \cdots & \cdots \\ I & P_{c-1,2} & P_{c-1,3} & \cdots & P_{c-1,m-1} \end{bmatrix} \tag{14.8}$$

where each submatrix $P_{i,j}$ is a permutation matrix. The power of P which determines $P_{i,j}$ (i.e., the position of the bit 1 the first column of $P_{i,j}$) can be found by using c_i^{j-1}, the ith element of the first base block in the class of blocks corresponding to the jth slope.

Similar parity check matrices have been obtained by several researchers using different approaches. The examples include Tanner's sparse difference codes [89], Blaum, Farrell, and Tilborg's [9] array codes, Eleftheriou and Olcer's [25] array LDPC codes for application in digital subscriber lines, and codes constructed by Kim et al. [39] with girth at least six.

Tanner graphs with large girth can be obtained by a judicious selection of sets of parallel lines included in the integer lattice 1-configuration. The resulting parity-check matrix is of the form of an array of circulant matrices. For example, for $c = 3$ it was shown in [92] that if the slope set represents an "arithmetically constrained" sequence, defined by Odlyzko and Stanley [68], then the resulting codes have girth at least eight. A generalization of this construction for higher girths is also straightforward. Other constructions of codes of large girth include Rosenthal and Vontobel's [77] construction based on an idea by Margulis [62], and the previously mentioned result by Kim et al. [39] based on work of Lazebnik and Ustimenko [49]. The problem of constructing designs with high girth appears to be a very difficult problem in general [5].

14.6   Application in the Partial Response (PR) Channels

As we already mentioned in the introduction for this chapter, the LDPC codes were originally constructed using an ensemble of random sparse parity-check matrices. Such random LDPC codes have been shown to achieve near-optimum performance when decoded by soft iterative decoding algorithms (e.g., message passing algorithm (MPA)), but due to the random structure of their parity check matrix large memory is required for their implementation. The complexity of implementation can be reduced by using structured LDPC codes constructed from algebraic and combinatoric objects described above. The generic channel architecture and decoding algorithms are similar to both random and structured LDPC codes, but due to the special properties of parity check matrices the implementation of the MPA algorithm for structured LDPC codes is much simpler than for the random LDPC codes.

The bit error rate (BER) and other characteristics of the described structured LDPC codes depend on a number of code parameters, such as rate, length, column weight of the parity check matrix, number of short cycles in the bipartite graph of the code, as well as the type of precoder, the type of the PR channel (i.e., equalization), amount of jitter and other factors. We present simulation results describing the effect of some of these factors in a perpendicular magnetic recording channel, but first let us describe the channel model used in simulations.

14.6.1   Channel Model and Signal Generation

Figure 14.6 shows a block diagram of a read/write channel using an LDPC code and an iterative detection scheme for decoding. As we can see from this figure, the user data are encoded by an outer ECC code (usually, Reed-Solomon (RS) code is used), then passed through a run-length limiting (RLL) encoder, and then go to the LDPC encoder. The output of an LDPC encoder can be also precoded before the coded sequence of bits is written on the medium. An analog part of the channel includes read/write heads, the storage medium, different filters, sampling and timing circuits. These components were modeled as described below. Due to electronic and media noise in the channel, the detector can not recover the original user data with arbitrary small error probability, and the ECC decoder is used to detect and correct as many channel errors as possible, decreasing the output BER and SFR of the channel to a level given by the

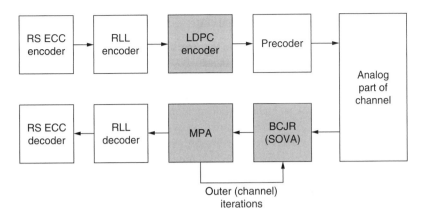

FIGURE 14.6 System view of the encoding and decoding operations in the read/write channel of a disk drive.

channel specifications. Combining RLL and LDPC codes is not a trivial problem which has some technical solutions, for example, described in [72, 104]. Here, we do not consider the effect of an outer ECC code, nor the effect of the RLL code, and in our simulations the user data are directly passed through an LDPC encoder.

Figure 14.7 shows how an input signal of the iterative detection scheme is generated. Formally, the readback signal $V(t)$ of a channel without jitter is defined as follows

$$V(t) = \sum_{k=-\infty}^{\infty} b_k g(t - kT_b) + V_n(t) \tag{14.9}$$

where $g(t)$ is the transition response of the channel, $b_k = \{0, 1, -1\}$ corresponding to no, positive or negative transitions, T_b is the bit interval, and $V_n(t)$ is Additive White Gaussian Noise (AWGN). The transition response of a perpendicular channel is modeled as

$$g(t) = \frac{1}{2} \left(\frac{\pi PW_{50}^2}{2 \ln 2} \right)^{1/4} erf \left(\frac{2\sqrt{2}}{PW_{50}} t \right) \tag{14.10}$$

where PW_{50} is the width of the perpendicular impulse response at 50% of its peak value, and $erf(\cdot)$ is an error function defined as

$$erf(x) = \frac{2}{\sqrt{\pi}} \int_0^x e^{-t^2} dt. \tag{14.11}$$

Here, the amplitude of the transition response is chosen in such a way that the energy of the impulse response is equal to one. In simulations, we defined SNR as $10 \log_{10} E_i / N_0$, where E is the energy of the impulse response which is assumed to be unity, and N_0 is the spectral density height of the AWGN. The linear density ND is defined as PW_{50} / T_b.

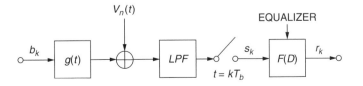

FIGURE 14.7 Signal generation in the magnetic recording channel.

14.6.2　Bit Error Rate (BER) of the Structured LDPC Codes Simulation Results

In this section, we present the BER performance of the structured LDPC codes in the perpendicular PR2 channel with target [121] at the normalized user density $ND = 2$ (channel ND is adjusted according to the code rate, and is equal to ND/R). The soft output Viterbi algorithm (SOVA, [33]) is used on the PRML trellis, and decoding is established by iterating between the inner SOVA decoder and the outer message passing algorithm (MPA, [31, 59]). That is, the soft information is extracted from a partial response channel using the SOVA operating on the channel trellis, and then used in the MPA for the LDPC decoding. The bit "likelihoods" obtained from the MPA are passed back to the SOVA as a priori probabilities and so on. The BER results are given below for the scheme in which the compound iteration "SOVA + 4 iterations of MPA" is performed four times before the final hard decision is taken. In other words, the LDPC decoder performs four internal (bit-to-check plus check-to-bit) iterations prior to suppling the inner decoder with extrinsic information. Larger numbers of compound iterations improve the BER, but as we can see from Figure 14.11 four compound iterations is a good compromise between the performance and increased latency and complexity [53].

　　The BER were evaluated by simulations for two groups of codes. The first group consists of three Kirkman codes with $c = 3$ and length $n = 2420, 4401,$ and 5430 (rates $R = 0.95, 0.963,$ and 0.966, respectively). The second group consists of lattice codes with $c = 4$ and length $n = 3481, 4489,$ and 7921 (rates $R = 0.933, 0.941,$ and 0.955, respectively). The BER of the first group of codes and the BER of the random LDPC code with column weight $c = 3$ are shown in Figure 14.8. As we can see from this figure, at the moderate SNR values the random LDPC code has a bit better BER, and in simulations of the size 10^9 bits does not show an error floor or a slope change. At the same time, the Kirkman LDPC codes exhibit a slope change at the BER level $10^{-6} - 10^{-7}$. The BER of lattice LDPC codes with the column weight $c = 4$ and the BER of the random LDPC code with column weight $c = 4$ are shown in Figure 14.9. In this case all BER are almost on the top of each other at the moderate SNR values, but again the lattice LDPC codes exhibit some signs of a slope change at the high SNR values. In Figure 14.10 we compare the BER of the structured LDPC codes with different column weights $c = 3, 4, 5$ and the BER of turbo product code with single parity checks

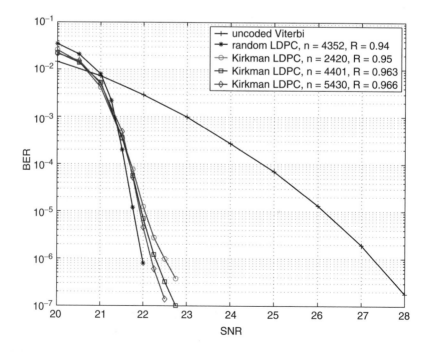

FIGURE 14.8　BER of different Kirkman LDPC codes with column weight 3 in the perpendicular PR2 channel.

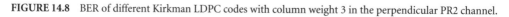

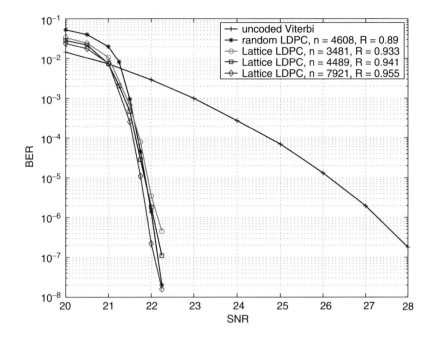

FIGURE 14.9   BER of different lattice LDPC codes with column weight 4 in the perpendicular PR2 channel.

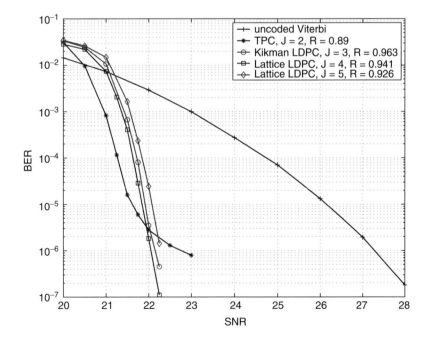

FIGURE 14.10   Comparison of the structured LDPC codes with different column weights of the parity check matrix.

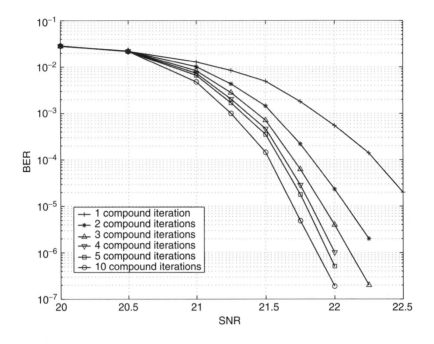

FIGURE 14.11   Effect of the number of iterations on the BER of the lattice LDPC code with $c = 4$.

(TPC/SPC, [51, 52]) considered as an LDPC code with column weight $c = 2$. In this figure, the BER of the TPC code is approximately 0.4 dB better at the BER level 10^{-4}, but has an error floor at the $SNR > 22$ dB, where all structured LDPC codes still are in the "waterfall" region. Finally, in Figure 14.11 we illustrate the typical dependence of the BER from a number of compound iterations G for the lattice and Kirkman LDPC codes. We can summarize the simulation results as follows:

- Kirkman and lattice LDPC codes show the BER that is close to the BER of random LDPC codes, but due to their mathematical structure can lend themselves to low complexity implementations.
- The tested structure LDPC codes with column weights $c = 3, 4$, and 5 exibit a slope change of the BER curves in the high SNR region, but not an error floor in contrast to the TPC/SPC codes.
- Larger numbers of compound iterations improve the BER, and in our simulations four compound iterations were already a good compromise between the performance and increased latency and complexity of the detector scheme.

14.7   Conclusion

We gave an overview of combinatorial constructions of high rate LDPC codes. The emphasis of the exposition was on codes that have column weight not less than three and girth at least six. We discussed BIBD codes obtained using constructions of cyclic difference families due to Bose, Netto and Buratti, and affine integer lattice code. We presented bounds on the minimum distance of the codes and determined the BER performance of these LDPC codes in PR magnetic recording channels by using computer simulations.

Acknowledgment

This work is supported by the NSF under grant CCR-0208597.

References

[1] I. Anderson, *Combinatorial Designs and Tournaments*, Oxford University Press, New York, 1997.

[2] E.F. Assmus and J.D. Key, *Designs and their Codes*, Cambridge University Press, London, 1994.

[3] L. R. Bahl, J. Cocke, F. Jelinek, and J. Raviv, Optimal decoding of linear codes for minimizing symbol error rate, *IEEE Trans. Info. Theory*, Vol. IT-20, pp. 284–287, 1974.

[4] G. Battail, M.C. Decouvelaere, and P. Godlewski, Replication decoding, *IEEE Trans. Info. Theory*, Vol.IT-25, pp. 332–345, May 1979.

[5] R.A. Beezer, The girth of a design, *J. Combinatorial Mathematics and Combinatorial Computing* [in press] (also http://buzzard.ups.edu/pubs.html).

[6] S. Benedetto, G. Montorsi, D. Divsalar, and F. Pollara, Serial Concatenation of Interleaved Codes: Performance Analysis, Design, and Iterative Decoding, *The Telecom. and Data Acquisition Progr. Rep.*, Vol. 42, pp. 1–26, August 1996.

[7] G. Berrou, A. Glavieux, and P. Thitimajshima, Near Shannon Limit Error-Correcting Coding and Decoding: Turbo-Codes, in *Proceedings of IEEE Internatinol Conference on Communications (ICC'93)*, Geneva, Switzerland, pp. 2.1064–2.1070, May 1993.

[8] T. Beth, D. Jungnickel, and H. Lenz, *Design Theory*, Cambridge University Press, London, 1986.

[9] M. Blaum, P. Farrell, and H. van Tilborg, Array codes, in *Handbook of Coding Theory*, V. Pless and W. Huffman, Eds., North-Holland, Elsevier, 1998.

[10] R.C. Bose, On the construction of balanced incomplete block designs, *Ann. Eugenics*, Vol. 9, pp. 353–399, 1939.

[11] R.C. Bose, An affine analogue of Singer's theorem, *Ind. Math. Soc.*, Vol. 6, pp. 1–5, 1942.

[12] R.A. Brualdi, *Introductory Combinatorics*, Prentice Hall, Upper Saddle River, New Jersey, 1999.

[13] M. Buratti, Construction of $(q, k, 1)$ difference families with q a prime power and $k = 4, 5$, *Discrete Math.*, 138, pp. 169–175, 1995.

[14] R.D. Cideciyan, F. Dolivo, R. Hermann, W. Hirt, and W. Schott, A PRML system for digital magnetic recording, *IEEE J. Sel. Areas Commn.*, Vol. 10, No. 1, pp. 38–56, January 1992.

[15] M.J. Colbourn and C.J. Colbourn, Recursive construction for cyclic block designs, *J. Stat. Plann. and Inference*, Vol. 10, pp. 97–103, 1984.

[16] C.J. Colbourn and J.H. Dinitz Eds., *The Handbook of Combinatorial Designs*, CRC Press, Boca Raton, 1996.

[17] C.J. Colbourn and A. Rosa, *Triple Systems*, Oxford University Press, Oxford, 1999.

[18] C. Colbourn and A. Rosa, *Steiner Systems*, Oxford University Press (Oxford Mathematical Monographs), London, 1999.

[19] C.J. Colbourn, J.H. Dinitz, and D.R. Stinson, *Applications of Combinatorial Designs to Communications, Cryptography, and Networking*, in "London Math. Soc. Lecture Note Ser.," vol. 267, pp. 37–100, Cambridge Univ. Press, Cambridge, UK, 1999.

[20] M.C. Davey and D.J.C. Mackay, Reliable communication over channels with insertions, deletions, and substitutions, *IEEE Trans. on Inform. Theory*, Vol. 47, No. 2, pp. 687–698, February 2001.

[21] A. Dholakia, E. Eleftheriou, and T. Mittelholzer, On Iterative Decoding for Magnetic Recording Channels, in *Proceedings of Second International Symposium on Turbo Codes*, Brest, France, September 4–7, 2000.

[22] C. Douillard, M. Jaezequel, C. Berrou, A. Picart, P. Didier, and A. Glavieux, Iterative correction of intersymbol interference: turbo equalization, *Eur. Trans. Telecommn.*, Vol. 6, pp. 507–511, September/October 1995.

[23] T.M. Duman and E. Kurtas, Comprehensive Performance Investigation of Turbo Codes over High Density Magnetic Recording Channels, in *Proceedings of IEEE GLOBECOM*, December 1999.

[24] T. M. Duman and E. Kurtas, Performance Bounds for High Rate Linear Codes over Partial Response Channels, in *Proceedings of IEEE International Symposium on Inform Theory*, Sorrento, Italy, p. 258, *IEEE*, June 2000.

[25] E. Eleftheriou and S. Olcerm, Low-density parity-check codes for digital subscriber lines, *IEEE Int. Conf. on Communications, ICC 2002*, Vol. 3, pp. 1752–1757, 2002.

[26] J.L. Fan, Array Codes as Low-Density Parity-Check Codes, in *Proceedings of 2nd International Symp. on Turbo Codes and Related Topics,* Brest, France, pp. 543–546, September 2000.

[27] J. Fan, A. Friedmann, E. Kurtas, and S.W. McLaughlin, Low Density Parity Check Codes for Partial Response Channels, *Allerton Conference on Communications, Control and Computing,* Urbana, IL, October 1999.

[28] J. Fan, Constrained Coding and Soft Iterative Decoding, in *Proceedings of 2001 IEEE Info. Theory Workshop,* Cairns Australia, pp. 18–20, September 2001.

[29] G.D. Forney, Jr., Codes on graphs: normal realizations, *IEEE Trans. on Info. Theory,* Vol. 47, No. 2, pp. 520–548, February 2001.

[30] G.D. Forney, Maximum-likelihood sequence estimation of digital sequences in the presence of intersymbol interference, *IEEE Trans. Info. Theory,* Vol. 18., No. 3, pp. 363–378, May 1972.

[31] B. J. Frey, *Graphical Models for Machine Learning and Digital Communication,* MIT Press, Cambridge, MA, 1998.

[32] R.G. Gallager, *Low-Density Parity-Check Codes,* MIT Press, Cambridge, MA, 1963.

[33] J. Hagenauer, E. Offer, and L. Papke, Iterative decoding of binary block and convolutional codes, *IEEE Trans. Info. Theory,* Vol 42., No. 2, pp. 439–446, March 1996.

[34] N. Hamada, On the p-rank of the incidence matrix of a balanced or partially balanced incomplete block design and its applications to error correcting codes, *Hiroshima Math. J.,* Vol. 3, pp. 153–226, 1973.

[35] C. Hartmann and L. Rudolph, An optimum symbol-by-symbol decoding rule for linear codes, *IEEE Trans. on Info. Theory,* Vol. 22, No. 5, pp. 514–517, September 1976.

[36] C. Heegard, Turbo Coding for Magnetic Recording, in *Proceedings of IEEE Information Theory Workshop,* San Diego, CA, USA, pp. 18–19, February 1998.

[37] S.J. Johnson and S.R. Weller, Regular Low-Density Parity-Check Codes from Combinatorial Designs, in *Proceedings of 2001 IEEE Information Theory Workshop,* pp. 90–92, 2001.

[38] P. Kabal and S. Pasupathy, Partial-response signaling, *IEEE Trans. Commn.,* Vol. COM-23, pp. 921–934, September 1975.

[39] Jon-Lark Kim, Uri N. Peled, Irina Perepelitsa, and Vera Pless, Explicit Construction of Families of LDPC Codes with Girth at least Six, in *Procedings of 40th Annual Allerton Conference on Communications, Control and Computing,* October 2002.

[40] T.P. Kirkman, Note on an unanswered prize question, *Cambridge and Dublin Mathematics Journal,* 5, pp. 255–262, 1850.

[41] H. Kobayashi, Correlative level coding and maximum-likelihood decoding, *IEEE Trans. Info. Theory,* Vol. IT-17, pp. 586–594, September 1971.

[42] Y. Kou, S. Lin, and M. Fossorier, Construction of Low Density Parity-Check Codes: a Geometric Approach, in *Proceedings of Second International Symposium on Turbo Codes,* Brest, France, September 4–7, 2000.

[43] Y. Kou, S. Lin, and M.P.C. Fossorier, Low density parity-check codes based on finite geometries: A rediscovery and new results, *IEEE Trans. Info. Theory,* 2001.

[44] F.R. Kschischang and B.J. Frey, Iterative decoding of compound codes by probability propagation in graphical models, *IEEE J. Select. Areas Commn.,* Vol. 16, pp. 219–230, February 1998.

[45] F.R. Kschischang, B.J. Frey, and H.-A. Loeliger, Factor graphs and the sum-product algorithm, *IEEE Trans. on Info. Theory,* Vol. 47, No. 2, pp. 498–519, February 2001.

[46] A.V. Kuznetsov and B.S. Tsybakov, On unreliable storage designed with unreliable components, In the book *Second International Symposium on Information Theory,* 1971, Tsahkadsor, Armenia. Publishing House of the Hungarian Academy of Sciences, pp. 206–217, 1973.

[47] A.V. Kuznetsov, On the Storage of Information in Memory Constructed from Unreliable Components, *Prob. Info. Transmission,* Vol. 9, No. 3, pp. 100–113, 1973 (translated by Plenum from Problemy Peredachi Informatsii).

[48] S. L. Lauritzen, *Graphical Models,* Oxford University Press, Oxford, 1996.

[49] F. Lazebnik and V.A. Ustimenko, Explicit construction of graphs with arbitrary large girth and of large size, *Discrete Appl. Math.*, Vol. 60, pp. 275–284, 1997.

[50] C.C. Lindner and C.A. Rodger, *Design Theory*, CRC Press, Boca Raton, 1997.

[51] J. Li, E. Kurtas, K.R. Narayanan, and C.N. Georghiades, On the performance of turbo product codes over partial response channels, *IEEE Trans. Magn.*, Vol. 37, No. 4, pp. 1932–1934, July 2001.

[52] J. Li, K.R. Narayanan, E. Kurtas, and C.N. Georghiades, On the performance of the high-rate TPC/SPC and LDPC codes over partial response channels, *IEEE Trans. Commn.*, Vol. 50, No. 5, pp. 723–735, May 2002.

[53] R. Lynch, E. Kurtas, A. Kuznetsov, E. Yeo, and B. Nikolic, The Search for a Practical Iterative Detector for Magnetic Recording, *TMRC 2003*, San Jose, August 2003.

[54] A.C.H. Ling and C.J. Colbourn, *Rosa triple systems*, in *Geometry, Combinatorial Designs and Related Structures*, Hirschfeld J.W.P., Magliveras S.S., de Resmini M.J., Eds., Cambridge University Press, London, pp. 149–159, 1997.

[55] J.H. van Lint and R.M. Wilson, *A Course in Combinatorics*, Cambridge University Press, London, 1992.

[56] S. Litsyn and V. Shevelev, On ensembles of low-density parity-check codes: asymptotic distance distributions, *IEEE Trans. Info. Theory*, Vol. 48, No. 4, pp. 887–908, April 2002.

[57] M.G. Luby, M. Mitzenmacher, M.A. Shokrollahi, and D.A. Spielman, Improved low-density parity-check codes using irregular graphs, *IEEE Trans. Info. Theory*, Vol. 47, No. 2, pp. 585–598, February 2001.

[58] D.J.C. MacKay and R.M. Neal, Good Codes Based on Very Sparse Matrices, in *Cryptography and Coding, 5th IMA Conference, in Lecture Notes in Computer Science*, C. Boyd, Ed., 1995, Vol. 1025, pp. 110–111.

[59] D.J.C. MacKay, Good error-correcting codes based on very sparse matrices, *IEEE Trans. Info. Theory*, Vol. 45, pp. 399–431, March 1999.

[60] D. MacKay and M. Davey, *Evaluation of Gallager Codes for Short Block Length and High Rate Applications*, http://www.cs.toronto.edu/ mackay/CodesRegular.html.

[61] F.J. MacWilliams and N.J. Sloane, *The Theory of Error-Correcting Codes*, North Holland, Amsterdam, 1977.

[62] M.A. Margulis, Explicit group-theoretic constructions for combinatorial designs with applications to expanders and concentrators, *Problemy Peredachi Informatsii*, Vol. 24, No. 1, pp. 51–60, 1988.

[63] R.J. McEliece, D.J.C. MacKay, and J.-F. Cheng, Turbo decoding as an instance of pearl's "Belief Propagation" algorithm, *IEEE J. Select. Areas Commn.*, Vol. 16, pp. 140–152, February 1998.

[64] R.J. McEliece, *Finite Fields for Computer Scientist and Engineers*, Kluwer, Boston, 1987.

[65] L.L. McPheters, S.W. McLaughlin, and K.R. Narayanan, Precoded PRML, Serial Concatenation and Iterative (Turbo) Decoding for Digital Magnetic Recording, in *IEEE International Magnetics Conference*, Kyongju, Korea, May 1999.

[66] E. Netto, "Zur Theorie der Tripelsysteme," *Math. Ann.*, 42, pp. 143–152, 1893.

[67] M. Oberg and P.H. Siegel, Performance analysis of turbo-equalized partial response channels, *IEEE Trans. Comm.*, Vol. 49, No. 3 , pp. 436–444, March 200.

[68] M. Odlyzko and R.P. Stanley, *Some Curious Sequences Constructed with the Greedy Algorithm*, Bell Labs Internal Memorandum, 1978.

[69] J. Pearl, *Probabilistic Reasoning in Intelligent Systems: Networks of Plausible Inference*, Morgan Kaufmann, San Mateo, CA, 1988.

[70] W.W. Peterson and E.J. Weldon, Jr., *Error-Correcting Codes*, MIT Press, Cambridge, MA, 1963.

[71] W. Pusch, D. Weinrichter, and M. Taferner, Turbo-Codes Matched to the 1 - D Partial Response Channel, in *Proceedings of IEEE International Symposium on Inform Theory*, Cambridge, MA, USA, p. 62, August 1998.

[72] W.E. Ryan, S.W. McLaughlin, K. Anim-Appiah, and M. Yang, Turbo, LDPC, and RLL codes in Magnetic Recording, in *Proceedings of Second International Symposium on Turbo Codes,* Brest, France, September 4–7, 2000.

[73] W.E. Ryan, L.L. McPheters, and S.W. McLaughlin, Combined Turbo Coding and Turbo Equalization for PR4-Equalized Lorentzian Channels, in *Proceedings of the Conference on Inform Sciences and Systems,* March 1998.

[74] W.E. Ryan, Performance of High Rate Turbo Codes on a PR4-Equalized Magnetic Recording Channel, in *Proceedings of IEEE International Conference on Communications.* Atlanta, GA, USA, pp. 947–951, June 1998.

[75] T. Richardson, A. Shokrollahi, and R. Urbanke, Design of Provably Good Low-Density Parity-Check Codes, in *International Symposium on Information Theory, 2000. Proceedings,* p. 199, June 2000.

[76] T.J. Richardson and R.L. Urbanke, The capacity of low-density parity-check codes under message-passing decoding, *IEEE Trans. Info. Theory,* Vol. 47, No. 2, pp. 599–618, February 2001.

[77] I.J. Rosenthal and P.O. Vontobel, "Construction of LDPC Codes using Ramanujan Graphs and Ideas from Margulis," in *Proceedings of 2001 IEEE International Symposium on Information Theory,* 2001.

[78] S. Sankaranarayanan, Erozan Kurtas, and Bane Vasic, Performance of Low-Density Generator Matrix Codes on Perpendicular Recording Channels, presented at *IEEE North American Perpendicular Magnetic Recording Conference,* January 6–8, 2003.

[79] S. Sankaranarayanan, B. Vasic, and E. Kurtas, A Systematic Construction of Capacity-Achieving Irregular Low-Density Parity-Check Codes, accepted for presentation in *IEEE International Magnetics Conference,* March 30–April 3, Boston, MA 2003.

[80] C.E. Shannon, A mathematical theory of communication, *Bell Syst. Tech. J.,* pp. 372–423, 623–656, 1948.

[81] J. Singer, a theorem in finite projective geometry and some applications to number theory, *AMS Trans.,* Vol. 43, pp. 377–385, 1938.

[82] H. Song, B.V.K. Kumar, E.M. Kurtas, Y. Yuan, L.L. McPheters, and S.W. McLaughlin, Iterative decoding for partial response (PR) equalized magneto-optical data storage channels, *IEEE J. Sel. Areas Commn.,* Vol. 19, No. 4, April 2001.

[83] H. Song, J. Liu, B.V.K. Kumar, and E.M. Kurtas, Iterative decoding for partial response (PR) equalized magneto-optical data storage channels, *IEEE Trans. Magn.,* Vol. 37, No. 2, March 2001.

[84] T. Souvignier, A. Friedman, M. Oberg, P.H. Siegel, R.E. Swanson, and J.K. Wolf, Turbo Codes for PR4: Parallel Versus Serial Concatenation, in *Proceedings of IEEE International Conference on Communications,* Vancouver, BC, Canada, pp. 1638–1642, June 1999.

[85] E. Spence and V.D. Tonchev, Extremal self-dual codes from symmetric designs, *Discrete Math.,* Vol. 110, pp. 165–268, 1992.

[86] M.G. Taylor, Reliable information storage in memories designed from unreliable components, *Bell Syst. Tech. J.,* Vol. 47, No. 10, 2299–2337, 1968.

[87] R.M. Tanner, A recursive approach to low complexity codes, *IEEE Trans. Infor. Theory,* Vol. IT-27, pp. 533–547, September 1981.

[88] R.M. Tanner, Minimum-distance bounds by graph analysis, *IEEE Trans. Info. Theory,* Vol. 47, No. 2, pp. 808–821, February 2001.

[89] R.M. Tanner, D. Sridhara, and T. Fuja, A Class of Group-Structured LDPC Codes, available at http://www.cse.ucsc.edu/tanner/pubs.html

[90] R. Townsend and E.J. Weldon, Self-orthogonal quasi-cyclic codes, *IEEE Trans. Info. Theory,* Vol. IT-13, No. 2, pp. 183–195, 1967.

[91] B. Vasic, Low Density Parity-Check Codes: Theory and Practice, National Storage Industry Consortium (NSIC) quarterly meeting, Monterey, CA June 25–28, 2000.

[92] B. Vasic, Combinatorial Constructions of Structured Low-Density Parity-Check Codes for Iterative Decoding, in *Proceedings 2001 IEEE Information Theory Workshop,* pp. 134, 2001.

[93] B. Vasic, Combinatorial Construction of Low-Density Parity-Check Codes, in *Proceedings of* 2002 *IEEE International Symposium on Information Theory,* p. 312, June 30–July 5, 2002, Lausanne, Switzerland.

[94] B. Vasic, E. Kurtas, and A. Kuznetsov, Lattice Low-Density Parity-Check Codes and Their Application in Partial Response Channels, in Proceedings 2002 *IEEE International Symposium on Information Theory,* p. 453–453, June 30–July 5, 2002, Lausanne, Switzerland.

[95] B. Vasic, Structured Iteratively Decodable Codes Based on Steiner Systems and Their Application in Magnetic Recording, in *Proceedings of Globecom* 2001, Vol. 5 pp. 2954–2960, San Antonio, Texas, November 26–29, 2001.

[96] B. Vasic, High-Rate Low-Density Parity-Check Codes Based on Anti-Pasch Affine Geometries, in *Proceedings ICC-2002,* Vol. 3, pp. 1332–1336, New York City, DC, April 28–May 2.

[97] B. Vasic, E. Kurtas, and A. Kuznetsov, LDPC codes based on mutually orthogonal latin rectangles and their application in perpendicular magnetic recording, *IEEE Trans. Magn.,* Vol. 38, No. 5, Part: 1, pp. 2346–2348, September 2002.

[98] B. Vasic, E. Kurtas, and A. Kuznetsov, Kirkman systems and their application in perpendicular magnetic recording, *IEEE Trans. Mag.,* Vol. 38, No. 4, Part: 1, pp. 1705–1710, July 2002.

[99] B. Vasic, K. Pedagani, and M. Ivkovic, "High-rate girth-eight low-density parity check codes on rectangular integer lattices," *IEEE Trans. Commn.* [in press].

[100] B. Vasic, A Class of Codes with Orthogonal Parity Checks and its Application for Partial Response Channels with Iterative Decoding, *Invited paper, 40th Annual Allerton Conference on Communication, Control, and Computing,* October 2–4, 2002.

[101] B. Vasic, E. Kurtas, and A. Kuznetsov, Regular Lattice LDPC Codes in Perpendicular Magnetic Recording, *Intermag 2002 Digest 151, Intermag 2002 Conference,* Amsterdam, The Netherlands, April 28–May 3, 2002.

[102] B. Vasic, High-Rate Low-Density Parity Check Codes Based on Anti-Pasch Affine Geometries, in *Proceedings of ICC-2002,* Vol. 3, pp. 1332–1336, New York City, NY, April 28–May 2.

[103] B. Vasic, E. Kurtas, and A. Kuznetsov, Kirkman Systems and their Application in Perpendicular Magnetic Recording, in *Proceedings 1st North American Perpendicular Magnetic Recording Conference* (NAPMRC), January 7–9th, 2002, Coral Gables, Florida.

[104] B. Vasic and K. Pedagani, Runlength Limited Low-Density Parity Check Codes Based on Deliberate Error Insertion, accepted for presentation at the *IEEE 2003 International Conference on Communications* (ICC 2003) 11–15 May, 2003 Anchorage, AK, USA.

[105] E.J. Weldon, Jr., Difference-set cyclic codes, *Bell Syst. Tech. J.,* 45, pp. 1045–1055, September 1966.

[106] S. Weller and S. Johnson, Iterative Decoding of Codes from Oval Designs, Defense Applications of Signal Processing, 2001 Workshop, Adelaide, Australia, September 2001.

[107] N. Wiberg, H.-A. Loeliger, and R. Ktter, Codes and iterative decoding on general graphs, *Eur. Trans. Telecomm.,* Vol. 6, pp. 513–525, September/October 1995.

[108] R.M. Wilson, Cyclotomy and difference families in elementary abelian groups, *J. Number Theory,* 4, pp. 17–47, 1972.

[109] V.V. Zyablov and M.S. Pinsker, Estimation of the error-correction complexity for Gallager low-density codes, *Probl. info. Transn.,* Vol. 11, pp. 18–28, 1975 (translated by Plenum from Problemy Peredachi Informatsii).

15

Turbo Coding for Multitrack Recording Channels

Zheng Zhang
Arizona State University
Tempe, AZ

Tolga M. Duman
Arizona State University
Tempe, AZ

Erozan M. Kurtas
Seagate Technology
Pittsburgh, PA

Zheng Zhang: received the B.E. degree with honors from Nanjing University of Aeronautics and Astronautics, Nanjing, China, in 1997, and the M.S. degree from Tsinghua University, Beijing, China, in 2000, both in electronic engineering. Currently, he is working toward the Ph. D. degree in electrical engineering at Arizona State University, Tempe, AZ.

His current research interests are digital communications, wireless/mobile communications, magnetic recording channels, information theory, channel coding, turbo codes and iterative decoding, MIMO systems and multi-user systems.

Tolga M. Duman: received the B.S. degree from Bilkent University in 1993, M.S. and Ph.D. degrees from Northeastern University, Boston, in 1995 and 1998, respectively, all in electrical engineering. Since August 1998, he has been with the Electrical Engineering Department Arizona State University first as an assistant professor (1998–2004), and currently as an associate professor. Dr. Duman's current research interests are in digital communications, wireless and mobile communications, channel coding, turbo codes, coding for recording channels, and coding for wireless communications.

Dr. Duman is the recipient of the National Science Foundation CAREER Award, IEEE Third Millennium medal, and IEEE Benelux Joint Chapter best paper award (1999). He is a senior member of IEEE, and an editor for IEEE Transactions on Wireless Communications.

Erozan M. Kurtas: received the B.Sc. degree from Bilkent University, Ankara, Turkey, in 1991 and M.Sc. and Ph.D. degrees from Northeastern University, Boston, MA, in 1993 and 1997, respectively.

His research interests cover the general field of digital communication and information theory with special emphasis on coding and detection for inter-symbol interference channels. He has published over 75 book chapters, journal and conference papers on the general fields of information theory, digital communications and data storage. Dr. Kurtas is the co-editor of the book *Coding and Signal Processing for Magnetic Recording System* (published by CRC press in 2004). He has seven pending patent applications. Dr. Kurtas is currently the Research Director of the Channels Department at the research division of Seagate Technology.

Abstract

Intertrack interference (ITI) is considered as one of the major factors that severely degrade the performance of the practical detectors, especially for narrow-track systems of future. As a result, multi-track systems, where data are written in a group of adjacent tracks simultaneously and read back by multiple heads in parallel, have received significant attention in recent years. In this chapter, we study the turbo coding and iterative decoding schemes suitable for the multitrack recording systems. We describe the maximum a posteriori detector for the multitrack channels in both cases of deterministic ITI and random ITI. We provide simulation results showing that this concatenated coding scheme works very well. In particular, its performance is only about 1 dB away from the information theoretical limits for the ideal partial response channels. It is also shown that the performance achieved by the multitrack system is much better than that of its single-track counterpart. Finally, we provide some results to illustrate the effects of the media noise on the system performance.

15.1   Introduction

The areal density of the hard disk drives has been doubled every 18 months over the last decade [1]. We can increase the areal density in two directions; the axial direction and the radial direction. In the axial direction, we write the information bits close to each other, thus decrease the pulse width which increases the *intersymbol interference* (ISI), while in the radial direction, we decrease the track width which increases the *intertrack interference* (ITI). ITI is considered as one of the major factors that severely degrade the performance of the practical detectors and limit the recording density, especially for narrow-track systems of future. In fact, even with wide tracks, ITI can be present due to possible head misalignment [2]. As a result, multitrack channels, where data are written in a group of adjacent tracks simultaneously and read back by multiple heads in parallel, have received great attention due to their capability in combating the problems caused by the ITI [2–6]. Compared to the single-track systems in the presence of ITI, by using a multitrack recording system, a significantly better performance can be achieved, which in turn results in a density increase for a given performance requirement. In [7, 8], the information theoretical results also show that multitrack systems have higher achievable information rates, which is also justified by the fact that multitrack systems are more robust than their single-track counterparts. Furthermore, we will show later in this chapter that the multitrack systems are not only robust against the ITI, but also to the media noise, which is a type of signal-dependent noise that exists in high-density recording channels.

Another motivation for employing the multitrack system lies in that it provides easier timing and gain control, since the timing and gain information can be derived from any track or any subset of tracks [9]. This brings the benefit of relaxing the synchronization constraints, such as the k constraint of the *run-length limited* (RLL) codes [10], which further increases the storage density. In order to make use of this advantage of multitrack systems, some work has already been performed on the design of the two-dimensional modulation or constrained codes [9, 11–13].

With all these advantages, multitrack systems have the potential to play an important role in the future storage industry. On the other hand, the practical difficulties and costs of developing multiple read/write heads, as well as the inevitable increase of the complexity of the detection and coding/decoding schemes, remain as the main obstacles.

Uncoded multitrack recording systems have been studied in [3, 6, 14–18] (see, also the references therein). These works mainly focus on the development of suitable equalization and detection algorithms. Channel coding techniques are used for the multitrack systems as well in [19, 20] to combat both the intersymbol and intertrack interference, where conventional block and convolutional codes are employed.

In 1993, turbo codes were proposed [21], where the message sequence is encoded by two parallel concatenated convolutional codes that are separated by an interleaver, and decoded by a suboptimal iterative decoding algorithm. The performance of the turbo codes is shown to approach the Shannon limit in the *additive white Gaussian noise* (AWGN) channel when the size of the interleaver is selected sufficiently large. Turbo coding schemes have also been proposed for the ISI channels, such as the magnetic recording channels [22, 23], and it is shown that they can achieve an excellent performance in this case as well.

In this chapter, we consider the application of turbo codes to the multitrack systems (see also [8, 24]). We use a turbo code, or just a single convolutional code with an interleaver, as the outer encoder, and map the coded bits to different groups and transmit them through adjacent tracks. At the receiver, we develop a modified *maximum a posteriori* (MAP) detector for the multitrack systems with deterministic or random ITI. The turbo equalization [25] and the iterative decoding can be performed by exchanging the soft information between the channel MAP detector and the outer soft-input soft-output decoder that corresponds to the outer encoder. We show that the resulting system performance is very close to the information theoretical limits obtained in [7, 8]. We also note that, other turbo-like codes, such as *low-density parity-check* (LDPC) codes [26] and the block turbo codes (or, turbo product codes) [27] can be applied to the multitrack recording systems as well.

The chapter is organized as follows. In Section 15.2, the multitrack recording channel model is given which is nothing but a *multi-input multi-output* (MIMO) ISI channel. In Section 15.3, the capacity and the achievable information rates over such channels with binary inputs are reviewed. These information theoretical limits are useful to evaluate the effectiveness of the specific turbo coding schemes to be proposed. In Section 15.4, a MIMO MAP detector is developed, and the turbo coding and iterative decoding scheme are presented. The performance of the turbo coded systems is then compared with the information theoretical limits developed in Section 15.3, for both multitrack and single-track systems when ITI is present. Finally, the conclusions are provided in Section 15.5.

15.2   Multitrack Recording Channels

Mathematically, magnetic recording channels can be modeled as ISI channels. In addition, in the narrow-track systems, which will be more and more popular in the future recording systems, the intertrack interference will also be present. Both the ISI and the ITI are illustrated in Figure 15.1, where the black block denotes the media track that records the desired signal and the gray ones represent the interfering sources. In the axial direction, the interference comes from the adjacent symbols, and in the radial direction, it results due to the adjacent tracks.

To deal with the increased ITI, we can write the signals to multiple tracks and use multiple heads to read them for joint detection. In an (N, M) multitrack system, for which there are N heads reading M tracks

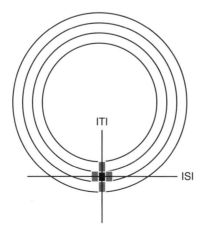

FIGURE 15.1   Multitrack system with intersymbol interference and intertrack interference.

simultaneously, the received signal for the nth head can be expressed as

$$r_n(t) = \sum_{m=1}^{M} a_{n,m} \sum_{k=-\infty}^{\infty} b_{m,k} \cdot p(t - kT) + w_n(t) \quad 1 \le n \le N \tag{15.1}$$

where $p(t)$ is the transition response of the channel, T is the symbol duration, $w_n(t)$ is the AWGN with two sided power spectral density of $N_0/2$, and $\{b_{m,k}\}$ is the transition symbol on the mth track with $b_{m,k} = x_{m,k} - x_{m,k-1}$ if $\{x_{m,k}\}$ is the transmitted binary signal.

Equivalently, the received signal is given by

$$r_n(t) = \sum_{m=1}^{M} a_{n,m} \sum_{k=-\infty}^{\infty} x_{m,k} \cdot h(t - kT) + w_n(t) \quad 1 \le n \le N \tag{15.2}$$

where $h(t)$ is the pulse response of the channel with $h(t) = p(t) - p(t - T)$. We assume that the pulse responses from different tracks are the same except for the amplitude varying with the distance between the track and the readhead, which is reflected by the coefficients $\{a_{n,m}\}$. This is a simplified model that is justified by the experimental measurements [2]. Other models can also be used, for which the coding/decoding schemes described later are still applicable.

For the sake of comparison, we also consider the single-track channels, which experience interference from adjacent tracks as well, but have only one read head. The receiver detects the desired signal from the corresponding track and considers the others as pure interference.

At the receiver, the output is passed through a matched filter, a sampler at the symbol rate and a noise-whitening filter [10]. The lth received symbol for the nth head is then given by

$$y_{n,l} = \sum_{m=1}^{M} a_{n,m} \sum_{k=0}^{K} x_{m,l-k} \cdot f_k + z_{n,l} \quad 1 \le n \le N \tag{15.3}$$

where $\{z_{n,l}\}$ is the white Gaussian noise sequence with variance $\sigma_z^2 = N_0/2$, and $\{f_l\}$ is the set of coefficients of the equivalent discrete-time ISI channel with memory K. We denote the ITI matrix as $A = [a_{n,m}]_{N \times M}$ and consider both cases of deterministic and random ITI.

In practice, the output signal of the magnetic recording channel is usually equalized to an appropriate *partial response* (PR) target using a linear equalizer. Therefore, for simplicity, the ideal PR channels are often used to model the magnetic recording channels, where the received signal can be expressed by Equation 15.3 with $\{f_l\}$ defined by the PR target. For the ideal normalized PR4 channel considered later in this chapter, we have $K = 2$ and $f_0 = 1/\sqrt{2}$, $f_1 = 0$ and $f_2 = -1/\sqrt{2}$.

As a more realistic example, we also use the longitudinal recording channel where the transition response $p(t)$ is modeled by the Lorentzian pulse

$$p(t) = \frac{1}{1 + (2t/PW_{50})^2} \tag{15.4}$$

where PW_{50} is the pulse width at the half height. We define the parameter $D_n = PW_{50}/T$ as the normalized density of the recording system.

In high-density magnetic recording channels, signal-dependent noise, called media noise exists [28, 29]. In this chapter, we only consider the fluctuation of the position of the transition pulse, called jitter noise, which is one of the main sources of the media noise. The received signal for the recording channel with jitter noise is then given by

$$r_n(t) = \sum_{m=1}^{M} a_{n,m} \sum_{k=-\infty}^{\infty} b_{m,k} \cdot p(t + j_{m,k} - kT) + w_n(t) \quad 1 \le n \le N \tag{15.5}$$

where $\{j_{m,k}\}$ is the independent zero-mean Gaussian noise term that reflects the amount of position jitter.

The *signal to noise ratio* (SNR) is defined as

$$\text{SNR} = \frac{C_1 \cdot C_2 \cdot E_s}{N_0} \tag{15.6}$$

where E_s is the transmitted symbol energy, and C_1 and C_2 are the normalization factors due to the received ITI and the channel response, respectively. For the deterministic ITI and the random ITI cases, C_1 can be expressed as $\frac{1}{N} \sum_{n=1}^{N} \sum_{m=1}^{M} a_{n,m}^2$ and $\frac{1}{N} \sum_{n=1}^{N} \sum_{m=1}^{M} E[a_{n,m}^2]$, respectively. For the longitudinal channel, $C_2 = \int_{-\infty}^{\infty} h^2(t)\, dt$, and for the normalized PR4 channel, $C_2 = 1$.

15.3 Information Theoretical Limits: Achievable Information Rates

Before we describe the turbo coding/decoding scheme for the multitrack recording channels, we present some results of the achievable information rates, which can be used as the ultimate theoretical limits of the coding/decoding schemes.

For *single-input single-output* (SISO) ISI channels, the capacity is derived in [30]. For multitrack systems with deterministic ITI, a lower bound on the capacity is derived in [31] following a similar approach as in [32]. In both cases, Gaussian inputs are assumed to be used. However, for practical digital communication systems, the inputs are constrained to be selected from a finite alphabet. For example, signals used in the magnetic recording systems are binary. In general, with the use of the constrained inputs, one cannot achieve the "unconstrained" capacity and the gap between the capacity and the achievable information rates under such input constraints is far from ignorable in high SNR regions, especially when the size of the alphabet is small. Therefore, it is also important to determine the information rates achievable under the specific input constraints.

Simulation-based methods have been recently proposed to estimate the achievable information rates for ISI channels with specific binary inputs [33–35]. The main idea is to use a simulation of the channel and employ the BCJR algorithm [36] to estimate the joint probability of the output sequence. Then, this estimate is used to compute the differential entropy of the output sequence and the mutual information between the input and the output, thus the achievable information rates over the noisy channel. By increasing the simulation complexity, one can easily obtain an accurate result. In [37], the maximization of the information rates is performed for the case of Markov inputs with a certain memory, as opposed to the use of *independent identically distributed* (i.i.d.) inputs.

We can extend these techniques to the case of MIMO ISI channels with deterministic and random ITI. The key step is still the computation of the entropy, and thus the probability of the output sequence for a given channel simulation. To accomplish this for the multitrack systems, we set up the channel trellis based on both the ISI and the ITI, instead of the one that only takes the ISI into account. The details can be found in [8].

To give specific examples in this chapter, we consider the two-track and two-head channel with the ITI matrix

$$A = \begin{bmatrix} 1 & \alpha \\ \alpha & 1 \end{bmatrix} \tag{15.7}$$

where the diagonal elements are set to 1, which represent the amplitudes of the desired signals to each readhead, and off-diagonal elements represent the amplitudes of the ITI from the other track. We also consider a single-track system with $M = 2$ and the ITI matrix $[1 \; \alpha]$ for comparison. For the case of deterministic ITI, the entries of the matrix are fixed. On the other hand, for the random ITI case, we model α as uniform random variables over $[0, 1]$, which are assumed to be independent spatially and temporally, and known to the receiver.

The information rates per track use for the multitrack and single-track systems with different values of deterministic α are shown in Figure 15.2 for the PR4 channel where the input is constrained to be binary and i.i.d. with equal probability of 1s and 0s. We observe that the improvement in the achievable information

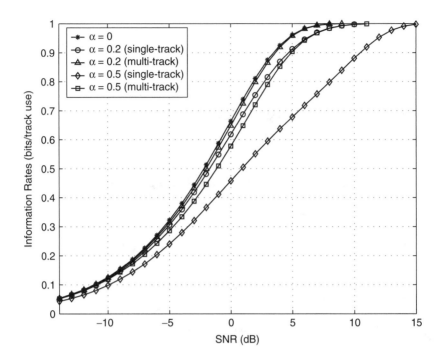

FIGURE 15.2   Information rates for the PR4 channel with deterministic α.

rates by adopting multitrack system is large, especially for high ITI levels. For example, to achieve a transmission rate of 16/17 bits per track use for the single-track system with arbitrarily low probability of error when $\alpha = 0.5$, about 11.6 dB is needed, but only 6.0 dB is enough for the multitrack system.

Figure 15.3 shows the information rates per track use for the PR4 channels with uniform ITI coefficients, which are known to the receiver only. We obtain similar conclusions by comparing the information rates for the multitrack and the single-track systems for this case as well.

In Figure 15.4, we maximize the information rates over all the Markov inputs with a memory $I = 2$ for the (2, 2) multitrack PR4 systems. We observe that in the low-to-medium SNR region, the information rate achieved by the optimized Markov inputs is obviously larger than the one with i.i.d. inputs. Particularly, to achieve a rate of 0.2 bits per track use, there is a gain of 2.6 dB when $\alpha = 0.5$. However, when the transmission rate considered is high, for example, 16/17 bits per track use, there is almost no difference in the required SNR to make the error probability arbitrarily small.

We also present several results for the longitudinal channels with a normalized density of 2.0. Figure 15.5 shows the information rates per track use for the channels with a deterministic α. We also observe the effectiveness of the multitrack recording compared to the single-track one. For example, when α is increased from 0 to 0.5, the SNR loss is about 6.6 dB for the single-track system, whereas it is only 1.7 dB for the multitrack case, for the transmission rate of 16/17 bits per track use.

15.4   Turbo Coding for Multitrack Recording Systems

In 1993, Berrou et al. introduced the turbo codes, which can achieve near Shannon-limit performance over the memoryless AWGN channel [21]. For this reason, turbo coding and decoding techniques have been comprehensively studied and applied to many other channels, such as the ISI channels [22, 23]. In this section, we extend these techniques to the multitrack recording system, which can be viewed as a special case of the general MIMO ISI channels. We first review the turbo codes and iterative decoding algorithm, as well as their application to the ISI channels for completeness.

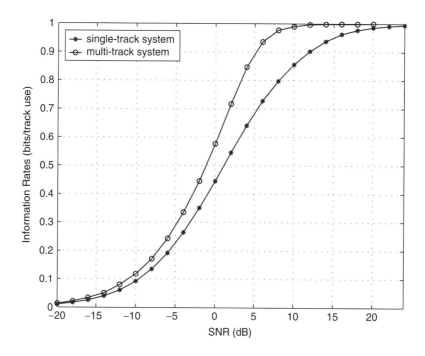

FIGURE 15.3 Information rates for the PR4 channel with uniform α.

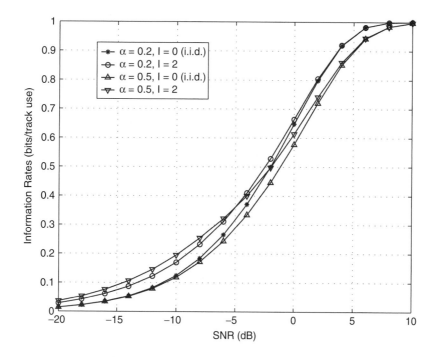

FIGURE 15.4 Maximized information rates of the (2, 2) multitrack PR4 system.

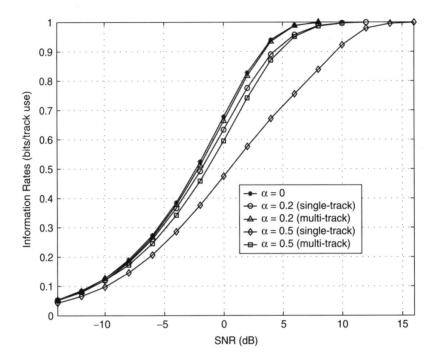

FIGURE 15.5   Information rates for longitudinal recording channels with deterministic $\alpha(D_n = 2.0)$.

15.4.1   Turbo Codes and Iterative Decoding Algorithm

The diagram of a turbo encoder is shown in Figure 15.6, where the uncoded message bits, denoted by **u**, are encoded by two parallel rate-1/2 *recursive systematic convolutional* (RSC) encoders that are separated by an interleaver. The codeword corresponding to the message **u** is then formed by the systematic information bits $\mathbf{x^s}$, and the parity bits, $\mathbf{x^{1p}}$ and $\mathbf{x^{2p}}$, generated by the two RSC encoders. This is an example of a rate 1/3 turbo code, however, by puncturing some of the parity bits, higher rate codes can be easily obtained [21].

The interleaver, which is used to permute the input bits, is very important in the construction of the turbo codes. For example, the average *bit error rate* (BER) over an AWGN channel is shown to be inversely proportional to the interleaver length when a uniform interleaver is used [38] (which is a probabilistic device that takes on each possible permutation with equal probability). This gain is referred as the interleaving gain.

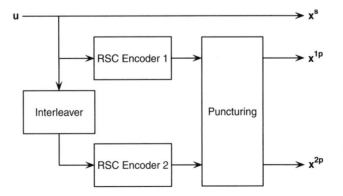

FIGURE 15.6   Turbo code block diagram.

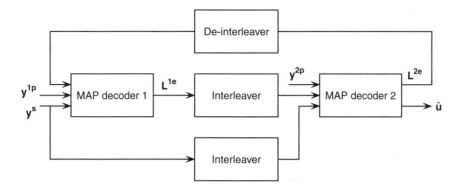

FIGURE 15.7 Block diagram of the turbo decoder.

Pseudo-random interleavers generally offer a good performance, therefore they are often used. However, well designed interleavers can outperform the average (uniform) interleavers significantly, especially in the high SNR region [39, 40].

The optimal decoder for the turbo code in the sense of minimizing the BER is the one that chooses the codeword with the maximum posterior probability. Due to the existence of the interleaver, such a decoder needs to perform an exhaustive search over the entire set of possible codewords. Therefore, its complexity is huge and it is not feasible, especially when the interleaver length is chosen to be relatively large to obtain a significant interleaving gain. A suboptimal decoding strategy is proposed in [21] to solve this problem. For the component convolutional codes, this decoder uses two MAP decoders that iteratively produce soft information about each bit, called extrinsic information, which is exchanged between the decoders. The diagram of the turbo decoder is shown in Figure 15.7, where $\mathbf{y}^s$, $\mathbf{y}^{1p}$ and $\mathbf{y}^{2p}$ are the observations corresponding to the systematic information bit sequence $\mathbf{x}^s$, and the two parity bit sequences, $\mathbf{x}^{1p}$ and $\mathbf{x}^{2p}$, respectively. $\mathbf{L}^{1e}$ and $\mathbf{L}^{2e}$ are the extrinsic information that is computed from the previous MAP decoder and fed to the next one. The turbo decoder works iteratively until the final hard decisions on the uncoded bits are made. For the details of the MAP algorithm and the iterative decoding for the component convolutional codes, the reader is referred to another chapter in this book [41].

15.4.2 Turbo Coding and Turbo Equalization for ISI Channels

After the invention of the turbo codes and the iterative decoding algorithm, many researches have proposed to apply these techniques for the transmission over other channels, including the ISI channels. For the case of ISI channels, instead of the *maximum likelihood sequence detector* (MLSD) [10, 42], it is necessary to use a soft-input soft-output detector or equalizer, which can generate soft reliability information about the coded symbols from the noisy channel observations. This information can then be passed to the turbo decoder to perform the iterative decoding. In addition, the extrinsic information generated by the turbo decoder can also be fed back to the channel detector to update the a priori probabilities. The concept of combining the channel equalizer and the turbo decoder in the iterative decoding scheme is called turbo equalization [25, 43].

It is clear that a MAP detector can be used as the soft-input soft-output detector for the ISI channel, where the soft information is computed based on the *log-likelihood ratio* (LLR) of the coded bit x_l that is transmitted over the ISI channel. We define the LLR as

$$\Lambda(x_l) = \log \frac{p(x_l = 1 \mid \mathbf{y}^n)}{p(x_l = 0 \mid \mathbf{y}^n)} \tag{15.8}$$

where $\mathbf{y}^n = \{y_1, \ldots, y_l, \ldots, y_n\}$ is assumed to be the channel output sequence or the observations with block length n.

To compute the *a posteriori probability* (APP) in Equation 15.8, we can use the BCJR algorithm [36] operating on the trellis of the ISI channel. Suppose the trellis state at the time instance l is S_l, then we have

$$\Lambda(x_l) = \log \frac{p(x_l = 1, \mathbf{y}^n)/p(\mathbf{y}^n)}{p(x_l = 0, \mathbf{y}^n)/p(\mathbf{y}^n)}$$

$$= \log \frac{\sum_{(i,j)\in\Omega(1)} p(S_{l-1} = i, S_l = j, \mathbf{y}^n)}{\sum_{(i,j)\in\Omega(0)} p(S_{l-1} = i, S_l = j, \mathbf{y}^n)} \tag{15.9}$$

where $\Omega(1)$ and $\Omega(0)$ are the sets of the valid state transitions where $x_l = 1$ and $x_l = 0$, respectively. We can further write $\Lambda(x_l)$ as [41]

$$\Lambda(x_l) = \log \frac{\sum_{(i,j)\in\Omega(1)} \alpha_{l-1}(i) \cdot \gamma_l(i,j) \cdot \beta_l(j)}{\sum_{(i,j)\in\Omega(0)} \alpha_{l-1}(i) \cdot \gamma_l(i,j) \cdot \beta_l(j)} \tag{15.10}$$

where $\alpha_l(j)$, $\beta_l(j)$ and $\gamma_l(i,j)$ are defined as

$$\alpha_l(j) = p(y_1, \ldots, y_l, S_l = j) \tag{15.11}$$

$$\beta_l(j) = p(y_{l+1}, \ldots, y_n \mid S_l = j) \tag{15.12}$$

and

$$\gamma_l(i,j) = p(S_l = j \mid S_{l-1} = i) \cdot p(y_l \mid S_{l-1} = i, S_l = j)$$

$$= p(x_l) \cdot p(y_l \mid S_{l-1} = i, S_l = j) \tag{15.13}$$

Here $p(x_l)$ is the a priori information about the coded bit x_l, and it can be updated iteratively using the extrinsic information generated by the turbo decoder. We can use the forward and backward recursions to compute $\alpha_l(j)$ and $\beta_l(j)$ in an efficient manner [36]. The details on the generation of the extrinsic information are presented in [41].

In the above discussion, we use the turbo code as the outer encoder, where two *parallel concatenated convolutional codes* (PCCC) separated by an interleaver are employed. As opposed to the PCCC scheme, a *single convolutional code* (SCC) can be employed instead, since the ISI channel can be viewed as a rate-one inner encoder. In this case, a precoder is necessary to make the channel (i.e., the inner code) recursive. Because there is only one component decoder in this case, the computational complexity is smaller than the one for the parallel concatenation, while the performance is shown to be comparable or, even better in some cases [23].

In addition, other turbo-like codes, including the LDPC codes and the block turbo codes, can also be used in place of the SCC or PCCC for the ISI channels [44, 45].

15.4.3   MAP Detector and Iterative Decoding for Multitrack Systems

We now extend the use of the code concatenation to the case of multitrack recording by designing the corresponding MAP detectors for different ITI scenarios. The block diagrams of the transmitter and receiver are shown in Figure 15.8. At the transmitter, we first encode the message bit sequence, denoted by $\mathbf{u}$, by using an outer encoder, such as the PCCC or SCC. After being passed through a random interleaver, the coded bits, represented by $\mathbf{x}$, are divided evenly into M groups, which are sent through the M transmitters or tracks. The N output sequences of the MIMO ISI channels, corrupted by the additive white Gaussian noise, constitute the received signal $\mathbf{y}$. At the receiver, the turbo equalization is used where a modified channel MAP detector takes the channel outputs and the extrinsic information fed back from the outer decoder as its inputs, and generates the soft information about the coded bits. This soft information about the inputs from different transmitters or tracks is deinterleaved and passed to the outer decoder. The outer decoder, that is, a turbo decoder or a MAP decoder for a single convolutional code, generates the extrinsic information, which is then fed back to the channel MAP detector after appropriate processing for the next iteration step. The LLR of the message bits are used to make hard decisions after a number of iterations.

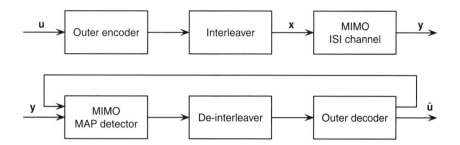

FIGURE 15.8 Block diagram of the turbo coding scheme for the multitrack system.

We use a (2, 2) multitrack system with the ISI memory of K as an example to illustrate the necessary modifications. First we should set up the channel trellis with 2^{2K} states according to the multiple inputs and multiple outputs, which has 2^{2K} states. Assume that, at time instance l, the two coded bits transmitted over the two tracks are $x_{1,l}$ and $x_{2,l}$, and the two output sequences with length n are $\mathbf{y}_1^n = \{y_{1,1}, \ldots, y_{1,l}, \ldots, y_{1,n}\}$ and $\mathbf{y}_2^n = \{y_{2,1}, \ldots, y_{2,l}, \ldots, y_{2,n}\}$. In order to compute the log-likelihood ratios for $x_{1,l}$ and $x_{2,l}$, we compute four probabilities: $P(x_{1,l} = 0, x_{2,l} = 0 \mid \mathbf{y}_1^n, \mathbf{y}_2^n)$, $P(x_{1,l} = 0, x_{2,l} = 1 \mid \mathbf{y}_1^n, \mathbf{y}_2^n)$, $P(x_{1,l} = 1, x_{2,l} = 0 \mid \mathbf{y}_1^n, \mathbf{y}_2^n)$ and $P(x_{1,l} = 1, x_{2,l} = 1 \mid \mathbf{y}_1^n, \mathbf{y}_2^n)$. Thus, the log-likelihood ratios of the two bits can be computed as

$$\Lambda(x_{1,l}) = \log \frac{P\left(x_{1,l} = 1, x_{2,l} = 0 \mid \mathbf{y}_1^n, \mathbf{y}_2^n\right) + P\left(x_{1,l} = 1, x_{2,l} = 1 \mid \mathbf{y}_1^n, \mathbf{y}_2^n\right)}{P\left(x_{1,l} = 0, x_{2,l} = 0 \mid \mathbf{y}_1^n, \mathbf{y}_2^n\right) + P\left(x_{1,l} = 0, x_{2,l} = 1 \mid \mathbf{y}_1^n, \mathbf{y}_2^n\right)} \qquad (15.14)$$

$$\Lambda(x_{2,l}) = \log \frac{P\left(x_{1,l} = 0, x_{2,l} = 1 \mid \mathbf{y}_1^n, \mathbf{y}_2^n\right) + P\left(x_{1,l} = 1, x_{2,l} = 1 \mid \mathbf{y}_1^n, \mathbf{y}_2^n\right)}{P\left(x_{1,l} = 0, x_{2,l} = 0 \mid \mathbf{y}_1^n, \mathbf{y}_2^n\right) + P\left(x_{1,l} = 1, x_{2,l} = 0 \mid \mathbf{y}_1^n, \mathbf{y}_2^n\right)} \qquad (15.15)$$

Similar to the single-input single-output case, we can employ the BCJR algorithm after a minor modification to compute the APPs. Suppose the trellis state at time instance l is S_l, then

$$\gamma_l(i, j) = P(S_l = j \mid S_{l-1} = i) \cdot P(y_{1,l}, y_{2,l} \mid S_{l-1} = i, S_l = j) \qquad (15.16)$$

$$= P(x_{1,l}, x_{2,l}) \cdot P(y_{1,l} \mid S_{l-1} = i, S_l = j) P(y_{2,l} \mid S_{l-1} = i, S_l = j) \qquad (15.17)$$

$$\approx P(x_{1,l}) P(x_{2,l}) \cdot P(y_{1,l} \mid S_{l-1} = i, S_l = j) P(y_{2,l} \mid S_{l-1} = i, S_l = j) \qquad (15.18)$$

where Equation 15.17 holds because the two observations are independent for a given state transition and the approximation in Equation 15.18 follows due to the use of the interleaver. When the ITI coefficients are random and known to the receiver, we just use them in the decoding as if they are constant for each symbol duration.

We can use a similar MAP detector and coding/decoding scheme for the single-track system, where there is only one received signal and one desired transmitted signal. Compared to the decoding for the multitrack systems, no detection is made for the other transmitted signal, which is considered as pure interference.

15.4.4 Examples

Figure 15.9 shows the performance of this turbo coding/decoding scheme for the (2, 2) PR4 channels with deterministic ITI where α is set to be 0.5. The component convolutional code used for the outer encoder is a (31, 33) code in octal form, the block length of the input to the outer encoder is 10016, the code rate is 16/17 and the iterative decoding algorithm with 15 iterations is used. We employ both SCC and PCCC and consider the use of a precoder, defined by $1/(1 \oplus D^2)$ where $\oplus$ indicates modulo-2 addition, as well. The performance of the uncoded system is obtained by passing the uncoded and unprecoded information bits through the channel and using the channel MAP decoder to detect them. The results show that SCC achieves a better performance compared to the PCCC, and its complexity is lower. The performance is only 1.2 dB away from the theoretical limit computed in Figure 15.2 for a code rate of 16/17 bits per track

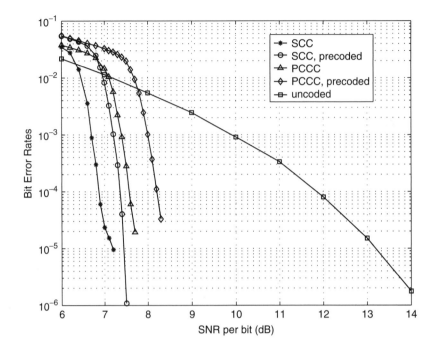

FIGURE 15.9   Performance of turbo coding for $(2, 2)$ multitrack PR4 channels with deterministic α.

use if we consider a BER of 10^{-5} as reliable transmission (where the rate loss of 0.26 dB is considered). This represents a coding gain of about 6 dB relative to the uncoded system. We also notice that, when the SCC is used, the BER curve has lower error floor with the appropriate use of precoding. Therefore, we only consider the combination of the SCC and the precoder in the following examples.

In Figure 15.10, the bit error rates of the iterative decoding with SCC for the multitrack and single-track systems with different values of α are shown. Same parameters are used as the previous example and a precoder of $1/(1 \oplus D^2)$ is still employed. Compared with the information rates shown in Figure 15.2, the distances from the theoretical limits are from 0.8 dB to 1.2 dB (where the rate loss is considered), respectively. To achieve a BER of 10^{-5} for the multitrack systems, there is an SNR gain of 1.6 dB and 5.3 dB over the single-track systems when $\alpha = 0.2$ and 0.5, respectively, which is in line with our expectations from the information theoretical results.

We consider the case with uniform α (known to the receiver) in Figure 15.11 by using the same coding/decoding scheme. The performance is about 1.5 dB away from the limits computed in Figure 15.3 for both multitrack and single-track cases. There is an SNR gain of about 8.2 dB for the multitrack system over the single-track one, which is also expected by comparing the achievable information rates for both systems.

In Figure 15.12, we show some results for the longitudinal channel with normalized density $D_n = 2.0$ and deterministic ITI, which is equalized to a $(2, 2)$ EPR4 channel $(1 + D - D^2 - D^3)$ with the same value of α. We use a single convolutional code as the outer encoder, together with the precoder $1/(1 \oplus D^2)$, and the code generators, the code rate and the block length are the same as in the earlier examples. We observe that, to achieve a BER of 10^{-5}, the required SNR per bit is 2.2 dB higher than the theoretical limit for $\alpha = 0.2$ and 3.0 dB higher for $\alpha = 0.5$. Although the performance is still very good, it is inferior to the previous examples. This may be due to two major factors. First, the equalizer target we used is based on the one-dimensional design and there should be better alternatives which take both the axial and radial interference into consideration. Also, the noise correlation due to the use of the equalizer is not taken into account in the decoding process.

We consider the effects of the jitter noise on the performance in Figure 15.13, where the detector ignores the existence of the jitter. We observe that the jitter degrades the system performance significantly. For

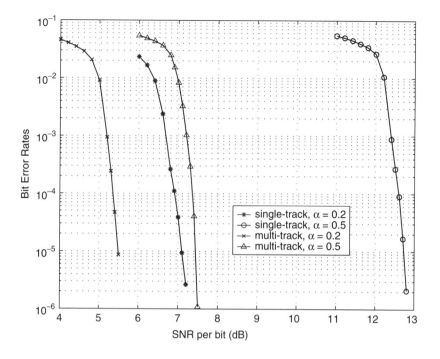

FIGURE 15.10 Performance of the SCC scheme for multitrack and single-track PR4 channels.

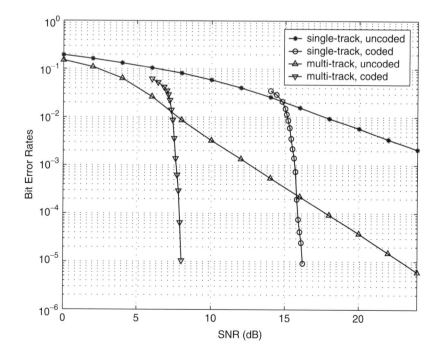

FIGURE 15.11 Performance of the SCC scheme for multitrack and single-track PR4 channels with uniform α.

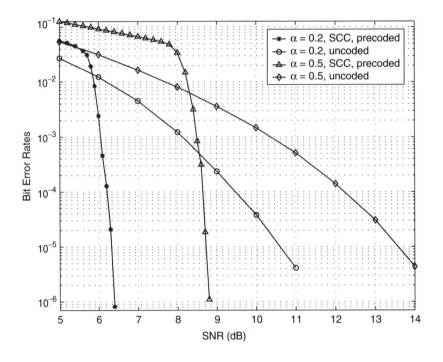

FIGURE 15.12   Performance of the SCC scheme for (2, 2) multitrack longitudinal channels.

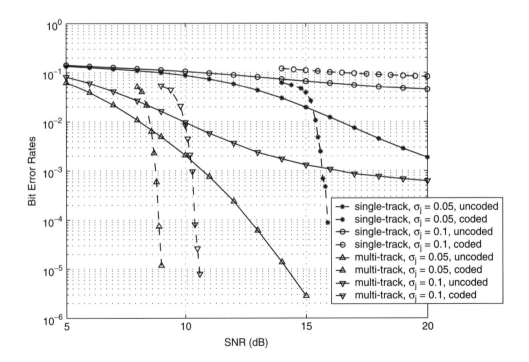

FIGURE 15.13   Performance of the SCC scheme for multitrack and single-track longitudinal channels with jitter noise.

the uncoded system, an error floor is observed, especially for the single-track case. Particularly, when $\sigma_j = 0.1$, the media noise is too high for the iterative decoding algorithm to correctly exchange the soft information in the single-track system, therefore the decoding algorithm almost fails. However, for the multitrack system, the turbo coding scheme still works well and very high coding gain can be obtained compared to the uncoded case.

15.5 Discussion

In order to increase the areal density of the magnetic recording systems, narrow tracks are used, which inevitably causes intertrack interference. We can employ multiple write and read heads to reduce the effects of the ITI. In this chapter, we considered the application of turbo codes to the multitrack recording systems, which are shown to achieve a very good performance (i.e., close to the information theoretical limits), for both cases with deterministic and random ITI. When compared with the single-track systems, the multitrack system has significant advantages in terms of the achievable information rates and the detection performance, which indicates that a multitrack system is a good solution to increase the areal densities of the future recording systems. In addition, we consider the effects of the media noise on the performance of the turbo coding scheme, and show that the multitrack systems are also more robust compared to the single-track ones even in the presence of media noise.

Since the design of the MAP detector for the multitrack systems is independent of the choice of the outer decoder, in addition to the turbo code (or, a single convolutional code), we can apply other turbo-like codes, such as the LDPC codes and block turbo codes, as well.

The MAP detector for the multitrack system can be considered as an optimal soft-output detector for the MIMO ISI channels. However, if a multitrack system with more tracks is considered, the computational complexity for this algorithm is high, as it is exponential in the number of tracks. Therefore, simplified schemes should be used instead, at the cost of some performance degradation. For example, we can give up the turbo equalization if the iterative decoding can be performed within the outer decoder (for the case of PCCC). In addition, we can use other soft-input soft-output detectors with lower complexities, such as the SOVA detector, or even other suboptimal detectors where the complexity increases linearly with the number of tracks.

We also note that, we did not consider the constrained modulation codes, such as the run-length limited codes for multitrack recording in this chapter. To make multitrack recording practical, new schemes combining both constrained coding and channel coding, as well as the iterative decoding algorithm should be employed [46].

References

[1] J. Moon, H. Thapar, B.V.K.V. Kumar, and K.A.S. Immink, (editorial) Signal processing for high density storage channels, *IEEE J. Select. Areas Commun.*, vol. 19, no. 4, pp. 577–581, April 2001.

[2] P.A. Voois and J.M. Cioffi, A decision feedback equalizer for multiple-head digital magnetic recording, in *Proceedings of IEEE International Conference on Communications (ICC)*, vol. 2, June 1991, pp. 815–819.

[3] L.C. Barbosa, Simultaneous detection of readback signals from interfering magnetic recording tracks using array heads, *IEEE Trans. Magn.*, vol. 26, no. 5, pp. 2163–2165, September 1990.

[4] P.A. Voois and J.M. Cioffi, Multichannel signal processing for multiple-head digital magnetic recording, *IEEE Trans. Magn.*, vol. 30, no. 6, pp. 5100–5114, November 1994.

[5] E. Soljanin and C.N. Georghiades, Multihead detection for multitrack recording channels, *IEEE Trans. Inform. Theory*, vol. 44, no. 7, pp. 2988–2997, November 1998.

[6] E.M. Kurtas, J.G. Proakis, and M. Salehi, Reduced complexity maximum likelihood sequence estimation for multitrack high-density magnetic recording channels, *IEEE Trans. Magn.*, vol. 35, no. 4, pp. 2187–2193, July 1999.

[7] Z. Zhang, T.M. Duman, and E.M. Kurtas, On information rates of single-track and multitrack recording channels with intertrack interference, in *Proceedings of IEEE International Symposium on Information Theory (ISIT)*, June-July 2002, p. 163.

[8] Z. Zhang and T.M. Duman, Information rates and coding for multitrack recording channels with deterministic and random intertrack interference, in *Proceedings of 40th Allerton Conference on Communications, Control and Computing (invited)*, Urbana, Illinois, October 2002, pp. 766–775.

[9] M.W. Marcellin and H.J. Weber, Two-dimensional modulation codes, *IEEE J. Select. Areas Commun.*, vol. 10, no. 1, pp. 254–266, January 1992.

[10] J.G. Proakis, *Digital Communications*, McGraw-Hill, Inc. New York, NY, 2001.

[11] J. Lee and V.K. Madisetti, Constrained multitrack RLL codes for the storage channel, *IEEE Trans. Magn.*, vol. 31, no. 3, pp. 2355–2364, May 1995.

[12] E. Soljanin and C.N. Georghiades, Coding for two-head recording systems, *IEEE Trans. Inform. Theory*, vol. 41, no. 3, pp. 747–755, May 1995.

[13] E.M. Kurtas, J.G. Proakis, and M. Salehi, Coding for multitrack magnetic recording systems, *IEEE Trans. Inform. Theory*, vol. 43, no. 6, pp. 2020–2023, November 1997.

[14] W.L. Abbott, J.M. Cioffi, and H.K. Thapar, Performance of digital magnetic recording with equalization and offtrack interference, *IEEE Trans. Magn.*, vol. 27, no. 1, pp. 705–716, January 1991.

[15] P.A. Voois and J.M. Cioffi, Multichannel digital magnetic recording, in *Proceedings of IEEE International Conference on Communications (ICC)*, vol. 1, June 1992, pp. 125–130.

[16] P.S. Kumar and S. Roy, Two-dimensional equalization: Theory and applications to high density magnetic recording, *IEEE Trans. Commun.*, vol. 42, no. 2/3/4, pp. 386–395, February/March/April 1994.

[17] M.P. Vea and J.M.F. Moura, Multichannel equalization for high track density magnetic recording, in *Proceedings of IEEE International Conference on Communications (ICC)*, vol. 2, May 1994, pp. 1221–1225.

[18] M.Z. Ahmed, P.J. Davey, T. Donnelly, and W.W. Clegg, Track squeeze using adaptive intertrack interference equalization, *IEEE Trans. Magn.*, vol. 38, no. 5, pp. 2331–2333, September 2002.

[19] A.M. Patel, Adaptive cross parity code for a high density magnetic tape subsystem, *IBM J. Res. Dev.*, vol. 29, pp. 546–562, November 1985.

[20] N. Kogo, N. Hirano, and R. Kohno, Convolutional code and 2-dimensional PRML class IV for multitrack magnetic recording system, in *Proceedings of IEEE International Symposium on Information Theory (ISIT)*, June 2000, p. 309.

[21] C. Berrou, A. Glavieux, and P. Thitimajshima, Near Shannon limit error-correcting coding and decoding: Turbo-codes, in *Proceedings of IEEE International Conference on Communications (ICC)*, vol. 2, May 1993, pp. 1064–1070.

[22] W.E. Ryan, Performance of high rate turbo codes on a PR4-equalized magnetic recording channel, in *Proceedings of IEEE International Conference on Communications (ICC)*, vol. 2, June 1998, pp. 947–951.

[23] T.V. Souvignier, M. Öberg, P.H. Siegel, R.E. Swanson, and J.K. Wolf, Turbo decoding for partial response channels, *IEEE Trans. Commun.*, vol. 48, no. 8, pp. 1297–1308, August 2000.

[24] E.M. Kurtas and T.M. Duman, Iterative decoders for multiuser ISI channels, in *Proceedings of Military Communications International Symposium (MILCOM)*, vol. 2, October-November 1999, pp. 1036–1040.

[25] D. Raphaeli and Y. Zarai, Combined turbo equalization and turbo decoding, in *Proceedings of IEEE Global Communications Conference (GLOBECOM)*, vol. 2, November 1997, pp. 639–643.

[26] R.G. Gallager, Low-density parity check codes, *IRE Trans. Inform. Theory*, vol. IT-8, pp. 21–28, January 1962.

[27] R. Pyndiah, A. Glavieux, A. Picart, and S. Jacq, Near optimum decoding of product codes, in *Proceedings of IEEE Global Communications Conference (GLOBECOM)*, vol. 1, November-December 1994, pp. 339–343.

[28] J.G. Zhu and H. Wang, Noise characteristics of interacting transitions in longitudinal thin film media, *IEEE Trans. Magn.*, vol. 31, no. 2, pp. 1065–1070, March 1995.

[29] T.R. Oenning and J. Moon, Modeling the Lorentzian magnetic recording channel with transition noise, *IEEE Trans. Magn.*, vol. 37, no. 1, pp. 583–591, January 2001.

[30] W. Hirt and J.L. Massey, Capacity of the discrete-time Gaussian channel with intersymbol interference, *IEEE Trans. Inform. Theory*, vol. 34, no. 3, pp. 380–388, May 1988.

[31] Z. Zhang, T.M. Duman, and E.M. Kurtas, Achievable information rates and coding for MIMO systems over ISI channels and frequency-selective fading channels, accepted for publication in the *IEEE Transactions on Communications*, 2004.

[32] H.E. Gamal, A.R. Hammons, Y. Liu, M.P. Fitz, and O.Y. Takeshita, On the design of space-time and space-frequency codes for MIMO frequency selective fading channels, *IEEE Trans. Inform. Theory*, vol. 49, no. 9, pp. 2277–2292, September 2003.

[33] D. Arnold and H.-A. Loeliger, On the information rate of binary-input channels with memory, in *Proceedings of IEEE International Conference on Communications (ICC)*, vol. 9, June 2001, pp. 2692–2695.

[34] H.D. Pfister, J.B. Soriaga, and P.H. Siegel, On the achievable information rates of finite state ISI channels, in *Proceedings of IEEE Global Communications Conference (GLOBECOM)*, vol. 5, November 2001, pp. 2992–2996.

[35] V. Sharma and S.K. Singh, Entropy and channel capacity in the regenerative setup with applications to Markov channels, in *Proceedings of IEEE International Symposium on Information Theory (ISIT)*, June 2001, p. 283.

[36] L.R. Bahl, J. Cocke, F. Jelinek, and J. Raviv, Optimal decoding of linear codes for minimizing symbol error rate, *IEEE Trans. Inform. Theory*, vol. 20, pp. 284–287, March 1974.

[37] A. Kavčić, On the capacity of Markov sources over noisy channels, in *Proceedings of IEEE Global Communications Conference (GLOBECOM)*, vol. 5, November 2001, pp. 2997–3001.

[38] S. Benedetto and G. Montorsi, Design of parallel concatenated convolutional codes, *IEEE Trans. Commun.*, vol. 44, no. 5, pp. 591–600, May 1996.

[39] T.M. Duman, Interleavers for serial and parallel concatenated (turbo) codes, *Wiley Encyclopedia of Telecommunications*, J.G. Proakis, Ed., December 2002, pp. 1141–1151.

[40] S. Dolinar and D. Divsalar, Weight distributions for turbo codes using random and non-random permutations, *TDA Progress Report*, vol. 42-122, pp. 56–65, August 1995.

[41] M.N. Kaynak, T.M. Duman, and E.M. Kurtas, "Turbo codes," *Coding and Recording Systems Handbook*, B. Vasic and E.M. Kurtas, Eds., CRC Press, Boca Raton, 2003.

[42] G.D. Forney, Maximum-likelihood sequence estimation of digital sequences in the presence of intersymbol interference, *IEEE Trans. Inform. Theory*, vol. 18, no. 3, pp. 363–378, May 1972.

[43] W.E. Ryan, L.L. McPheters, and S.W. McLaughlin, Combined turbo coding and turbo equalization for PR4-equalized Lorentzian channels, in *Proceedings of Conference on Information Sciences and Systems*, Princeton, NJ, March 1998, pp. 489–493.

[44] J.L. Fan, E.M. Kurtas, A. Friedmann, and S.W. McLaughlin, Low density parity check codes for partial response channels, in *Proceedings of Allerton Conference on Communications, Control and Computing*, Urbana, IL, October 1999.

[45] H. Song and J.R. Cruz, Block turbo codes for magnetic recording channels, in *Proceedings of IEEE International Conference on Communications (ICC)*, vol. 1, June 2000, pp. 85–88.

[46] J.L. Fan, Constrained Coding and Soft Iterative Decoding for Storage, Ph.D. dissertation, Stanford University, Stanford, CA, 1999.

Index